Studies in Human Ecology and Adaptation

Volume 13

Series Editors

Daniel G. Bates, City University of New York, Hunter College, New York, NY, USA

Ludomir R. Lozny, City University of New York, Hunter College, New York, NY, USA

This series reflects a growing trend in the social sciences towards interdisciplinary study and research spanning a large number of disciplines, such as archaeology, anthropology, environmental sociology, geography and demography. It focuses on the adaptation of humans to their environment throughout the world. Within the environmental sciences, the field of human ecology is the study of the ways in which human social behavior is affected by environmental factors and events, including but not limited to natural resources, anthropogenic environmental changes and the interactions of competing or co-operating human groups. It encompasses a broad perspective that views the biological, environmental, demographic, and technological aspects of human existence as interrelated. Because the approach is multidisciplinary, it brings new and often unexpected insights to many topical issues. This series publishes cutting-edge and critical research in the field of human adaptation to their environments. Both edited volumes and monographs are welcome.

E. N. Anderson • Raymond Pierotti

Respect and Responsibility in Pacific Coast Indigenous Nations

The World Raven Makes

Springer

E. N. Anderson
Department of Anthropology
University of California
Riverside, CA, USA

Raymond Pierotti
Department of Ecology and Evolutionary Biology
University of Kansas
Lawrence, KS, USA

ISSN 1574-0501
Studies in Human Ecology and Adaptation
ISBN 978-3-031-15588-8   ISBN 978-3-031-15586-4   (eBook)
https://doi.org/10.1007/978-3-031-15586-4

© The Editor(s) (if applicable) and The Author(s), under exclusive license to Springer Nature Switzerland AG 2022

This work is subject to copyright. All rights are solely and exclusively licensed by the Publisher, whether the whole or part of the material is concerned, specifically the rights of translation, reprinting, reuse of illustrations, recitation, broadcasting, reproduction on microfilms or in any other physical way, and transmission or information storage and retrieval, electronic adaptation, computer software, or by similar or dissimilar methodology now known or hereafter developed.

The use of general descriptive names, registered names, trademarks, service marks, etc. in this publication does not imply, even in the absence of a specific statement, that such names are exempt from the relevant protective laws and regulations and therefore free for general use.

The publisher, the authors, and the editors are safe to assume that the advice and information in this book are believed to be true and accurate at the date of publication. Neither the publisher nor the authors or the editors give a warranty, expressed or implied, with respect to the material contained herein or for any errors or omissions that may have been made. The publisher remains neutral with regard to jurisdictional claims in published maps and institutional affiliations.

This Springer imprint is published by the registered company Springer Nature Switzerland AG
The registered company address is: Gewerbestrasse 11, 6330 Cham, Switzerland

The Song of the Sky
I will sing the song of the sky. This is the song of the tired—the salmon panting as they swim up the swift current.
I go around where the water runs into whirlpools.
They talk quickly, as if they are in a hurry. The sky is turning over. They call me. Traditional Tsimshian song as sung by Tralahaet (Tsimshian chief); translated by Benjamin Munroe and William Beynon, early 20th century. Recall that the salmon lay eggs and then die; this poem is about sacrificing one's life for the rising generations. (Garfield et al. 1950: 132.)

*To the survival of plants and animals on the Northwest Coast and elsewhere
And to the cause of land claims and management rights for Indigenous people.*

Preface and Acknowledgments

This book discusses resource conservation on the Pacific coast of North America, primarily the Northwest Coast: Oregon, Washington, British Columbia, and the southern parts of Alaska. We also consider California in an appendix; the original idea was to include a separate book-length study of it, but considerations of space, and the existence of excellent reference works that cover resource management, led us to confine our efforts to the heart of our project: documenting conservation and management ideology.

This book is unique in drawing on a wide range of Indigenous materials: writings, field records, our own research, and the vast amounts of textual materials that now reside in relative obscurity in old collections from the nineteenth through the twentieth centuries. Most of these collections have not been previously investigated for research on conservation and resource management. What has been done has concentrated on plants and to some extent on fish; we expand to take major account of land animals and hunting.

We are grateful to a wide range of people. Anderson owes a special debt to my coworker in the 1980s Evelyn Pinkerton, who is not only the finest ethnographer I have ever seen at work but a superb scholar and human being. He has also had special help and support on the Northwest Coast from Chelsey Armstrong, Fikret Berkes, Sara Breslow, Karen Capuder, Marnie Duff, Leslie Main Johnson, Joyce LeCompte-Mastenbrook, Dana Lepofsky, Robin Ridington, Nancy Turner, and many other scholars. Anderson acknowledges, in particular, Guujaw (Gary Edenshaw), Ki-?e-in (Ron Hamilton), Nelson Keitlah, the late Ray Seitcher, and several other First Nations friends and consultants. For particularly valuable advice in the field work years and since, I am grateful to Ronald and Marianne Ignace. For California, deepest thanks to M. Kat Anderson, Paul Apodaca, Jeanne Arnold, Robert Bettinger, Thomas Blackburn, Lowell Bean, Christopher Chase-Dunn, Deborah Dozier, Catherine Fowler, Lynn Gamble, Robert Heizer, Sandy Lynch, Malcolm Margolin, Daniel McCarthy, George Phillips, Katherine Saubel, Mark Sutton, Jan Timbrook, Robert Yohe, and many more.

Pierotti is also grateful to a number of individuals with whom he has worked: a number of Indigenous colleagues who have shared stories or discussed the manifold ways in which Indigenous people perceive and deal with the natural world, including Thom Alcoze (Cherokee), Joseph Bruchac (Abenaki), Vine Deloria Jr. (Lakota), Wilfred Denetclaw (Dine), Robin Kimmerer (Pottawatomie), Brent and Lori Learned (Arapaho), Henry Lickers (Seneca), Merlin Little Thunder (Cheyenne), Oren Lyons (Onondaga), Henrietta Mann (Northern Cheyenne), James Peshlakai (Dine), Enrique Salmon (Tarahumara), Dean Smith (Mohawk), Earl Swift Hawk (Lakota), Joseph Brophy Toledo (Jemez Pueblo), Albert Whitehat (Lakota), and the eminent Indigenous writers Joseph Marshall III (Lakota), and the late Louis Owens (Choctaw/Cherokee),

We also thank many people whose names have been withheld for reasons of privacy.

Riverside, CA, USA
Lawrence, KS, USA

E. N. Anderson
Raymond Pierotti

Contents

1 Commons and Management 1
2 Looking to the Sea: Economics and Ecology in the Pacific Northwest 13
3 Looking to the Land: Terrestrial Ecology 39
4 Traditional Cultural Areas 55
5 Social and Cultural-Ecological Dynamics 65
6 Traditional Resource Management 89
7 White Settler Contact and Its Consequences 127
8 The Ideology Behind It All 151
9 Animism and Rationality: North vs "West" 189
10 Respect and Its Corollaries 203
11 Teachings and Stories 227
12 The Visual Art 251
13 Conclusions 261
14 Appendix 1: Indigenous California 265
15 Appendix 2: Wider Connections 287
16 Appendix 3: The "Wasteful" Native Debunked 307

About the Authors

E. N. Anderson is Professor Emeritus of Anthropology at the University of California, Riverside. He has lived in California almost all his life. He conducted field research on the Northwest Coast during 1983-87, primarily in 1984 and 1985. He lived in Seattle from 2006 through 2009. His research was rather varied in nature, and in 1985-87 focused largely on the writings of Wilson Duff, an anthropologist active in the mid-twentieth century (Duff 1996). He did field work on ecological and environmental issues for about several months in 1984 and another three in 1985, largely among the Nuu-chah-nulth and Haida. (The Nuu-chah-nulth were formerly called, or miscalled, "Nootka" in the literature; the new name was coined some years ago, and means something like "people on our side of the mountains," referring to the west coast of Vancouver Island. The name is correctly transcribed *Nuučaańuł*, but the syllabic spelling is now established and hard to displace.) Anderson draws comparative material from field work conducted among the Yucatec Maya, who maintain beliefs and ontologies similar to the ones described herein. Travels in Mongolia and Kazakhstan added direct experience with northern societies in those locales. His books include *The East Asian World-system* (with Springer, 2019), *Caring for Place* (2014), and *Ecologies of the Heart* (1996). Both of us have traveled widely in the region, from northern Alaska south throughout southern British Columbia, Washington, Oregon, and California.

Raymond Pierotti is Associate Professor in the Department of Ecology and Evolutionary Biology and in the Indigenous Nations Study Program at the University of Kansas. He draws on his experiences with Plains Indigenous groups for general perspectives. He draws upon 50 years as a trained ecologist, and scholar of Native stories and writing. His concentration has been on peoples of US Midwest and Southwest, having spent 30 years as professor at the University of Kansas, and having grown up largely in Texas and New Mexico. His books include *Indigenous*

Knowledge, Ecology, and Evolutionary Biology (2011) and (with Brandy Fogg) *The First Domestication* (2019). Both of us have traveled widely in the region, from northern Alaska south throughout southern British Columbia, Washington, Oregon, and California.

Chapter 1
Commons and Management

Common-Pool Resource Management

Humanity has a long record of managing resources with highly variable degrees of success. Current worldwide management is collapsing fast, with incredible waste added to other types of overconsumption. Some argue that humans are incapable of managing natural resources, or at least commonly held ones. This flies in the face of tens of thousands of years of reasonably sustainable use of many resources and environments, often through common property management and reliance on traditional Indigenous ways of thinking.

Current failures have been blamed on human nature, capitalism, consumerism, industrial civilization, the concept of private individual ownership, and many other factors. We have both argued at length elsewhere (Anderson, 2010, 2014; Pierotti, 2011) that blame lies with a sense of "nature" as enemy: something to conquer and exploit, rather than to coexist with. This attitude goes back at least to ancient Mesopotamia, is reflected in the Epic of Gilgamesh and the first chapter of Genesis, and has been most dramatically stated in Mao Zedong's phrase "Struggle against nature." In a worldwide context, it is a historic exception. In fact, it was a new idea for China when Mao adopted it, which reflects the argument made by Pierotti and Wildcat (2000) that capitalism and communism are simply two sides of the same Eurocentric coin.

Other forms of resource management lack this feature, depending on quite different conceptions of the wild, of resources, of management, and of humanity. Among management systems that have succeeded in maintaining large human populations sustainably over centuries, the Indigenous system of the Northwest Coast of North America stands out. It was not perfect. It did not support a very dense population, although they did enjoy a high standard of living. Yet it did support large numbers of people for thousands of years in a harsh and demanding

© The Author(s), under exclusive license to Springer Nature Switzerland AG 2022
E. N. Anderson, R. Pierotti, *Respect and Responsibility in Pacific Coast Indigenous Nations*, Studies in Human Ecology and Adaptation 13,
https://doi.org/10.1007/978-3-031-15586-4_1

natural environment, where the currency involved was survival, rather than amassing wealth.

The pervasive disaster of Western economic management, especially in colonial and settler situations, led many biologists, economists, and others to assume that no humans could manage resources well, especially if the resources were held in common-pool status. This problem was raised by John Locke (1975 [1697]), who inaccurately saw Native Americans as being held down by holding all things common. He and many others saw the cure in private ownership, having failed to read about the problems of private ownership in settler societies like the British colonies in America. Locke, instead, was channeling Thomas Hobbes, who believed that "the life of man in his natural state was solitary, poore, nasty, brutish and short" (Hobbes, 1950 [1657]), and that only life under a king could cure this perpetual state of "warre."

This view was most famously stated in the twentieth century by Garrett Hardin (1968), whose paper "The Tragedy of the Commons" became one of the most cited papers in the history of science. Hardin expressed the views of most biologists of the time that humans were destroying the world's resources at ever-increasing rates, and blamed unrestricted use of common-pool resources as the greatest problem, because all too frequently, people can draw from commons without restricting their take without paying immediate costs, leading to overexploitation. He advocated privatization of the world's commons.

This argument provoked a strong reaction within the field of anthropology, leading to an American Anthropological Association conference section that became a book, *The Question of the Commons* (McCay & Acheson, 1987). Many participants studied fisheries, which are supposed to be notoriously hard to manage because of the problem identified by Hardin and classically described by Arthur McEvoy in *The Fisherman's Problem* (1986).

It turned out that some fisheries self-regulated with varying degrees of success. McEvoy, and others such as Evelyn Pinkerton (Pinkerton & Weinstein, 1995), remarked on the problem of fishermen knowing the fishery was in trouble, but "it won't hurt to take just one." More serious was another part of the same problem: Chronically overoptimistic estimates by regulators of how many fish were out there resulted in the Maximum Sustainable Yield (MSY) approach that become popular in the 1950s (Pierotti, 2011). "Western economic models assume relatively constant conditions, which is why they set ... (MSY) at approximately 50 percent of the theoretical carrying capacity (K). Because Euro-Americans assume the environment is stable, they believe that carrying capacity is a real and constant value, rather than fluctuating and unpredictable" (Wilson, 2005). In the real world, there is no constant K and environmental conditions change regularly in unpredictable fashion... the variable nature of the fish populations means that (commercial fishing) is fated to be a boom-and-bust industry (Pierotti, 2011: 65–66). Often, governments interfered with their own scientists; forcing them to set extremely optimistic limits. This led, among other things, to the destruction of the Canadian cod fishery. Similar results are well known in game hunting, wildfowl shooting, plant collecting, and indeed in management of every resource. Chronic erring on the optimistic side, "taking just one (often several), just this time," inevitably leads to destruction.

Another problem is unexpected risk. Salmon runs suddenly fail without warning. Late blizzards decimate plant flowering. Traditional societies, such as those of the Northwest Coast, spread the risk (Cronk et al., 2020; Suttles, 1987; Trosper, 2009). Social mechanisms such as wide-flung kin relations provide security.

Soon after these studies of fisheries and other commons became available, Elinor Ostrom produced a short but extremely influential book, *Governing the Commons* (Ostrom, 1990), which refuted Hardin's "tragedy" view of the English and other commons. Several design features allowed people to manage common-property resources effectively, including clear boundaries, clear lines of management, community involvement, measurable levels of offtake, and similar guidelines. She later added several features (Ostrom, 2005, 2009). The book led to several of us coming together with Elinor Ostrom to found the International Association for the Study of Common Property, which had a major influence in providing an alternative to Hardin's views. Hardin, an older but open-minded scientist, acknowledged this effort. ENA met him shortly before he died, at a conference on common property regulation. He generously acknowledged our findings and wrote several end-of-life papers correcting his original view. One of his last papers was "The Tragedy of the *Unmanaged* Commons" (Hardin, 1991; his emphasis).

Since that time, literally, thousands of studies of common property management have appeared, chronicling the successes of traditional and modern people all over the world. Many skeptical biologists went to the field expecting to find tragedies and found enlightenment instead (e.g., Aswani & Sheppard, 2003). On the other hand, thousands of other studies have found that not only modern industrial society but also many traditional societies in all parts of the world are conspicuous failures at commons management (e.g., Alvard, 1995; Kay & Simmons, 2002). A complicating factor is that many traditional societies simply do not live at high enough densities, or use sophisticated enough equipment, to decimate their environments (Beckerman et al., 2002); a circumstance sometimes called "epiconservation" (Lozny & McGovern, 2020). This is the case, for instance, with the Matsigenka people, who do not practice conservation as such but do not overhunt their game because they have no incentive to do so (Alvard, 1995; Beckerman et al., 2002; Allen Johnson, pers. comm. to ENA). A similar case of open access without self-conscious management exists in the Lake Chad grasslands, which can be grazed only briefly in the year and are open to small, highly nomadic populations of grazers (Moritz et al., 2020). Under those conditions, the resource is more than adequate for all comers.

These various findings led to continual elaboration and refinement of Ostrom's design principles. One particularly thorough and useful summary elaborated considerably on Ostrom's principles (Agrawal, 2002). Agrawal's final lists can be generalized as "smaller is easier," "clearer is easier" (boundaries, counting, evaluating), and enforcement is served by socially mandated equality and fairness, graduated sanctions, clear lines of authority, and other fairly obvious dimensions. In short, the simpler and clearer everything is, the easier it is to enforce rules. This has been incorporated in at least one standard book on conservation (Borgerhoff Mulder &

Coppolillo, 2005). Ostrom added several elaborations, mostly in the direction of more specification and feedback (Ostrom, 2005). She also clarified the concept of rules: "Rules are ATTRIBUTES of participants who are OBLIGED, FORBIDDEN, OR PERMITTED to ACT (or AFFECT an outcome) under specified CONDITIONS, OR ELSE" (Ostrom, 2005: 187). Ronald Trosper quotes and elaborates on this passage, noting that without the "or else" clause, the rest becomes a mere strategy statement. The "or else" makes it a rule or law (Trosper, 2009: 72). These rules are enforced in traditional societies by local means, ranging from disapproval and ostracism, to actual killing of offenders.

Traditional societies have serious problems with some of these vectors. The degree of specifying, counting, measuring, evaluating, and modeling that Agrawal's principles would require is difficult without modern scientific equipment and backup. Traditional hierarchies can interfere with equality. In practice, as Agrawal realized, it is difficult even in modern industrial societies to satisfy his requirements.

As a result, traditional societies go in the other direction. They usually embed good management in spiritual traditions. These are often characterized in Western scholars as "religion," and therefore ignored, because "religion" is supposed to contrast with "science." However, highly emotional, value-laden, community-accepted ideological systems can be based on fact and can function to conserve resources, as pointed out by ourselves and others (Berkes & Turner, 2006; Anderson, 1996, 2014; Pierotti, 2011). Good management is represented as a moral duty, a part of caring for all relatives and community members, both human and nonhuman (Pierotti, 2011). This emotional and social management of commons, and indeed of all resources contrasts with the strictly rational method implied by Agrawal. The cultural construction of management is a thoroughly social undertaking, based on mutual responsibility.

David Hume wrote: "Reason is, and ought only to be the slave of the passions, and can never pretend to any other office than to serve and obey them" (Hume, 1969 [1739–1740]: 462). This is factually true: humans will not act without a motivation—either emotional or need-based. We are rational when rewarded for it, or when emotion gives us an urgent reason to deploy good sense. Our reason is also limited by many heuristics and shortcuts, inborn, culturally imposed, or self-imposed. A deep and intense emotional framework is needed for effective conservation (Anderson, 1996: 2014).

In the Northwest, insecurity about the food supply, including genuine concerns about possible starvation, underlies the recognition of a need to manage. By itself, this would be powerful, but might not provide sufficient guidance. What guides it is recognition of the need for mutual consideration, mutual respect, and mutual responsibility both among humans and among all species. Indigenous peoples are much more aware of the naturally occurring variation in environmental conditions, both over a few years and even at longer scales (Anderson, 2014; Pierotti, 2011). Western thought tends to emphasize balance and stability, and gets away with this because technological industrial agriculture buffers individuals and small communities from having to worry about starvation, because supply chains ensure reasonable availability of food. The possibility of imminent, visible shortages of key sources of food, which are, after all, other species, is a different matter. It causes humans to care about these other life forms, because human survival is recognized as being dependent on their presence in adequate numbers and this is codified through ritual and ceremony in which all of the people participate (Pierotti, 2011, Chap. 5). This induces Indigenous people to show respect and reciprocity, thus generating more positive emotions, and directing people to care for the world around them as they care for their families—largely because they know their families depend on such care for all beings.

In Indigenous cultures, as exemplified by the First Nations of the Northwest Coast and California, the realm of "people" does not stop with human society. All beings, at the least all animals, are persons. Many Indigenous cultural traditions consider plants and even mountains to be persons as well (Turner, 2005; Pierotti, 2011). In the moral systems of the groups in question, animals and other beings are thus owed respect, care, consideration, and decent treatment, even if they must be killed at times to provide food and materials for clothing and shelter.

This works powerfully against the fatal line "It can't hurt to take just one," so endemic to "maximum sustainable yield" management and rationally calculated game and foraging limits. People need to be aware of the actual state of wild populations; if a population's numbers are falling or are low, there should be no taking of any individuals and certain individuals are completely off limits at all times (Pierotti, 2011). This means that people must be intimately familiar with the ecological circumstances facing all of the members of their community on a consistent basis. If one species is undergoing the breeding season, or at low numbers, humans should turn to other species for sustenance.

Just as the line about "taking just one" is invoked by almost every small-scale poacher, it can be extended to include the inevitably overoptimistic calculations that have bedeviled MSY style management. It is always easy—too easy—to assume this just happens to be a bad year, and to base yield targets on the best year of record, in spite of clear records that the fishery has been declining for decades (Holt, 1975; Larkin, 1977; Myers & Worm, 2004). Logging is less easy to ignore, but the fond hope that a beautiful, diverse, old-growth Douglas fir forest will magically reappear soon has led to wildly overoptimistic tree-farming. In contrast, if a cultural tradition is to treat animals and trees with respect and care, its members will attend to their needs and quickly recognize their problems—especially when guided by thousands of years of experience with the local place and the species that the humans share it with (Pierotti, 2011).

The Pacific Coast of North America is of interest for many reasons, not least being the careful management of plant and animal resources by its Indigenous peoples. Thanks to care backed by an ideology of respect, the Native people supported dense populations with rich artistic traditions and complex stratified societies (Johnson, 2019; Deur & Turner, 2005).

Reciprocity, including feasting and potlatching, led to interaction that stressed moral values and kept everyone up-to-date on conditions, both environmental and social (Trosper, 2009). The environment was not just sustainably managed; it was managed to get better every time that conditions allowed and to minimize harm during lean times. Controlled burning to increase berry and herb production, stocking fish, cultivating root crops, and other techniques exemplified in real time the visionary goal of "sustainable development" that seems so hard to imagine in today's world. Most of the existing literature on this, e.g., Nancy Turner's numerous publications about Coastal nations, concentrate on plants, so in this book we devote more attention to animals.

The reason we are able to generalize about these societies is straightforward: the Indigenous peoples of North America, and their Siberian relatives developed and broadly shared a long-standing cultural ideology based on respect, reciprocity, relatedness, and responsibility, the four R's of Indigenous spiritual and cultural traditions. These approaches generated sustainable management and use of resources because the resources were not being exploited for profit, but to maximize the chances of survival of both human and ecological communities (Anderson, 1996; Pierotti, 2011).

Religion was intimately connected with links to the environment and to other species, i.e., ecology. The artificial separation of "economy" or "subsistence" from "religion" in early ethnographies led to misunderstanding. Social science, including theorists of religion such as Emile Durkheim (1995 [1912]), separated "the sacred" from the everyday. To scholars from this tradition, Native American religions can be surprising: ordinary activities like cutting up fish or preparing basket materials have vast spiritual importance, with the most high-flown and remote of spiritual teachings applied to such apparently minor practical issues.

The world could obviously benefit from employing such an ideology today. We are likely to find it impossible simply to recycle the old Indigenous views, because too much knowledge had been lost as a consequence of 300 years of colonial attitude and exploitation. Nevertheless, we can be inspired by these ways of thinking and living. What is ironic is that as ENA pointed out in *Ecologies of the Heart* (1996), aspects of this knowledge may have been lost, but the attitude located in the 4R's (above) can be regained by any culture, or individuals within that culture, who are willing to make the effort (Pierotti, 2011). At the very least, they prove that intensive use of natural resources does not automatically lead to depletion. It is possible for humans to be reasonable and responsible. As we have written previously, Western technology has the knowledge and tools to solve every environmental problem that humans face, what we lack in the technological world is the will to employ our knowledge (Anderson, 1996), especially if it is feared that this will somehow hurt "The Economy" and lead to loss of "jobs." If people refuse to make such changes today, it is their choice, not some Hobbesian imperative. The world and all its residents, both human and nonhuman, will pay the price. This book can be regarded as our extended commentary on that realization. Thus, environmental management should be regarded as a part of the cultural construction of emotional relationships within society. In other words, we humans will have to learn to care about the environment, and its myriads of other species, as much as we care about ourselves.

The present book is unique in one important way: it draws on the accumulated wealth of knowledge found in early collections of texts. Myths, tales, speeches, and other documents were recorded by early anthropologists, linguists, and—not least—

Native people preserving their own records. These texts are laden with vast amounts of ecological teachings, mostly concerned with dealing with animals and plants by teaching respect and cautious, caring use and management. Religion and spirituality were the drivers, bringing the deepest human emotions to bear (Anderson, 1996). Considerable use of myths and texts has been made in regard to plant use (Anderson, 2005; Turner, 2005), but animals, from insects and fish to elk and bear, have not been subjected to as much analysis. Some amazing treasures of literature are found in these texts, from the wonderful recordings of the Tsimshian ethnographer William Beynon to the soaring epic poetry of Kato nation elder Bill Ray, dutifully recorded in detail by the linguist Pliny Earle Goddard (1909) at the beginning of the twentieth century. These old texts are widely dismissed, and the records of the time disregarded, but they turn out to be filled with lessons we all need today. The central focus of this book is bringing those teachings back, with applications to the present time spelled out.

There are many similarities between Northwest Coast views and those of other hunting and fishing peoples of the northern hemisphere. In this book, we trace resemblances and relationships—more suggestively than systematically, but at least with some grounding in data and traditional stories.

One topic that must be addressed is the issue of whether or not Native peoples really think in terms of "resources" that have to be "managed," and if such concepts really can be applied to people who think of themselves as being "related" to the nonhuman aspects of their world. The Onondaga scholar and elder, Oren Lyons, once addressed a mixed gathering of tribal citizens and resource managers of European ancestry by asking "Why is it whenever we gather, we always want to talk about relatives, and you want to talk about resources?" The message Lyons intended to convey was that the approach could be very different if the individuals involved assumed they were developing tactics for addressing how to deal with beings with whom they assumed they had a personal relationship, or whether they were simply interested in maximizing the productivity of certain species to improve human hunting or harvesting (see Anderson, 1996 for a detailed discussion of this topic).

The Pacific Coast of North America

This book is a belated fulfillment of a promise: to help Native people get fair treatment in their land claims. Aboriginal title to land and rights to manage resources were not recognized in Canada, especially in British Columbia, which did not join Canada until after Confederation in 1867, and as a consequence was not subject to the British North America Act, and thus had no treaties of any kind with its First Nations as they are now known in Canada.

These First Nations were finally given some standing in the late 1990s, as a consequence of the Delgamuukw Decision by the Supreme Court of Canada, which established that Indigenous accounts had legal standing when presented in courtroom proceedings and were given some control over how their traditional lands could be used and exploited (Glavin, 1998).

Use and management rights are still being debated in Canada, but there has been progress to allow greater First Nations input and control in the twenty-first century. In the United States, Native peoples were recognized as "dependent nations" from the earliest years of the republic, but interpretations of tribal rights have fluctuated, depending on whether "dependent" or "nations" was being foregrounded. Treaty rights to fisheries were disregarded in Washington state, for instance, until Judge Boldt in 1978 ruled that they had to be enforced and that Native peoples were entitled to a significant portion of the annual harvest of salmonids. Title to land was recognized, but was signed away (often at gun point), through unequal treaties written in a language not spoken or understood by tribal members, or was simply stolen or acquired by duplicity. Various claims and compensation have resulted, but in both Canada and the United States there is a long way to go. In 1985 Anderson promised to Haida friends that he would write what could be useful. He has since written several items about Northwest Coast human ecology, notably his introduction to Wilson Duff's writings (1996), and in relevant sections in my books *Ecologies of the Heart* (1996) and *Caring for Place* (2014). The time has come to bring all these ideas together.

Earl Maquinna George, Nuu-chah-nulth anthropologist, has written: "The work of a non-native is colored by the inability of the outsider to experience the context of the information collected…there are many parts of native life that have never come out in their work" (George, 2003: 38). We are painfully aware of this issue, and not really positioned to do any better as a person of largely European ancestry (ENA) and a person of mixed blood from the Numic (Shoshonean) people. Fortunately, we can rely on quoting George and other First Nations writers, and view them from a very different perspective than have other scholars. Many Indigenous persons are now leading scholars of their traditions, and we will have much occasion to refer to them below.

There is a serious issue with making use of traditional stories recorded in collections of texts. Many of these were recorded with full permission, for publication, and were told by people who had the rights to tell them. However, many others were not. We make an effort to confine my citations of traditional stories to material in the first category, but to do anything like a proper job, at times, we must cite materials that may be in the second. We have, however, consciously tried to avoid using distorted, "edited," bowdlerized, or otherwise "settler"-altered materials, and materials clearly recorded from shaky sources or published without even the pretense of permission. We ask forgiveness for errors. The traditional teachings of the Northwest Coast are so important that they should be made available to the wider world, even if they are already out there in published form.

In usage, we prefer the term "Native American," though it has drawbacks. Most of the people in question call themselves "Indians," but prefer to be referred to as "Native Americans" in formal contexts. In Canada "First Nations" is now the polite term of reference. "Indigenous" is a vexed term, but has a solid meaning in the ecological context, and is difficult to avoid. "Local" and "traditional" have their usual vague meanings (see Pierotti, 2011, and Chap. 1, below, for a more detailed discussion of this topic). We remain aware that while some traditions are thousands of years old, others are only a generation old; if their bearers think of them as "traditions," we are not going to object. As one of us has written previously,

> What is ignored in many discussions of traditional knowledge, ecological or otherwise, is that the stories, the oral tradition, are only the foundation of the actual knowledge base, and functionally equivalent to basic concepts or premises, ... Thus, Indigenous stories function (to) allow a philosophical framework to develop... Traditional Ecological Knowledge can be rebuilt and changed by each generation in the same way that each new generation of graduates in ecology or evolutionary biology go out and make new discoveries that add on to the base of empirical knowledge that currently exists.
>
> The tendency of Indigenous societies to incorporate new information, and new technologies, is crucial... adding new ideas, and especially new technologies, has led to debate over whether or not use of modern technology, such as guns or outboard motors, removes hunting and fishing practices from the category of "traditional" in terms of treaty arrangements. Indigenous people invariably contend that the attitude and philosophy involved, rather than the technology, are what make a practice traditional. In contrast, many of European heritage argue that traditional practices can employ only primitive technologies, such as bows and arrows, presumably using stone arrowheads, or canoes, presumably dugout or manufactured from birchbark.
>
> Following such logic to an absurd extreme, it might be questioned whether the buffalo cultures of the Plains peoples of North America, ... can be considered as traditional. Hunting from horseback was a major part of this tradition, yet it is well established that Europeans introduced domestic horses into North America in the sixteenth century.... Prior to this the only domestic animal of plains peoples was the wolf (or dog, *Canis lupus*)... most people of European ancestry assume that *traditional* describes only those conditions that existed when a tribe was initially visited ... by Europeans. This contains the tacit assumption that Indigenous peoples remained essentially unchanged and uninfluenced by any other cultures prior to European contact (Pierotti, 2011: 13–14).

Similarly, we are stuck with "supernatural" to refer to spirit beings, though they are considered perfectly natural on the Northwest Coast, and include the immortal essences of ordinary animals, these having been "supernaturals" in ancestral times. We try to avoid making irrelevant discriminations such as "natural" and "supernatural" (see Miller, 1999), but most sources use the word, and better recent coinages like "supramundane" are not current usage.

Finally, the term "authentic" has become so polarized in Northwest Coast studies that we avoid it altogether. It has been especially problematic in art, where modern Native American works have often been criticized for not being identical to those of a hundred or more years earlier, these latter being assumed to be somehow more "authentic." We apologize for any offense given by particular usages.

The general characteristics of the Indigenous Northwest are well described in so many books that it would be tedious to go into detail; see the classic cultural summaries in the Smithsonian Institution's *Handbook of North American Indians, Northwest Coast* volume, edited by Wayne Suttles (1990). Defining the region we are covering, however, is necessary. The Smithsonian Institution Handbook is divided into volumes by culture areas. Our "Northwest Coast" overlaps their Plateau (Walker, 1998) and Subarctic (Helm, 1981), and even makes incursions into California (Heizer, 1978). This may be convenient; however, Indigenous peoples did not divide themselves into neat culture areas. What has not been covered, however, is the philosophical underpinnings of these cultures, i.e., how they dealt with the natural world in order to survive and even thrive for more than 10,000 years before Europeans arrived without overusing resources or driving any species to extinction.

References

Agrawal, A. (2002). Common resources and institutional stability. In E. Ostrom, T. Dietz, N. Dolsak, P. C. Stern, S. Stonich, & E. U. Weber (Eds.), *The drama of the commons*. National Academy Press.

Alvard, M. (1995). Interspecific prey choice by Amazonian hunters. *Current Anthropology, 36*, 789–818.

Anderson, E. N. (1996). *Ecologies of the heart*. Oxford University Press.

Anderson, M. K. (2005). *Tending the wild: Native American knowledge and the management of California's natural resources*. University of California Press.

Anderson, E. N. (2010). *The pursuit of ecotopia: Lessons from indigenous and traditional societies for the human ecology of our modern world*. Praeger (imprint of ABC-Clio).

Anderson, E. N. (2014). *Caring for place*. Left Coast Press.

Aswani, S., & Sheppard, P. (2003). The archaeology and ethnohistory of exchange in precolonial and colonial Roviana: Gifts, commodities, and inalienable possession. *Current Anthropology, 44*, S51–S78.

Beckerman, S., Valentine, P., & Eller, E. (2002). Conservation and native Amazonians: Why some do and some don't. *Antropologica, 96*, 31–51.

Berkes, F., & Turner, N. (2006). Knowledge, learning, and the evolution of conservation practice for social-ecological system resilience. *Human Ecology, 34*, 479–494.

Borgerhoff Mulder, M., & Coppolillo, P. (2005). *Conservation: Linking ecology, economics and culture*. Princeton University Press.

Cronk, L., Berbesque, C., Conte, T., Gervais, M., Iyer, P., McCarthy, B., Sonkoi, D., Townsend, C., & Aktipis, A. (2020). Managing risk through cooperation: Need-based transfers and risk pooling among the societies of the human generosity project. In L. R. Lozny & T. H. McGovern (Eds.), *Global perspectives on long term community resource management* (pp. 41–75). Springer.

Deur, D., & Turner, N. (Eds.). (2005). *Keeping it living: Traditions of plant use and cultivation on the northwest coast of North America*. University of British Columbia Press.

Duff, W. (1996). *Bird of paradox*. Hancock House.

Durkheim, E. (1995). *The elementary forms of religious life*. Trans. by Karen E. Fields. Free Press.

George, E. (2003). *Living on the edge: Nuu-Chah-Nulth history from an Ahousaht Chief's perspective*. Sono Nis Press.

Glavin, T. (1998). *A death feast in Dimlahamid*. New Star Books.

Goddard, P. E. (1909). Kato Texts. *Berkeley: University of California Publications in American Arcaheology and Ethnology, 5*(3), 65–238.

Hardin, G. (1968). The tragedy of the commons. *Science, 162*, 1243–1248.

Hardin, G. (1991). The tragedy of the *unmanaged* commons: Population and the disguises of providence. In R. V. Andelson (Ed.), *Commons without tragedy* (pp. 162–185). Barnes & Noble.

Heizer, R. (Ed.). (1978). *California. Handbook of North American Indians* (Vol. 8). Smithsonian Institution Press.

Helm, J. (Ed.). (1981). *Subarctic. Handbook of North American Indians* (Vol. 6). Smithsonian Institution Press.

Hobbes, T. (1950 [1657]). *Leviathan*. E. P. Dutton.

Holt, S. J. (1975). The concept of maximum sustainable yield (MSY) and its application to whaling. In *Fao/UN scientific consultation on marine mammals*. Document ACMRR/MM/SC/4.

References

Hume, D. (1969 [1739–1740]). *A treatise of human nature*. Penguin.

Johnson, L. M. (Ed.). (2019). *Wisdom engaged: Traditional knowledge for northern community well-being*. University of Alberta Press.

Kay, C. E., & Simmons, R. T. (Eds.). (2002). *Wilderness and political ecology: Aboriginal influences and the original state of nature*. University of Utah Press.

Larkin, P. A. (1977). An epitaph for the concept of MSY. *Transaction of the American Fisheries Society, 106*, 1–11.

Locke, J. (1975 [1697]). *An essay concerning human understanding*. Oxford University Press.

Lozny, L. R., & McGovern, T. H. (Eds.). (2020). *Global perspectives on long term community resource management*. Springer.

McCay, B., & Acheson, J. (Eds.). (1987). *The question of the commons*. University of Arizona Press.

McEvoy, A. F. (1986). *The Fisherman's problem: Ecology and law in the California fisheries, 1850–1980*. Cambridge University Press.

Miller, J. (1999). *Lushootseed culture and the shamanic odyssey: An anchored radiance*. University of Nebraska Press.

Moritz, M., Scholte, P., Hamilton, I. M., & Kari, S. (2020). Open access, open systems: Pastoral resource management in the Chad Basin. In L. R. Lozny & T. H. McGovern (Eds.), *Global perspectives on long term community resource management* (pp. 165–187). Springer.

Myers, R. A., & Worm, B. (2004). Rapid worldwide depletion of predatory fish communities. *Nature, 423*, 283–290.

Ostrom, E. (1990). *Governing the commons: The evolution of institutions for collective action*. Cambridge University Press.

Ostrom, E. (2005). *Understanding institutional diversity*. Princeton University Press.

Ostrom, E. (2009). A general framework for analyzing sustainability of social-ecological systems. *Science, 325*, 419–422.

Pierotti, R. (2011). *Indigenous knowledge, ecology, and evolutionary biology*. Routledge.

Pierotti, R., & Wildcat, D. (2000). Traditional ecological knowledge: The third alternative. *Ecological Applications, 10*, 1333–1340.

Pinkerton, E., & Weinstein, M. (1995). *Fisheries that work: Sustainability through community-based management*. David Suzuki Foundation.

Suttles, W. (1987). *Coast Salish essays*. University of Washington Press.

Suttles, W. (Ed.). (1990). *Northwest coast. Handbook of north American Indians* (Vol. 7). Smithsonian Institution.

Trosper, R. L. (2009). *Resilience, reciprocity, and ecological economics: Northwest coast sustainability*. Routledge.

Turner, N. J. (2005). *The Earth's blanket*. University of Washington Press.

Walker, D. E., Jr. (Ed.). (1998). *Plateau. Handbook of North American Indians* (Vol. 12). Smithsonian Institution.

Wilson, K. (2005). MSY, no net loss and the future of Fraser River sockeye. In *Proceedings from the World Summit on Salmon, Chapter 10*. University of Utah Press.

Chapter 2
Looking to the Sea: Economics and Ecology in the Pacific Northwest

Defining the Region

To begin, it is crucial to envision the habitat encountered by the first humans to move down the Pacific coast of North America, at least 15,000 years ago. They almost certainly did this by boat along the coast, rather than through the "ice-free" corridor so beloved to certain paleoecologists and archaeologists (e.g., Martin, 2005; Krech, 1999). They had to move along the coast, crow-hopping along a series of coastal ice-free refugia where they could extract a living from the sea, especially by taking marine mammals. Their success belies the idea of Pleistocene Overkill, because they survived by taking the largest megafauna of all, baleen whales, for thousands of years, without causing any significant decline in the numbers of their prey.

The northwest corner of North America is a land of vast forests and high mountains. At first glance, it seems to be a difficult country for making a living, especially for hunter-gatherers. More importantly, it is also a land bordering possibly the richest marine environment in the world. On land, most of the biomass is tied up in wood, and the coniferous trees are living chemical factories, producing compounds that discourage insects and other herbivores. Thus, compared to the deciduous forests of eastern North America or tropical rainforests, Northwest rainforests are typically quite poor in wildlife. One can hike for miles without encountering anything more than a squirrel or two and a flock of chickadees and nuthatches. Old-growth mixed forests develop a richer fauna, but much of the region is covered with Douglas fir, famously well protected by its hardness and its chemical defenses against herbivores of all types. Almost nothing consumes it, making it valuable as lumber but poor for hunters.

In contrast, the marine environment is full of large packages of protein, especially prior to the arrival of European invaders: Russians, Spanish, and eventually British and Americans in the late eighteenth century. The North Pacific has the highest diversity of marine mammals on the planet, with 17 species of cetacean (8 baleen

© The Author(s), under exclusive license to Springer Nature Switzerland AG 2022
E. N. Anderson, R. Pierotti, *Respect and Responsibility in Pacific Coast Indigenous Nations*, Studies in Human Ecology and Adaptation 13,
https://doi.org/10.1007/978-3-031-15586-4_2

whales, at least 9 toothed whales), 11 species of pinniped (5 true seals, 2 fur seals, 2 sea lions, and walrus), sea otters, and, until the eighteenth century, the only non-tropical sirenian, Steller's Sea Cow, *Hydrodamalis gigas*. In addition, there are over 20 species of marine birds, and more than 10 species of sea duck and loons, many of which are most available in winter, because they are non-breeding migrants which swarm to these productive waters from their breeding grounds in the Arctic or the interior. Rivers and lakes also teem with these birds.

Most importantly, there are fish, or there were until the advent of commercial fishing by Europeans and Japanese in the nineteenth century. The main commercial fisheries in the Northeast Pacific are the species of Pacific salmon (Chinook, Sockeye, Coho, Pink, Chum, and Steelhead: the anadromous form of Rainbow Trout), flatfishes (including halibut), cod, Alaska pollock, hake, lingcod, mackerel, Pacific herring, and various crabs, shrimps, clams, squid, and oysters. Salmon are particularly important because they can be found in both fresh and salt water. They are most often taken by Indigenous fishers as they migrate up the numerous rivers that resulted from the melting of glacial ice, especially after about 12,000 years ago.

What we describe as the "Northwest" comprises southern and southeastern Alaska; all of British Columbia, along with Washington and Oregon, and northwestern California, i.e., the belt of coastal wet forests. This is a land of fog and rain. It is the land of Raven, the large omnivorous bird that chooses to associate with humans, regarded as both a trickster and a Creator figure. Wherever humans go, there is Raven, waiting for them, showing them where to find food (Pierotti, 2011: vii). The large salmon runs in Alaska connect the marine and inland waters through the huge rivers. These salmon runs also include, or once included, most of Idaho, Oregon, Washington, and even western Montana, where the rivers run to the Pacific. There, the Indigenous people of the western interior also adapted to other resources, functioning partly as hunter-gatherers.

The classic coastal Northwestern human community is defined by large plank group houses (typically made from cedar), distinctive style of visual art, ranked societies with chiefs, commoners, and enslaved individuals (largely captured in war or raid), dependence on fish and marine mammals for subsistence, and heavy use of wood and wood products, which was the primary link to the land in this thickly-forested environment.

The southern interior part of the Northwest is the Columbian Plateau. It depends heavily on riverine fish (anadromous and otherwise), large mammal hunting, especially deer, elk, and antelope, heavy use of roots dug in semidesert lithic plains, less strongly ranked societies, and smaller settlements. Plateau peoples are the most varied ecologically, ranging from the Shuswap Lakes in the northern forests to the Yakima on the sagebrush deserts of interior Washington and British Columbia. They lived on a varied diet of fish, game, roots, berries, and other plant foods. Their usual houses were substantial pit houses, in which the foundation was dug out two or three feet deep into the ground, large logs were set up in a square, and substantial timbers were put over these to make a domed roof; the whole was then covered with sod. These large, comfortable, well-insulated houses were ideal for the climate and hardly changed for 5000 years. The Plateau people were highly mobile, at first in canoes, later on horseback, and their mobility patterns have locally persisted to this day; they may suddenly disappear from one place and reappear in another they supposedly "abandoned" decades before, to the surprise of Euro-American settlers (Ackerman, 2005).

The northern interior was a cold, mountainous, difficult land of highly dispersed hunting groups, living largely on mammals, though fish were important, and in many areas the dominant food. Their homes were variations on the "tipi" or "wigwam" theme. Social groups were mobile, though reuniting often at especially productive fishing and gathering sites. Berries were a vital supplemental food, their abundance increased by selective burning.

Obviously, there are border groups that partake of two or three cultural areas—sharing one trait with one area, other traits with another. The Tsetsaut, Wetsuweten, and other Athapaskan-speaking groups shared varying amounts of Northwest culture while being Subarctic in other ways. The Wasco and Wishram along the Columbia mediated between Coast and Plateau, as did some other groups. The little-known Kalapuya and Molalla of Oregon do not seem to fit comfortably into any category. The Yurok, Karok, and their neighbors in California neatly bridge the gap between Northwest and California. The Klamath and Modoc might be considered Northwest Coast, Plateau, California, or even Great Basin. The spatial classifications are modern constructs, ultimately arbitrary as to boundaries.

Given this ambiguity, we draw examples and principles from the farthest extent possible. Any group in a drainage basin whose outlet is on the Pacific is within our purview, allowing us to go far inland up the Columbia, Fraser, Nass, Skeena, and even Yukon rivers. At times, we will take examples from even farther afield, where they are relevant, largely in the case of Athapaskan groups that currently inhabit other drainages, but are similar in culture to the ones in our area of focus, but also from eastern Siberia, whose cultural similarities to the Northwest have long been known. As we argue, Indigenous peoples show a number of philosophical links and concepts that extend these connections well beyond narrowly defined geographical areas.

River drainages are physically, ecologically, and biologically real and important entities, unlike culture areas. This is particularly true in the Pacific Northwest, where most of the inland areas were under ice until around 11,000 years ago. As the glacial ice melted, the combination of rapid melting and high rates of runoff opened up new habitats that could be used by humans and nonhumans alike, as they followed the retreating ice.

Rivers and other bodies of water are also assumed to be persons by Indigenous tradition (Pierotti, 2011). For many of the creator figures of these cultures, one thing that is paramount is their capacity for shapeshifting, an attribute well represented in the wide variety of tales concerning Raven, who can assume the forms of a small piece of dirt, a pine needle, a rock, a human child, a grown human, and various other shapes. While occupying these various shapes Raven is apparently involved in serious hydrological activity. A Tlingit story concerning the Birth of Raven involves two Raven characters, Raven at the head of the Nass (*Nas-caki-yel*) and his nephew Raven (*Itcaku*) (Swanton, 1909: 119–120). The uncle initially tries to kill his nephew out of jealousy. When he cannot accomplish this, Nas-caki-yel becomes frustrated and releases heavy rains, leading to major flooding, in which many people and their communities are damaged.

A more blatant bit of hydrological tampering by Raven involves the story of the *Theft of Fresh Water*.

> At this time there was no fresh water in the world, except rain, all other water was salt. Raven visited some people and asked them for water to drink. They said, "We have none, water is very scarce. We get a mouthful sometimes from the man who owns it." Raven asked the name of the man and where he lived. They told him that the man's name was Kan'ugu (Ganuk) and pointed out where he lived. (After some trickery Raven convinces Kan'ugu to leave him alone with the spring that is the source of fresh water). Raven then goes and drinks up almost all the fresh water. Raven escapes and lets water fall from his mouth to make all the large rivers, saying "Henceforth water will run here and there all over the country and everyone will have plenty of water" (Swanton, 1909: 4)

These stories suggest that Raven is considered to function as the entity who changed the world, releasing the water from its frozen form. This also reveals that these peoples were present at the melting of the glaciers and that prior to this time, fresh water was scarce in this area.

A more explicit story about a similar trickster/creator figure comes from the Blackfoot:

> "When Naapi encounters other beings, he often...emerges from or takes refuge in a source of water...he returns to the water element...water is thus the method of transportation as a way of transformation... and can include fluid, transient, inconsistent, highly variable conditions..." (Howe, 2019: 26). Howe follows this statement with a list of "examples of how Naapi's actions shape and make places" providing a series of names of lakes, creeks, and rivers that carry his name. In Indigenous ways of understanding, a creator/trickster figure can represent as water, which is known to be the most changeable element. A shape-shifter can thus assume the forms of water: mist, ice (Including glaciers), hail, snow, rain, creeks, rivers, ponds, and lakes. Such accounts are metaphors for instantaneous changes that can take place in the environment. Some forms of water, i.e., mist, sleet, snow, and fog, change constantly. Thus, the trickster can be considered an archetype that represents change and adaptability over untold spans of time (Howe, 2019). This stands in contrast to Western thinking, with its "Balance of Nature" and equilibrium-based thinking derived from economics (Pierotti, 2011).

For humans, the release of all this solid water into its liquid form was probably a disaster at first. These events may have lasted for centuries, as floods resulting from melting glaciers carved out the Columbia and Fraser River systems, as well as many of the rivers along the coasts of contemporary British Columbia and Alaska. This flooding and landslides almost certainly led to many deaths and even more discomfort. Once things stabilized, however, possibly over a period of millennia, humans found flowing fresh water to be a valuable addition to their environment, as did the salmon who quickly evolved in response to this changed environment. As the coastal forests expanded into these empty habitats out of coastal refugia, the salmon obviously followed, finding newly available, extensive inland spawning areas, safe from the abundant predators of the sea. Humans may have speeded the process by transplanting them to newly-available streams.

Humans benefited not only from the increased abundance of salmon but also because the rivers and lakes functioned as trade and communication corridors. Northwest Coast trading was long-range and extensive, with routes leading inland for hundreds of miles. Heavy-laden canoes went up and down the coast, and there was even regular canoe traffic across the dangerous waters of Hecate Strait between Haida Gwaii and the mainland. Rivers unified peoples, serving as highways into the interior. Individuals who spoke totally unrelated languages could, and frequently did, meet at major fishing areas, trade, exchange ideas, and frequently intermarry. It was evidently rather rare, in anything close to a border zone, to speak only one language. Many people grew up with at least two languages at home.

A related theme is that the links between Asian Indigenous peoples, e.g., the Ainu, Sahka, and Chukchi peoples, and Pacific Coast Indigenous Americans have complex cultural relationships that are not well resolved (Glavin, 2000). Since Franz Boas, anthropologists have realized that the northeast Siberian Indigenous peoples were culturally similar to the Northwest Coast ones, though the languages were different (except for Yuit, shared across Bering Strait). As a result, study of the two sides of the North Pacific as one linked area has been pursued by numerous anthropologists (e.g., Fitzhugh & Crowell, 1988; Colombi & Brooks, 2012). Connecting links to the Siberian societies proves a continuous cultural pattern, not independent invention, extending as far as the Great Plains (Schlesier, 1987).

Among people of the Arctic, similar skis/snowshoes, dogsleds, complex harpoons, skin tipis, and particular forms of sewn skin clothing prove conclusively that there was a "circumpolar" cultural area, defined by the Russian ethnographer Vladimir Bogoras (1929). Similar patterns of belief and practice in regard to hunting and animals prevalent throughout this area, allow the assumption that there has been at least some sort of ongoing contact, however indirect, qualified, and tentative. It is also true that many Indigenous cultures derive similar approaches to conservation because of the need to generate sustainable ways of life in harsh and changeable environments (Pierotti, 2011). We can define an "Old Northern worldview," based on these patterns and on Bogoras' article (see Chap. 15). Generalizing on this scale may lose specific details of cultural context, local interpretation, personal and family traditions, and the richness, variety, and dynamism of the cultures in question. Nonetheless, both of us have established that there are conceptual philosophical similarities rooted in philosophical traditions that apply across a range of cultures (Anderson, 1996; Pierotti, 2011).

> Indigenous peoples have not only used the resources around them, but have maintained and enhanced them... (Turner, 2005: 14). *One of our intentions* is to establish principles that underlie Indigenous knowledge and make it comparable to Western *concepts such as Conservation Biology and Population Dynamics*... (We hope to) establish Indigenous approaches as *superior* to those of Western Conservation because they are different in approach... It is important to understand that the knowledge of Indigenous peoples emerges

from a different philosophical tradition than what is described in the Western tradition as scientific knowledge. To understand this, it is essential to consider differences in worldview... (The) the concept of worldview *emerges from* the answers to the following questions: (1) What is real? (metaphysics); (2) What can we understand? (epistemology); (3) How should we behave? (ethics); (4) What is pleasing to the senses? (aesthetics); and (5) What are the patterns upon which we can rely? (logic). Indigenous traditions would answer each of those questions differently than would the European tradition. (emphasis and parenthetical elements modified from Pierotti, 2011: 7–8).

We must emphasize that *all cultural groups within this vast area are different*; even families and individuals within them differ from each other. Generalizing at the level found in this volume contains some risk, because it blurs major differences and minor nuances. Our major concern is with establishing broad principles of environmental management. These can be safely generalized, if they are kept at a high level of abstraction. In what follows, we make distinctions where necessary, but readers should remember that *extensive quotes from an individual fully apply, in the last analysis, only to that individual* and, hopefully, to his or her ethnic group. These quotations serve to illustrate very broad and general points about northwestern North America or even Native American worldviews in general, but they also reveal nuances and specific details relevant to the quoted individual's own life and experience.

As Pierotti has written previously:

North American Indigenous worldviews obviously vary in the details of their belief systems, but much less so in the principles that underlie their philosophical beliefs.... As the Canadian Salish/Metis writer Lee Maracle states, "We come from our own specific places, but we have a commonality and a common dream"... The influence of local places upon cultures and the corresponding diversity of peoples attached to those places guarantee the existence of variation in ... Indigenous worldviews. Despite this spatial variation in ecology and physical space there appear to exist a fundamental shared way of thinking and a concept of community common to Indigenous peoples of North America, (Pierotti, 2011: 5)

There are some quite specific similarities across the entire North Pacific. The reverence for salmon stretches from Hokkaido to Monterey Bay. Personification of moose (elk in European usage) is done by the Yukaghir of Siberia and the Dunne-Za of Canada (Ridington, 1988; Willerslev, 2007). Nonetheless, we do not agree with the ascription of "shamanism" to all Indigenous peoples, the myth of the Noble Savage (Ellingson, 2001), or the stereotype of the "primitives in harmony with nature" (Pierotti, 2011: 34). We remain conscious of both the differences between groups and of the extreme danger of essentializing and overgeneralizing. We explore particular themes that are both widespread and valuable from a maze of cultural specifics, similar to the way that Anderson (1996) emphasized "caring for nature" across a wide range of cultural traditions, and Pierotti (2011) discussed Indigenous Knowledge in relation to Western Science. Readers should not take anything we say as essentializing some mystic unity among these groups, but more as evidence of convergence in important concepts and traditions, which allowed various human cultures to survive, and even thrive, for millennia without the aid of modern technology. People learned over time, shared their knowledge, and good ideas spread.

People in similar environments may come to similar conclusions. The Haida have been compared for over 100 years with the Vikings, which has been put on a systematic footing by Ling et al. (2018). These three authors define a type of society, the maritime chiefdom, based on seafaring, trade, and slaving, combined with some on-shore production of goods and food. This description also fits well also with the Makah Nation, one of our major examples (Coté, 2010; Reid, 2015). Ling et al. (2018) observe that societies from the Haida to the Solomon Islanders display similar patterns: chiefdoms consisting of fiercely independent people, united behind leaders of descent groups. These in turn are compared with nomadic herding peoples, who often have similar patterns of high mobility and extensive trading.

On the Northwest Coast, the Haida are certainly a good example of such a culture. Some of the Kwakwala-speaking groups, such as the Lekwiltok, come close. There is, in fact, a smooth gradation from the Haida through these groups to the Salish peoples, perhaps more often targets of slaving than of functioning as slavers themselves, and much more local in their journeys. Interior peoples often traded and had some enslaved captives, but did not take long sea journeys.

Northwest Coast people share certain concepts and cultural themes with Indigenous Alaskans and Siberians, just as westerners share ideas of individualism and individual rationality without losing cultural identity. Such groups also share much more specific things: details of spiritual (often shamanic) practices and vision quests, specific folktales and songs, specific traits of material culture, even games and clothing styles. Within North America, they also maintained important trade kept links; Angelo Anastasio (1972: 169) provided a list of 53 classes of goods regularly traded in the Plateau, from abalone shells to yew wood for bows. Nancy Turner (2014: 137ff) similarly discusses the effects of trade on language; words were borrowed freely all over the region. The cultural links are real and multifarious.

Some Northwest Coast ethnographic collaborations between academic anthropologists, bilingually educated local Indigenous people, and First Nations are particularly impressive. The Tsimshian chief William Beynon collected materials for several ethnographers. He never got authorship credit for this but did most of the work. His first major publication under his own name came out in 2000, 42 years after his death (Beynon, 2000), although he was in fact the real ethnographer in many earlier cases (Halpin, 1978). Also incredibly impressive, and much better acknowledged, was the work of Edward Sapir with Tom Sayachapis and other Nuu-chah-nulth elders, of Franz Boas with George Hunt, and of James Teit with many interior Salish. Carrying on this tradition were Eugene Hunn, working with James Selam, a Yakima elder (Hunn & Selam, 1990), and Nancy Turner, coauthoring with many First Nations elders, importantly bringing the female perspective into these accounts.

Particularly noteworthy is the contribution of Charlotte Coté, a member of the Tseshaht community of the Nuu-chal-nulth First Nation, and Professor of American Indian Studies at the University of Washington. Her monograph, *Spirits of our Whaling Ancestors* (2010), is a personalized account of Makah and Nuu-chal-nulth life and spiritual beliefs, told by a person who is participating directly in efforts to revitalize First Nation traditions. It often goes unrecognized that women, both as scholars and sources, have contributed a great deal to understanding how cultural traditions function. In addition, many Native Americans from Canada and the USA, such as Richard Atleo, Earl George, Ki-Ke-In, Simon Ortiz, Bea Medicine, and Viola Cordova, Dan Wildcat, Don Fixico, and Michael Yellowbird, have become professional anthropologists, historians, and sociologists.

Fisheries

From a human perspective, food and foraging concentrate along the rivers and coasts in the Pacific Northwest; even local bears and wolves, not to mention Ravens and Glaucous-Winged Gulls (*Larus glaucescens*), rely on salmon as major food sources. The fish resources are legendary, to the extent that some Western Scholars have assumed they were constantly productive and inexhaustible. Even now, some must be seen to be believed. The major fish resource is anadromous salmonids of seven species in the genus *Oncorhynchus* ("swollen snout"). Five species are called salmon (after a distant Atlantic relative) by Anglo-Americans, two others are "trout." The "salmon" are—in rough order of popularity as food—the chinook or king (*Oncorhynchus tschawytscha*), sockeye (*O. nerka*), pink (*O. gorbuscha*), coho (*O. kisutch*), and chum or keta (*O. keta*). The "trout" are the rainbow, which when sea-running is called "steelhead" or "salmon-trout" (*O. mykiss*), and the cutthroat (*O. clarki*), which locally sea-runs as well. These seven species are much more closely related to each other than to the "salmon" and "trout" of the Atlantic drainage and the Old World. Several other trout and char species occur, and a few of them occasionally sea-run.

    Important books on Pacific salmonids include Thomas Quinn's *The Behavior and Ecology of Pacific Salmon and Trout* (2nd ed. 2017), and Mark Kurlansky's *Salmon: A Fish, the Earth, and the History of their Common Fate* (2021), both of which are of interest for scholars interested in Northwest ecology. (On fish in the Pacific Northwest, see Hart, 1973; McPhail, 2007; Wydoski & Whitney, 2003). An excellent, accessible account of upriver fish in the Columbia drainage and how the Nez Perce Nation practices fishing and management can be found in Landeeen and Pinkham (1999). Ethnographies that provide good accounts of salmon and Native salmon fisheries are Stephen Grabowski's study of salmonids in the Columbia (2015), Eugene Hunn and James Selam's *Nchi'wana, the Big River* (1990), and John Ross' *The Spokan Indians* (2011) for the Columbia. A review of their importance and traditional management is provided by Courtney Carothers (2012). The effects of salmon on streams and riparian environments have been reviewed by Walsh et al. (2020). An odd account from the Oowekeno suggests salmon may be latecomers, people having eaten flounders (more likely halibut) before salmon were created (Walkus, 1982: 36–46). This may recall a time before salmon moved up the coast after the Ice Age, which seems reasonable, since some of these rivers did not exist in anything like their current form until after the end of the Ice Age around 11,000 years ago. If glaciers covered most of their current habitat in the Northwest, and they had to reinvade after the great melt, perhaps humans helped them, by stocking streams, as Native people did in historic times.

There is a myth in older anthropological literature that salmon runs always produced a surfeit and there was little to no variation in salmon runs from Alaska to at least the Columbia River. This is incorrect; runs vary enormously from place to place and year to year.

As Pierotti has written:

> (M)odes of relationship within an animistic worldview take on two main forms: reciprocity and predation. Some argue that reciprocity "implies the *need for a total balance* in the exchange of substance between human and nonhuman beings," whereas predation presumably involves "*aggressive exploitation* of nonhuman beings as are useful" (Harkin, 2007: 217; emphasis added). Harkin goes on to argue that humans "never repay their debts to animals and thus never achieve genuine reciprocity... humans appear almost exclusively as predators rather than as equal partners" (Harkin, 2007: 218)... Harkin seems to completely ignore the idea that ceremonies and rituals designed to placate the animals taken are as close to full reciprocity as any predator can get... Following Harkin's logic, no predator exists in a co-evolutionary or egalitarian relationship with prey species, which, although it ignores evolutionary ecology (and) is a position regularly occupied by scholars who argue that Indigenous people do not really understand how to maintain a sustainable relationship with nature... More importantly, ... Harkin does not seem to understand the nature of reciprocity in natural systems... "parental care doesn't need to favor one's own offspring (and) assistance given to others doesn't require the actor to know if, when, and how they will benefit from it" (de Waal, 2009) (Pierotti, 2011: 63).

Large rivers, such as the Fraser, Columbia, Yukon, and even the Amur, may have several runs of the major species, each run coming at a different time and showing slight, but important genetic differences from other runs of the same species in the same river. Both spring and fall runs of many or most species occur in the larger rivers. At least in chinook, these seasonal runs are differentiated only by a single two-gene region on Chromosome 28, which influences maturation time and fat storage (among other things); otherwise, the spring and fall chinooks in a given river are much the same and can interbreed, except that they are temporally isolated (Thompson et al., 2020). The spring-run fish are vitally important to organisms ranging from bears to water plants and were also important to Native Americans. They have been hit even harder than fall fish by dams, diversion, and the other usual problems (Thompson et al., 2020), and thus, many spring runs are now lost, and with them large parts of their ecosystems.

Salmon use chemoreception to smell and taste their way home. The young imprint on the chemical fingerprint of their place of rearing. They can detect the scent of their native river far out to sea. Once in the river, they smell their way up to their origin stream. Magnetic sense also helps them to navigate. They evidently recognize bodies and currents of water. They often do, however, pioneer new homes, especially if—as often happens—landslides or other events. e.g., dams and clearcutting of headwaters have made old homes unreachable or uninhabitable.

Salmon are famous for their ability to cross many obstacles—swimming up major waterfalls—but they cannot deal with a really high waterfall. Also, making a river harder to negotiate may cause them to starve before they reach home. Each run has evolved to store exactly the amount of fat it needs to swim upriver to its birthplace (Quinn, 2017). Most species do not feed after they enter the rivers and metabolize everything but muscle and gonads. Modern dam, channelization, railroad and highway bridge construction that make rivers faster, rockier, and more barrier-laden, or have raised culverts, lead to the elimination of runs.

The quantity of fish was enormous in the old days. The Fraser River sockeye run was estimated to be around 160,000,000 fish in 1901, though in 1904 it was down to 6,500,000 due to cyclic variations and overfishing. The Skeena run was about 1.8 million fish a year in the early 1900s; it is down to 470,000 now, with all 13 of its runs declining, some by 90% (Ogden, 2019). Runs are subject to disastrous accidents that depress that year's run for a long time without affecting other years' runs. The total Fraser salmon run may have produced 3 million to 60 million kg of fish (figures from Michael Kew as summarized by Ignace et al., 2017: 515, and set lower than the 1901 estimate). The Columbia and Yukon systems had many more. Smaller but still enormous runs negotiated the other rivers. Even small creeks had their own small runs. The large rivers had runs of all six species. Smaller streams usually had fewer species. Steelhead were particularly adaptable (even tiny California creeks have steelhead runs). Tales of streams that appear "more fish than water" abound from the old days, and a photograph in Thomas Quinn's book (p. 434) shows that this is only a slight exaggeration for some streams even today.

This allowed heavy takes by fishers. Estimates of consumption by the Tlingit alone range from 6,000,000 to 48,000,000 pounds of fish (Arnold, 2008: 24), the latter being more in line with earlier population levels.

> Recent studies that analyze bottom sediment from coastal waters of British Columbia and southeast Alaska have shown that far from being consistently predictable, there has been enormous variation in the availability of salmon over the last two millennia .... *Year-to-year fluctuations in salmon runs typically show relatively little variation, but over 20 to 30-year periods the availability of salmon approximates a wavelike pattern during which salmon populations peaked, declined, remained in a trough for several years, came back, and then diminished again after crests that lasted five to ten years.* In the early 16th century there were big runs, followed by a steady fall to the middle of that century. In the early eighteenth century populations were low, but they recovered near the end of that century *so that when Europeans arrived there appeared to be a time of abundance for salmon, which the Europeans apparently assumed was the normal state.* There was another peak in the late 19th century that seems to coincide with the beginning of commercial salmon fishing. Starting in the 1950s until the present there has been a consistent low in salmon numbers, which corresponds with a decline in the salmon fishery (Pierotti, 2011: 65, emphasis added).

Pacific salmon are semelparous, which means they breed only once, and die after spawning. This is crucial, because their decaying bodies fertilize the water and the surrounding forest, which provides necessary nutrients for the newly hatched young. This sacrifice of life for the newborn is widely noticed and emotionally appreciated by Native peoples, who—in most areas—quickly learned to return all bones and innards to the water to sustain the runs. Bears, wolves, and birds take enormous quantities of salmon and eat them on the banks, often dragging them off into the surrounding forest to consume, thus fertilizing the forest as well as the water. This allows lush vegetation to thrive along the rivers. Closing the cycle, some of that vegetation has evolved to disperse by bears—typically, the bears eat the fruit, usually berries, and then excrete the seeds packaged in fertilizer. By managing berry and root resources, humans took this natural cycle to a new level of sophistication.

The odd-sounding scientific species-specific names of the salmon are derived from vernacular Siberian-Russian names for the species. Common names are more local, as an example, "Sockeye" (*Sau-ki* or "chief of fish" from coastal peoples, or *suk-kegh* "red fish" from Salish) is a straight borrowing from Indigenous languages. All the Native languages have different terms for each species. They may also have a general term for "salmon" (*me* in Kwakwala and *xāt* and cognates in Tlingit, for instance), but it is used only when speaking very generally of the total resource, not for fish of one species. Many Indigenous people find it ridiculous that Whites lump such dissimilar fish under one term. Outsiders who come to the Northwest soon acculturate, and refer to "pinks," "sockeyes," and so on, never using the word "salmon" except in broad economic-statistical contexts. As a result, not only Indigenous but also settler Northwest Coasters are apt to laugh at outsiders naïve enough to refer to a given fish as a "salmon." References to the "Salmon Wars," over whether or not the tribes could take fish in traditional areas, as guaranteed in treaties, use the term in its broad economic sense.

These species are ecologically differentiated. Chinooks used to become huge, especially in the Columbia River drainage, where early-running "springs" were known as "June hogs" from their enormous size. They run far up the rivers. Sockeye are smaller, spawn in small streams draining into lakes, and the young fish grow up in the relatively predator-free lakes before moving to sea. Some populations have evolved to spend all their lives in lakes, never growing large or going to sea; these are known as "kokanee salmon." Kokanees are usually found in lakes far from the ocean, e.g., in Idaho. Pinks run into the lower courses of rivers in enormous numbers. Coho, usually rather small, run very far up into small tributaries and streams as far south as Central California. This species requires particularly cold and clean water. Coho are rapidly dying out in areas affected by global warming and by clearcutting, which destroys cover and increases runoff. Chum salmon stay in wide lower river courses and estuaries. Steelhead run into even smaller and more marginal streams than coho, and were the widest of all in distribution, with runs all the way to southern California, where they would go up quite small streams draining the Coast range. Steelhead can handle rougher water and more variable streams than others, and thus can spawn upstream of the competition (Grabowski, 2015) and in smaller watercourses. Steelhead are the only iteroparous (multiple-breeding) "salmon." They do not necessarily die after spawning and can repeat the pattern of running to sea and back upriver to spawn, in subsequent years.

The earliest currently known use of salmon by humans in North America occurred in central Alaska some 11,500 years ago; the salmon were chum (Halffman et al., 2015). It is interesting that salmon had already reached that northern drainage by then; it had been totally frozen and unusable not many thousand years before. As Terry Glavin describes it:

There are several important differences between aboriginal salmon fisheries that persisted for so many centuries throughout the Northwest Coast and the industrial (white) fisheries that emerged in the late 1800s. *A key feature of aboriginal salmon fisheries was that fishing was spread out along the salmon's migratory routes*...within the coast's river systems where trap-and-weir complex were the central mode of harvesting...(Very) few salmon of any species were caught in the ocean...apart from the (Georgia) Straits Salish reef-net fisheries, *it appears unlikely that there were any fisheries directed at sockeye salmon, in salt water, anywhere on America's west coast*... Because of the vast cultural and spatial diversity of the Northwest Coast... (Indigenous) salmon-management regimes... contribute(d) to maintenance of the genetics and spatial diversity necessary for the survival of salmon runs...*productive trap-and-weir fisheries were directed at specific runs... managed and regulated according to customary laws that required adequate escapement for fisheries upriver, and for spawning returns*... (these) tended to be stock and run specific...within the. Sustainable limits of familiar salmon runs... *(and) were managed according to strict rules arising from a variety of myths within the aboriginal oral tradition, in a local, decentralized way...a simple management rule prevailed: You screw up, you starve to death* (Glavin, 2000: 93, parenthetical elements and emphasis added.)

Another major issue is that, in the 1980s, the Atlantic salmon (*Salmo salar*) became a ranched or farmed fish in the Northwest, with locally disastrous consequences; having escaped locally and competing with the native fish, but far more serious are the diseases and pests it brought with it, which have locally wiped out the pinks, and decimated other species, in the areas of farming, primarily in the Strait of Georgia between Vancouver Island and southern British Columbia. Farming has now been reduced, to save the native species.

One thing the carnivore world, from bears to humans to eagles to wolves to ravens, seems to agree on is the superior eating quality of salmonids. They are the height of gourmet dining more or less everywhere they occur, including areas where they have been recently introduced—notably the southern hemisphere, which has few indigenous salmonids but which has taken to farming them on a huge scale.

A connoisseurship of salmon exists on the Northwest Coast, among both Native people and settler societies. True salmon gourmets can not only tell the species, they can tell which river a salmon came from, and sometimes even claim to be able to tell which run in a particular river was the source of their dinner. One learns to tell the rich, subtle flavor of chinook from the deep, oily meatiness of sockeye, and the ethereal, evanescent fragrance of fresh-caught pink. Coho and farmed Atlantic salmon are more modest but still excellent food, though true Northwesterners tend to abstain religiously from farmed salmon. The Puyallup activist Billy Frank once told Pierotti that "hatchery reared fish don't taste right," and ENA has heard similar statements. Chum or dog salmon is less flavorful, as the names imply; it was saved for use as chum (bits of fish thrown out to lure other fish to be caught) or for feeding the dogs. Recent attempts to rehabilitate it as "keta salmon" have not met with much success. Still, it was highly valued by groups like the Nuu-chah-nulth because it was abundant and easy to dry. It is said to develop a fine flavor when properly dried, in the wonderful account of Tlingit foods, from salmon—always put first—to berries, with details on quality and preparation (Jacobs & Jacobs, 1982).

Salmon and herring roe are delicacies; salmon roe was often buried, producing "stink eggs" and "Indian cheese," fermented commodities reminiscent of the buried salmon of Scandinavia (*gravlax*). It can develop malignant bacteria that kill those who consume it, but this is very rare in carefully prepared eggs (Jacobs & Jacobs, 1982).

Related to the salmon, but restricted to fresh water, are whitefish (genera *Coregonus, Prosopium,* and *Stenodus*), with many species in each genus. These are lake fish, largely found high in the northern hemisphere where they are important sources of protein. Trout include not only the cutthroat and rainbow trout in *Oncorhynchus*, but the char group, *Salvelinus* (*S. confluentus,* bull trout; *S. malma*, Dolly Varden, locally very important; *S. namaycush*, Lake Trout, native to the northeast corner of our region and widely introduced recently); the Arctic Char (*S. alpinus*) in the very far north; and Brook Trout (*S. fontinalis*), widely introduced. The closely related Arctic Grayling (*Thymallus arcticus*) occurs in the far north. All of these species were locally important foods throughout the Arctic and Subarctic.

Other riverine resources included sturgeon, which, like the salmon, are anadromous. The white (*Acipenser transmontanus*, family Acipenseridae) and green (*A. medirostris*) occur. Both are large; the white can be mammoth, reaching a length of more than six meters and a mass exceeding 500 kg). Both have suffered from overfishing and dams even more than most river fish.

Among smelts (Osmeridae), the candlefish (*Thaleichthys pacificus*), a smelt so rich in oil that dried ones can literally be used as candles, was a staple food, specifically an oil source. It is known on the coast as "eulachon," pronounced and frequently spelled by whites as "ooligan," from Tsimshian *halimootxw*, "savior"— the same term now used for Jesus (Daly, 2005: 113, 193). Getting oil from it originally involved letting the fish decay in cold water till the water could be heated to cook the fat out. Properly aged ooligan grease is an acquired taste, somewhat reminiscent of the medicinal cod liver oil once given to children. Ooligans occur throughout the region, running up rivers, but were most important in northern British Columbia. In Washington State, it is an important food for marine mammals and birds in Gray's Harbor.

Several fully marine fish were at least as important as salmon for the more maritime groups. Herring (*Clupea harengus*) were especially abundant, gathering to spawn in bays. The roe was almost as valuable a resource as the herring themselves. Herring have been drastically overfished, locally impacted by pollution, and reduced to a tiny fraction of their former abundance. Since they were basic to the entire marine food chain, their demise has led to collapse all the way up, through salmon and seabirds to whales and sea lions. Halibut (*Hippoglossus stenolepis*) which could weigh from about five to 100 kg, was a staple food of ocean-fishing groups such as the Haida and Makah. Rockfish (genus *Sebastes* with many species, family Scorpaenidae), "cod" (a general term for fish, mostly not true cod of the family Gadidae), and other marine fish were important. Octopuses, which grow very large in the north Pacific, were fairly important. The Haida, living on islands with small and generally poor salmon streams, depended more on marine fish than on anadromous ones.

Other marine resources were extremely important. The Northwest Coast is a paradise for shellfishing. Rocky shores provide attachment and shelter for abalone, mussels, chitons, limpets and other gastropods. Chitons, now a little-noted resource, were very popular because of their excellent flavor (Croes, 2015; ENA can confirm both their popularity and their excellent flavor in Haida Gwaii). Sandy beaches and, above all, the vast mudflats of the bays and estuaries provide major habitat for clams (including geoducks, *Panopea generosa*) and cockles. Shifting bars in bays and estuaries were ideal for oysters, which require this environment. The cold, nutrient-rich waters moved by currents carry enormous amounts of nutrients to these relatively passive filter-feeders (see Ellis & Swan, 1981 for a thorough account of Nuu-chah-nulth shellfish use).

Shellfish do not offer many calories, with the possible exception of geoducks, which weigh from 0.5 to as much as 7 kg The name geoduck is derived from a Nisqually word, gʷídəq, either meaning "genitals" (referring to the shape of the fleshy part of the clam) or a phrase meaning "dig deep"; perhaps both, as a double entendre. Shellfish were so common on coastlines that they provided much nutrient-rich food. Vast shell middens dot the coastline, and shellfish are still an enormously important subsistence and commercial resource in the Pacific Northwest.

Seaweeds were also major resources, though less well known because they do not preserve archaeologically and have only occasionally been well documented as foods. (On these types of sea resources, see Jacobs & Jacobs, 1982; Lamb & Hanby, 2005 for fine photographs and general accounts).

Another important freshwater resource, especially in the Columbia River drainage, were, "eels"—actually lampreys, *Lampreta* or *Entosphenus*, formerly called *Petromyzon,* family Petromyzontidae. A particularly detailed Native account of catching and use is given by John Hudson (Jacobs, 1945: 24–25), with an excellent anthropological account provided by Miller (2012). Especially important is the Pacific lamprey *E. tridentata*. These too are anadromous. They have proved less popular with settler societies, possibly because of their strange appearance, but they are said to be superb eating. Many Native groups still relish them and want to manage them for recovery, though they are notorious parasites of salmon, to which they attach, feeding on flesh. They have suffered even more than salmon from dams and pollution, and are rare in much of their former range. Like all animals (including fish), they warranted respect. Miller quotes Patricia Phillips on Oregon usage: "Night eels were only supposed to be cut with a knife made from freshwater mussel shell, otherwise the eels would feel insulted and fishermen might not catch any more of them." Lampreys caught at night were supposed to be better and healthier than day-caught ones (Miller, 2014: 130).

Many freshwater fish species occur, with suckers (Catostomidae, genus *Catostomus*) being locally very important, especially well inland where anadromous fish are available only during major runs. Now avoided because of their numerous small bones, suckers were extremely abundant and extremely popular, supporting huge Indigenous fisheries (see Landeen & Pinkham, 1999 for the suckers of the Columbia drainage). Some suckers have huge spawning runs of their own, into shallows or spring areas, and these runs may be critically important to local groups. The Lost River Sucker of the Oregon-California border, now acutely endangered, was so important to the Modoc that it was of religious importance. It was confined to a single river; as its name implies, the Lost River ends in a closed lake basin, from which the fish used to run upriver in incredible numbers to spawn. The Achomawi similarly revere and enjoy the suckers of Big Lake in northeastern California, and they have been able to get some protection for this run (Floyd Buckskin, pers. comm. to ENA).

Other freshwater fish of lesser importance were minnows and pikeminnows (Cyprinidae), catfish (Ictaluridae), sculpins (Cottidae), sticklebacks (Gasterosteidae), sunfish (Centrarchidae), perch (Percidae), and a few others. Good accounts of all the above fishes are found especially in J. D. McPhail's great work *The Freshwater Fishes of British Columbia* (2007, see also Landeen & Pinkham, 1999). The enormous wealth and variety of fish in the rivers allowed people to live on fish even when salmon runs failed.

The riverine world is unimaginable now; we have some idea of how many salmon occurred, but only recently are archaeological studies revealing the enormous amount and importance of suckers, whitefish, lampreys, and other little-remarked resources. In the Columbia River drainage, salmon were far from the only important group of fishes (Butler & Martin, 2013), being the most important single resource, but, locally, the total biomass of other species was greater. In the Pend Oreille drainage, for instance, a wide range of local trout, whitefish and other species was important (Lyons, 2015).

On rivers, there were choke points where falls and rapids delayed salmon migration and concentrated the fish ascending the river; the most famous ones were The Dalles and Kettle Falls on the Columbia, but these were flooded by dam building for irrigation and for hydropower to support Washington's aluminum industry. Other, less dramatically productive, cascades in most large rivers may also have been impacted by dams. Only Hells Gate on the lower Fraser River seems to have gone undammed.

Marine Mammals

Cetaceans

It is difficult to discuss the population sizes of marine mammals prior to the mid-eighteenth century, when Europeans arrived and began keeping records of the number of individuals they were taking. What we do know is that almost all marine mammal species underwent precipitous declines during the nineteenth century, because of whaling and sealing by Europeans, although some species, like gray whales, fur seals, and sea otters were also being heavily exploited by Indigenous people as well, who were either trading with Europeans, or enslaved by them.

Some Northwest Coast peoples conducted extensive whaling, from large open dugout canoes, thousands of years before Europeans engaged in such activities (Colson, 1953; Coté, 2010; Reid, 2015; Sapir, 2004; Sepez, 2008). The importance of sea mammals has been underestimated, because heavy whaling, sealing, and sea otter hunting by Settler societies depleted these resources before anthropologists got to the area. Despite the absence of anthropologists, local Europeans were aware of this activity because The Makah and Nuu-chal-nulth were trading whale oil and fur seal skins in Victoria and Port Townsend practically as soon as these communities were established (Reid, 2015).

It is unclear which whales were taken by Indigenous Peoples, but the Pacific Coast has several species, only a few of which were the targets of Indigenous whalers. It is very unlikely that blue whales, *Balaenoptera musculus*, and fin whales, *B. physalis*, were ever taken, or even pursued. These species are so large and fast that even European whalers did not try to take them in the age of sail. It was too difficult and dangerous to take them using oar-driven whaleboats. It is also claimed they took sperm whales, *Physeter catodon*, but this deep-diving and aggressive species does not seem a likely prey for canoe-based hunting.

They focused on gray whales, *Eschrictius robustus*, which is relatively small, swims slowly, and migrates close to the coast. When gray whale numbers declined precipitously in the early twentieth century, the Makah and Nu-Chah-nulth voluntarily imposed their own moratorium of hunting this species (Coté, 2010), which European whalers did not. (They exterminated the Atlantic form of this species.) Gray whales have made a major comeback in the last 50 years, since the passage of the Marine Mammal Protection Act. Their populations have recovered to pre-exploitation levels, and many now frequent the sounds of western Vancouver Island, Hecate Strait, and Coastal Alaska into the Bering and Chukchi Seas. Unfortunately, global warming, and overfishing in their feeding grounds, now threatens them; they are in danger of losing all their recent gains.

Another species known to have been taken was the Humpback Whale, *Megaptera novaeangliae*, which was greatly depleted by European overexploitation. It has made a significant recovery as well and can be observed regularly from Monterey Bay to SE Alaska and up into the Bering and Chukchi Seas. Whaling could pay well under such circumstances, though it was still difficult enough to demand high levels of ceremony meant to show respect for the lives of the whalers and their prey (Sapir, 2004).

Numerous smaller, toothed cetaceans inhabit the seas of the Pacific Northwest; however, these are cetaceans with teeth and high intelligence. They are not easily taken by people operating from canoes. The largest species of this group is the killer whale *Orcinus orca*, the wolf of the sea, which lives in family groups that never separate. Their families are referred to as "pods"—humans seem incapable of describing any social organizations but those of humans as families, or communities (Pierotti, 2011). Unlike wolves, however, the offspring never leave their parents and siblings. Even grown males live with their mothers and sisters, aunts, and cousins, for their entire lives. *Orcinus* has the tightest, most cohesive, family groups on the planet, and they are the largest predators on the planet today, even great white sharks being on their menu. They also have strong cultural traditions, with families and local populations showing extreme specialization of diet and of the dialects they use in their communication. The fish-eating population of the Salish Sea and surrounding area is behaviorally and apparently somewhat genetically distinct from other orcas, and is acutely threatened by overfishing of its food source.

The Lummi Nation of NW Wasington is currently engaged in a campaign and legal battle for the rights of orcas. In addition, The Mowachaat-Muchalaht First Nation in British Columbia has succeeded in postponing the capture of the now-famous orca whale known as Tsuxit or Luna. The tribe says the whale is the spirit of their late tribal chief, who wished to reside with the orcas and passed away shortly before Tsuxit showed up in Nootka Sound 3 years ago (https://www.culturalsurvival.org/news/canada-first-nation-wins-first-round-orca). The Haida First Nation has numerous stories concerning the role of orcas in their creation stories, including the idea that orcas are closely related to wolves, and even to the Wascos (or *wasgo*), the giant wolves of the sea that catch huge whales and carry them in their tail curls—their vast tails curl up like sled dogs' tails.

Pinnipeds

One important point about understanding the difference between marine and terrestrial ecosystems is that on land both the largest species and the most common species are typically herbivores, whereas in marine systems the most abundant forms are herbivores, but the largest are always carnivores. This is because in marine systems plants are tiny, and underlie very long food chains, whereas on land plants are often as large or larger than the species that feed on them, and in the case of trees are the largest organisms in their ecological communities. In addition, large herbivorous ungulates often live in large groups, e.g., bison, elk, pronghorn.

In terms of hunting by Indigenous peoples, this produces patterns surprising to settler societies. Pinnipeds (seals, sea lions, and walrus), which actually are carnivores in their evolutionary history, are treated more like ungulates in terms of exploitation. Pinnipeds tend to live like, and hunt, schooling fish in large groups. In addition, they are amphibious, rather than being full aquatic, like cetaceans. This means that when they haul out on land in groups, they are vulnerable to terrestrial predators. They could avoid bears, wolves, and big cats by hauling out and breeding on offshore islands, but humans hunting from canoes could not be avoided. As a consequence, to the Indigenous peoples of the Pacific Northwest who depended on marine hunting, baleen whales, pinnipeds (especially sea lions and fur seals), and large fish became the ecological equivalents of bison, elk, antelope, and deer.

As a consequence, sealing by Indigenous peoples was extensive enough that it seemed to scholars of European ancestry that the seal resource base was reduced even before European contact (Braje & Rick, 2011). A range of opinions can be found with regard to negative impacts of Indigenous peoples when hunting pinnipeds. The species involved are primarily Otariid pinnipeds, i.e., eared seals: (1) Steller Sea Lion (*Eumetopias jubatus*), (2) Northern Fur Seal (*Callorhinus ursinus*), (3) California Sea Lion (*Zalophus californianus*), and (4) Guadalupe Fur Seal (*Arctocephalus townsendi*). In addition, there was some taking of Phocid pinnipeds (Harbor Seal *Phoca vitulina* and Northern Elephant Seal *Mirounga angustirostris*). Further north in the Bering Sea Pacific Walrus (*Odobenus rosmarus*) were also exploited. Pierotti participated in studies involving all these species except *Arctocephalus* and *Odobenus*. My study areas focused on Southeast Farallon, Año Nuevo, and Santa Barbara Islands, and the Monterey Bay area in general, which provided insight and experience concerning handling and hunting pinnipeds.

With the exception of *Phoca* these species are strongly sexually dimorphic, with males weighing from 300 (*Callorhinus*) to 2000 kg (*Mirounga*), which means they are as large, or larger than any ungulate species hunted by Indigenous Americans over the last 10,000 years. Such hunting can generate a lot of meat, however, there can be risk involved hunting males, who are most vulnerable, but also most dangerous, during reproduction, when they are defending territories for breeding (Otariids) or groups of females (*Mirounga*). Breeding seasons occur during spring and early summer except for *Mirounga*, which breeds in winter. As a result, winter seas must be dealt with to exploit elephant seals. It is unsurprising that elephant seals, especially males, are taken least often in Coastal California.

Another species taken rarely was *Eumetopias*, a species where males weigh up to a 1000 kg, making them one of the largest mammalian carnivores to ever exist. *Eumetopias* bulls have canine teeth the size of grizzly bear fangs. They also display a protective thick mane which gives them their name and can be aggressive toward humans. Pierotti observed a human dummy tossed in front of a bull *Eumetopias* being grabbed, thrown into the ocean, and torn apart.

Climate change and accompanying variation in environmental conditions over historical time should be considered. Crockford and Frederick (2011) describe hunting regimes associated with ice cover, which came down as far as the Aleutians in recent times, which would have prevented *Callorhinus* from breeding on the Pribilof Islands, their major breeding colonies today. These islands would have been ice-bound in spring and summer, during the Neoglacial period (4700–2500 ybp).

Crockford believes that climates change constantly and that current variations might be considered natural variation in global conditions (Pierotti, 2011: 33). She contends that colder climates caused the expansion of sea ice in the North Pacific and Bering Sea. This expansion would have led to changes in distribution and life history features of pagophilic pinnipeds and that more temperate pinniped species, like sea lions and fur seals, were effectively excluded from this environment for a considerable period of time, which in turn had a major impact on patterns of human exploitation during this period (Pierotti, 2011: 33).

Ecologists employed a more comprehensive approach to examining a 4500-year time series involving how otariid and sea otter were exploited by Indigenous Alaskans in the western Gulf of Alaska (Betts et al., 2011). They argue that a combination of climatic variation and human exploitation on both the local and metapopulation level explain observed fluctuations, which over 2000 years appear to have evolved into a sort of natural predator/prey equilibrium. This suggests that human predation could have had a significant impact upon otariid populations, especially in *Callorhinus*, where pelagic hunting of fur seals by the Makah was significant during the late eighteenth and early nineteenth centuries when they were hunting to trade pelts with Europeans in Victoria and Vancouver (Reid, 2015). Betts et al. (2011) argue that pinnipeds evolved significant behavioral responses which minimized negative impacts. Exploitation increased when fur seal populations increased under colder, more productive conditions, and declined during warm less productive conditions.

Betts et al. (2011) rely upon methods developed by Lyman (2011), and are critical of the Krechian perspective (see Chap. 16), arguing that it is hard to estimate historical population sizes based solely upon taphonomic remains (Pierotti, 2011: 34). Lyman contends that many archaeologists struggle to identify marine mammal remains to species level; also that few museums have collections of marine mammal skeletal material adequate for comparative purposes. As an example, the skeletal material of *Odobenus* that vary in antiquity shows up all over the North Pacific, even as far south as California, whereas walrus are absent from these areas today. It may help help to be skeptical of modern biogeography when identifying recent remains.

As an example, two species of fur seal (Northern: *Callorhinus*, and Guadalupe: *Arctocephalus*) have complex histories over the last few centuries. *Callorhinus* breed today primarily in subarctic waters and probably consisted of two or more distinct forms, possibly even distinct species, ranging from coastal Washington southward. *Arctocephalus*, as implied by its common name, today breeds exclusively south of the US/Mexico border. Historically, however, it ranged at least as far north as the Farallon Islands near San Francisco where it was extirpated by Euroamerican sealers in the nineteenth century (Busch, 1987; Pierotti, 2011: 34). Lyman (2011) reports earlier specimens from coastal Washington.

Some scholars have attempted to apply Optimal Foraging Theory (OFT) to the exploitation of pinnipeds, making the argument that, "The prey choice model predicts that as the availability of large-bodied taxa decline, predation of smaller bodied and lower-ranked taxa increases" (Veronica et al., 2011: 111 The prey choice theorem was developed by watching Great Tits choose mealworms off a conveyer belt. As a result, search and handling times were not assessed (Pierotti, 2011: 34). Such a simple prey choice model may not be applicable to whether human foragers preferentially select large Otariids as opposed to medium-sized phocids or 30 kg sea otters as food (Pierotti, 2011).

This point may seem minor; however, it is crucial to understanding the Krechian mindset. Individuals following such beliefs employ models from foraging theory to explain how Indigenous hunters are not typical predators and are thus, according to their thinking, not ecological (Pierotti, 2011: Chap. 8, 2011: 34; see also our Chap. 16). Of the Northwest Pacific marine mammals, sea otters are the least palatable, being bony and low-fat mustelids, whose pelts are valuable in trade. Sea lions and fur seals are more formidable prey, with teeth and jaws the size and strength of their ursid relatives, especially *Eumetopias*. As a result, many Indigenous hunters probably concentrated on the medium-sized, sexually monomorphic *Phoca*, which are less aggressive and dangerous than otariids, and more palatable than otters (Pierotti, 2011: 34).

Different insights arise from isotopic analysis, which can reveal whether food is primarily obtained from terrestrial or marine environments, because Carbon/

Nitrogen ratios are markedly different between these habitats (Pierotti, 2011: 34). Some isotopic studies reveal that pre-contact humans, especially along the Strait of Georgia, took primarily terrestrial ungulates (deer, moose), even though they lived in a coastal environment (Drucker, 1951). Data reveal that pinnipeds and otters (they include river otters as marine prey; river otters often hunt in shallow ocean waters) were important prey items only on the outer coast of Vancouver Island and that most of take was smaller species like harbor seals, northern fur seals, and otters (McKechnie & Wigan, 2011). One of their most compelling findings of this study is that as Europeans invaded their ranges, the First Nations peoples apparently abandoned taking marine mammals. This could be either because their own populations were decimated by disease or because of European exploitation patterns, which extirpated populations of fur seals and sea otters (Pierotti, 2011: 35).

A study of human exploitation of pinniped exploitation in the estuaries of southern Washington and northern Oregon (Moss & Losev, 2011), critiques the use of foraging theory models, pointing out that size alone may not be an important component of prey choice, especially when accessibility is factored in. This use of OFT is more sophisticated, because it incorporates search and handling times into prey selection (Pierotti, 2011: 35). They point out that smaller *Phoca*, being year-round residents that use regular haul-outs, and are slow on land, are the most consistently available prey (Moss & Losev, 2011). Along with *Enhydra*, also year-round residents, these two species make up the preponderance of the prey taken in these areas (Moss & Losev, 2011).

Whitaker and Hildebrandt (2011) argue that Indigenous hunters had little impact upon fur seal populations in Northern California. However, these populations were subsequently extirpated by Europeans (Busch, 1987), but Whitaker and Hildebrandt contend that this was because exploitation of fur seals was an example of "prestige economy." Male *Callorhinus* were targeted because taking males conferred higher status on hunters who took males; because their teeth could be used as amulets. In traditional conservation practices, subadult males are exploited because females are protected to maintain high population numbers.

Hildebrandt and Jones (2002) refer to an earlier study in which they argued that Indigenous Californians created a "tragedy of the commons" that led to over-exploitation of pinniped populations, even though there is little evidence that a collapse of these populations ever took place. Hildebrandt also referred to similar ambivalence in a volume of decidedly Krechian cast (Kay & Simmons, 2002) where he argued that even though Aboriginal hunters had eliminated mainland rookeries of pinnipeds, they may not have had much of a negative impact because of the ability of these populations to establish offshore rookeries rendered them harder to access (Pierotti, 2011: 35). This way of thinking ignores the likelihood of the original problem for mainland rookeries having been the presence of wolves and grizzly bears. These predators were themselves extirpated by European invaders in the 1600s. As a result, recent evidence of mainland breeding colonies may not be related to human exploitation (Pierotti, 2011: 35). This is typical of the idea that aboriginal human exploitation is to blame for any identified problem, while ignoring (1) the presence and possible impacts of nonhuman predators, and (2) that any evidence of "overexploitation" is played up, even when there is no evidence of a population decline of the exploited prey (White, 2000; Pierotti, 2011: 170, 2011: 35).

The history of fur seals around Central California, including Monterey Bay, was assessed, and these populations were probably established because the breeding colonies in the Bering Sea and surrounding waters may not have been available during the Neoglacial as discussed above (Gifford-Gonzalez, 2011; Crockford & Frederick, 2011). In consequence, the disappearance of these colonies was likely caused by climatic factors, not human exploitation. There is some evidence of aboriginal exploitation on Año Nuevo Island, which was connected to the mainland until at least the nineteenth century (Pierotti, 2011: 35). The global metapopulation of northern fur seals seems to have been quite stable over the last few thousand years. Only the advent of the European and Japanese sealing industries led to the complete extirpation of populations until the establishment of the North Pacific Fur Seal Commission (Pierotti, 2011: 35).

There is also an evaluation of the behavior and ecology of these mammals (Rick et al., 2011). This study reveals how archaeologists and marine mammal ecologists can work together to produce genuine insights into long-term processes. These authors show that pre-contact human hunting had some impact on pinniped populations. They make it clear, however, that pinniped populations on San Miguel Island remained large until recent commercial exploitation by European invaders, despite continued hunting pressure by Indigenous hunters over the last ten thousand years.

Another point is that the predominant species found in middens throughout much of the Channel Islands was *Arctocephalus*, which is the only species found year-round in these islands in contemporary times. Other supposedly vulnerable species, like northern fur seal, are found only rarely and may represent scavenging events. This is important, because as DeLong points out, *Mirounga* weanlings are left unprotected by adults for several months following the breeding season from February through May, yet they are not found in large numbers in the archaeological sites.

There is little overall evidence that Holocene exploitation had serious negative impacts upon marine mammal populations, except for some local impacts and possibly forcing rookeries onto smaller less accessible islands and rocks (Pierotti, 2011: 36). T issue of predation by grizzly bears goes unaddressed. The overall consensus is that a complex mix of human impacts and climate/environmental changes have shaped marine carnivore populations along the Pacific Coast of North America over the last several millennia (Pierotti, 2011: 36).

A somewhat related issue involves the tendency of harbor seals and male sea lions to swim up rivers to feed on various species of aquatic vertebrates, including salmon and lampreys. Seals and sea lions apparently used to come far inland along the rivers. Here they are alleged to have eaten a great deal of fish, although complaints of this nature are invariably exaggerated, especially by Europeans, who blame everyone but themselves for declines in fish abundance, even if such declines rarely occurred prior to their arrival. White settlers have tried to exterminate them from the rivers, but they are protected under the 1972 Marine Mammal Protection Act, which makes killing them problematic from a legal perspective. One case in the Rogue River involved a young male Steller sea lion that was alleged to be damaging salmon runs. When the animal was shot its gut contents proved to be nothing but lampreys, so it was actually benefitting salmon. If the salmon were returned to their original abundance this would be a non-issue. Restoring Native hunting practices would be a good way to handle the situation.

Sea Otter

The last species of marine mammal that was hunted by the Coastal First Nations was the sea otter, *Enhydra lutris*. The cultural significance of this species is hard to assess because it seems to have been important only after contact with Europeans, primarily as a valuable commodity traded with them (Reid, 2015). First Russians, followed shortly thereafter in the late eighteenth and early nineteenth century by Spanish, English, and Americans, placed great value on sea otter pelts because of the lush, dense fur (125,000 hairs/sq. cm), which were prized for their luxury and their ability to keep humans warm in cold climates. Jones et al. (2011) reviewed the prehistory of the southern sea otter, which was almost completely extirpated by Europeans starting with the Russian fur trade in the 1600s. They argue that even though hunting of sea otters, especially females, was intense, it did not depress populations to such a degree that subsequent commercial exploitation by Europeans was precluded, which seems to be cold comfort indeed from an ecological perspective.

Sea otters are primarily specialists in benthic marine invertebrates, including sea urchins, abalone, snails, clams, and mussels. They have been implicated in the "destruction" of fisheries Europeans like to harvest, like Red Abalone and Pismo Clams, although as always seems to be the case, what otters actually do is simply depress local populations to the degree where it becomes difficult to exploit them commercially. As an example, in the 1970s it was claimed that otters had destroyed the Red Abalone populations along the Monterey coastline. However, ecological study revealed that sea otters picked off the abalone that were not inhabiting rock crevices, their true preferred habitat. This meant that commercial abalone divers would have actually had to work to harvest abalone, rather than simply picking off the ones that were exposed, and helping destroy kelp bed ecosystems. Also, disease and parasites were impacting the Reds. Governor Ronald Reagan initially sympathized with abalone divers until conservationists managed to get signatures supporting sea otter restoration of prominent wealthy citizens who lived along the 17 Mile Drive in Pacific Grove, and golfed at Monterey's famous golf courses, at which point, Reagan melted like California snow.

References

Ackerman, L. A. (2005). Residential mobility among Indians of the Colville reservation. *Journal of Northwest Anthropology, 39*, 21–31.

Anastasio, A. (1972). The southern plateau: an ecological analysis of intergroup relations. *Northwest Anthropological Research Notes, 6*, 109–229.

Anderson, E. N. (1996). *Ecologies of the heart*. Oxford University Press.

Arnold, D. F. (2008). *The Fisherman's frontier: People and Salmon in Southeast Alaska*. University of Washington Press.

Betts, M. W., Maschner, H. D. C., & Lech, V. (2011). A 4500-year time series of Otariid abundance on Sanak Island, Western Gulf of Alaska. In T. Braje & T. Rick (Eds.), *Human impact on seals, sea lions and sea otters: Integrating archaeology and ecology in the Northeast Pacific*. University of California Press. https://doi.org/10.1525/california/9780520267268.001.0001

Beynon, W. (2000). *Potlatch at Gitsegukla: William Beynon's 1945 field notebooks*. University of British Columbia Press.

Bogoras, W. (1929). Elements of culture of the circumpolar zone. *American Anthropologist, 31*, 597–601.

Braje, T. J., & Rick, T. C. (Eds.). (2011). *Human impacts on seals, sea lions, and sea otters: Integrating archaeology and ecology in the Northeast Pacific*. University of California Press.

Busch, B. C. (1987). *The war against the seals: A history of the north American seal fishery*. McGill/Queens' University Press.

Butler, V. L., & Martin, M. A. (2013). Aboriginal fisheries of the lower Columbia River. In R. T. Boyd, K. M. Ames, & T. A. Johnson (Eds.), *Chinookan peoples of the lower Columbia* (pp. 80–105). University of Washington Press.

Carothers, C. (2012). Enduring ties: Salmon and the Alutiiq/Sugpiaq peoples of the Kodiak archipelago, Alaska. In B. J. Colombi & J. F. Brooks (Eds.), *Keystone nations: Indigenous peoples and Salmon across the North Pacific* (pp. 133–160). School of American Research Press.

Colombi, B. J., & Brooks, J. F. (Eds.). (2012). *Keystone nations: Indigenous peoples and Salmon across the North Pacific*. School of American Research Press.

Colson, E. (1953). *The Makah Indians: A study of an Indian tribe in modern American society*. University of Minnesota Press.

Coté, C. (2010). *Spirits of our whaling ancestors: Revitalizing Makah and Nuu-chah-nulth traditions*. University of Washington Press.

Crockford, S. J., & Frederick, G. (2011). Neoglacial Sea ice and life history flexibility in ringed and fur seals. In T. Braje & T. Rick (Eds.), *Human impact on seals, sea lions and sea otters: Integrating archaeology and ecology in the Northeast Pacific*. University of California Press. https://doi.org/10.1525/california/9780520267268.003.0004

Croes, D. R. (2015). The undervalued black Katy chitons (*Katharina tunicata*) as a shellfish resource on the northwest coast of North America. *Journal of Northwest Anthropology, 49*, 13–26.

Daly, R. (2005). *Our box was full: An ethnography for the Delgamuukw plaintiffs*. University of British Columbia Press.

De Waal, F. (2009). *The age of empathy: Nature's lessons for a kinder society*. Harmony Books.

Drucker, P. (1951). *The northern and central Nootkan tribes*. Smithsonian Institution, Bureau of American Ethnology, Bulletin 144.

Ellingson, T. (2001). *The myth of the Noble savage*. University of California Press.

Ellis, D. W., & Swan, L. (1981). *Teachings of the tides: Uses of marine invertebrates by the manhusat peoples*. Theytus Books.

Fitzhugh, W. W., & Crowell, A. (1988). *Crossroads of continents: Cultures of Siberia and Alaska*. Smithsonian Institution Press.

Gifford-Gonzalez, D. (2011). Holocene Monterey Bay fur seals. In T. Braje & T. Rick (Eds.), *Human impact on seals, sea lions and sea otters: Integrating archaeology and ecology in the northeast Pacific*. University of California Press. https://doi.org/10.1525/california/ 9780520267268.003.0010

Glavin, T. (2000). *The last Great Sea: A voyage through the human and natural history of the North Pacific Ocean*. Douglas & McIntyre.

Grabowski, S. (2015). Structure of a resource: Biology and ecology of Pacific Salmon in the Columbia River basin. In P. Yu (Ed.), *Rivers, fish, and the people: Tradition, science, and historical ecology of fisheries in the American west* (pp. 12–41). University of Utah Press.

Halffman, C., Potter, B. A., McKinney, H. J., Finney, B. P., Rodrigues, A. T., Yang, D., & Kemp, B. M. (2015). Early human use of anadromous Salmon in North America at 11,500 years ago. *Proceedings of the National Academy of Sciences, 112*, 12344–12348.

Halpin, M. (1978). William Beynon, ethnographer, Tsimshian, 1888-1958. In M. Liberty (Ed.), *American Indian intellectuals* (pp. 140–156). West Publishing Co.

Harkin, M. E. (2007). Swallowing wealth: Northwest coast beliefs and ecological practices. In M. E. Harkin & D. R. Lewis (Eds.), *Native Americans and the environment: Perspectives on the ecological Indian* (pp. 211–232). University of Nebraska Press.

Hart, J. L. (1973). *Pacific fishes of Canada*. Fisheries Research Board of Canada.

Hildebrandt, W. R., & Jones, T. L. (2002). Depletion of prehistoric pinniped populations along the California and Oregon coasts: Were humans the cause? In C. Kay & R. T. Simmons (Eds.), *Wilderness and political ecology: Aboriginal influences and the original state of nature* (pp. 72–110). University of Utah Press.

Howe, N. (2019). *Retelling trickster in Naapi's Language*. University Press of Colorado.

Hunn, E., & Selam, J. (1990). *Nch'i-Wana, the big river*. University of Washington Press.

Ignace, M., Ignace, R., & Chief. (2017). *Secwépemc people, land and Laws*. McGill-Queen's University Press.

Jacobs, M. (1945). *Kalapuya texts* (p. 11). University of Washington. Publications in Anthropology.

Jacobs, M., Jr., & Jacobs, M., Sr. (1982). Southeast Alaska native foods. In A. Hope III (Ed.), *Raven's bones* (pp. 112–130). Sitka Community Association.

Jones, T. L., et al. (2011). Toward a prehistory of the Southern Sea otter (*Enhydra lutris nereis*). In T. Braje & T. Rick (Eds.), *Human impact on seals, sea lions and sea otters: Integrating archaeology and ecology in the Northeast Pacific*. University of California Press. https://doi.org/10.1525/j.ctt1pntkp.14

Kay, C. E., & Simmons, R. T. (Eds.). (2002). *Wilderness and political ecology: Aboriginal influences and the original state of nature*. University of Utah Press.

Krech, S. (1999). *The ecological Indian: Myth and reality*. W. W. Norton.

Kurlansky, M. (2021). *Salmon: A fish, the earth, and the history of their common fate*. Oneworld Publications.

Lamb, A., & Hanby, B. P. (2005). *Marine life of the Pacific northwest*. Harbour Press.

Landeen, D., & Pinkham, A. (1999). *Salmon and his people: Fish and fishing in Nez Perce culture*. Confluence Press.

Ling, J., Earle, T., & Kristiansen, K. (2018). Maritime mode of production: Raiding and trading in seafaring chiefdoms. *Current Anthropology, 59*, 488–524.

Lyman, R. L. (2011). A history of paleoecological research on sea otters and pinnipeds of the eastern Pacific Rim. In T. Braje & T. Rick (Eds.), *Human impact on seals, sea lions and sea otters: Integrating archaeology and ecology in the Northeast Pacific*. University of California Press. https://doi.org/10.1525/california/9780520267268.003.0002

Lyons, K. (2015). Recognizing the archaeological signatures of resident fisheries. In P.-L. Yu (Ed.), *Rivers, fish, and the people: Tradition, science, and historical ecology of fisheries in the American west* (pp. 96–126). University of Utah Press.

Martin, P. S. (2005). *Twilight of the mammoths: Ice age extinctions and the rewilding of America*. University of California Press.

McKechnie, I., & Wigen, R. J. (2011). Toward a historical archaeology of pinniped and sea otter hunting traditions on the coast of southern British Columbia. In T. Braje & T. Rick (Eds.), *Human impact on seals, sea lions and sea otters: Integrating archaeology and ecology in the Northeast Pacific*. University of California Press. https://doi.org/10.1525/california/ 9780520267268.003.0007

References

McPhail, J. D. (2007). *The freshwater fishes of British Columbia*. University of Alberta Press.

Miller, J. (2012). Lamprey 'eels' in the greater northwest: A survey of tribal sources, experiences, and sciences. *Journal of Norhwest Anthropology, 46*, 65–84.

Miller, J. (2014). *Rescues, rants, and researches: A review of Jay Miller's writings on northwest Indian cultures*. Northwest Anthropology, Memoir 9.

Moss, M., & Losev, R. J. (2011). Native American use of seals, sea lions and sea otters in estuaries of northern Oregon and southern Washington. In T. Braje & T. Rick (Eds.), *Human impact on seals, sea lions and sea otters: Integrating archaeology and ecology in the Northeast Pacific*. University of California Press. https://doi.org/10.1525/california/9780520267268.003.0008

Ogden, L. E. (2019). Salmon-smeared notebooks reveal fisheries [sic] past bounty. *Science, 365*, 733.

Pierotti, R. (2011). *Indigenous knowledge, ecology, and evolutionary biology*. Routledge.

Quinn, T. (2017). *The behavior and ecology of Pacific Salmon and Trout* (2nd ed.). University of Washington Press in association with American Fisheries Society.

Reid, J. L. (2015). *The sea is my country: The maritime world of the Makah*. Yale University Press.

Rick, T. C., Braje, T., & DeLong, R. L. (2011). People, pinnipeds, and sea otters of the Northeast Pacific. In T. Braje & T. Rick (Eds.), *Human impact on seals, sea lions and sea otters: Integrating archaeology and ecology in the Northeast Pacific*. University of California Press. https://doi.org/10.1525/j.ctt1pntkp.4

Ridington, R. (1988). *Trail to heaven: Knowledge and narrative in a northern native community*. University of Iowa Press.

Ross, J. A. (2011). *The Spokan Indians*. Michael J. Ross.

Sapir, E. (2004). *The whaling Indians: West coast legends and stories, legendary hunters*. Canadian Museum of Civilization.

Schlesier, K. H. (1987). *The wolves of heaven: Cheyenne shamanism, ceremonies, and prehistoric origins*. University of Oklahoma Press.

Sepez, J. (2008). Historical ecology of Makah subsistence foraging patterns. *Journal of Ethnobiology, 28*, 110–133.

Swanton, J. R. (1909). *Tlingit myths and texts*. Bureau of American Ethnology, Bulletin 39.

Thompson, N. E., Anderson, E. C., Clemento, A. J., Campbell, M. A., Pearse, D. E., Hearsey, J. W., Kinziger, A. P., & Garza, J. G. (2020). A complex phenotype in Salmon controlled by a simple change in migratory timing. *Science, 370*, 609–613.

Turner, N. J. (2005). *The Earth's blanket*. Douglas & McIntyre.

Turner, N. J. (2014). *Ancient pathways, ancestral knowledge: Ethnobotany and ecological wisdom of indigenous peoples of northwestern North America* (p. 2). McGill-Queen's University Press.

Veronica, L., Betts, M. W., & Maschner, H. D. G. (2011). An analysis of seal, sea lion, and sea otter consumption patterns on Sanak Island, Alaska: An 1800-year record of aleut consumption behavior. In T. Braje & T. Rick (Eds.), *Human impact on seals, sea lions and sea otters: Integrating archaeology and ecology in the Northeast Pacific*. University of California Press. https://doi.org/10.1525/california/9780520267268.003.0007

Walkus, S. (1982). *Oowekeeno Oral traditions as told by the late Chief Simon Walkus Sr*. National Museums of Canada, National Museum of Man, Mercury Series 84.

Walsh, J., Pendray, J. E., Godwin, S. C., Artelle, K. A., Kindsvater, H. K., Field, R. D., Harding, J. N., Swain, N. R., & Reynolds, J. D. (2020). Relationships between Pacific Salmon and Aquatic and terrestrial ecosystems: Implications for ecosystem-based management. *Ecology, 101*(9), e3060. https://doi.org/10.1002/ecy3060

Whitaker, A. R., & Hildebrandt, W. (2011). Why were northern fur seals spared in northern California? A cultural and archaeological explanation. In T. Braje & T. Rick (Eds.), *Human impact on seals, sea lions and sea otters: Integrating archaeology and ecology in the Northeast Pacific*. University of California Press. https://doi.org/10.1525/california/9780520267268.003.0013

White, R. (2000, June 24). Review of *The Ecological Indian* by Shepard Krech. New Republic.

Willerslev, R. (2007). *Soul Hunters: Hunting, animism, and personhood among the Siberian Yukaghirs*. University of California Press.

Wydoski, R. S., & Whitney, R. R. (2003). *Inland fishes of Washington* (2nd ed.). University of Washington Press.

Chapter 3
Looking to the Land: Terrestrial Ecology

Mammals

Land mammals of many species were numerous and remain so to this day. Moose were very local in northern regions before recent times; and have expanded into coastal areas quite recently, because logging and other activities have opened up the conifer forests, allowing browse to flourish. Deer, caribou, elk, mountain goat, mountain sheep, and smaller animals such as marmots (locally "whistlers" or "groundhogs"), porcupines, martens, fishers, and squirrels occurred. The deep conifer forests near the coast were generally rather poor in game, but many other areas were rich: lush mountain meadows, lowland prairies, riparian strips, and wetlands. Interior groups hunted extensively, knowing where these rich habitat patches were located, and concentrating on them. These animals supplied not only meat, but also skins and wool for clothing and other material resources.

The bond between humans and wolves is always complex, but it is also very deep in many cultures (Fogg et al., 2015; Pierotti & Fogg, 2017). Wolves have always been the closest companions of humans, and they maintain this relationship to this day in the form of the descendants of wolves, what we call domestic dogs. The delicate, but profound essence of this relationship can be seen in the following story:

> Once a man caught a young wolf, and raised him as a dog. He took good care of him, and gave him the best of meat to eat. When he went out hunting, and saw sheep or caribou, he showed them to his wolf-dog, who chased them to the bottom of the hills, where he killed them one after another. The man followed him, and opened and skinned the animals as fast as possible. distance away, waiting to be fed. As soon as he cut up an animal, he gave some of the best meat and fat to the wolf... through the aid of his dog, the hunter always had an abundance of meat on hand (Teit, 1919: 249–250).

Unfortunately, the man's relatives are jealous and ask to borrow his hunting companion. He agrees and they treat the wolf in a disrespectful manner, so it leaves. The man seeks to recover his companion, who has returned to his pack, and is told that because of the disrespect shown by his relatives his companion will never return (Teit, 1919: 250).

Wolves are the greatest hunters of both Siberia and North America. For this reason, they were honored and respected. The way wolves lived in monogamous family groups also impressed Indigenous peoples. This combination of devotion and the ability to kill effectively was admired and humans wanted to emulate this, which is why they constructed ceremonies that were intended to generate similar qualities in the people themselves. The purpose of the Kluckwalle and related ceremonies is to create warriors worthy of being called "wolves."

In contrast, Europeans feared and hated wolves, so they were determined to exterminate them (Pierotti, 2011, Chap. 3; Pierotti & Fogg, 2017). This was frequently couched in their mythology as removing a threat to their own lives, but attacks by American wolves on humans are extremely rare. No Indigenous American culture describes wolves as threats to humans, even though they are honored for their ability to hunt and kill. The difference appears to arise because Indigenous peoples recognized themselves as predators, whereas Europeans have a long history of regarding themselves as prey, creating cultural traditions and stories that reinforce this image (Pierotti, 2011, Chap. 3). The attitude towards wolves is more likely a result of the Europeans' dependence on domestic ungulates as part of their tradition, and recognition that wolves could be predators upon these dependent, relative helpless animals, bred to function as prey for the humans themselves (Pierotti, 2011). This identification with livestock actually reveals a root cause of the difference in worldviews. North and west Europeans used to treat wolves the way First Nations did, until medieval times, when concentrated stock rearing became important.

A close relative of wolves, the coyote (*Canis latrans*) is another important figure in Indigenous spiritual and ecological traditions. This species is absent (or was until recently) from the coastal Northwest, where Raven takes its place as the combination of trickster and creator figure. In, contrast, in the inland areas and California, Coyote is a major cultural "hero." On the Plateau, where both were common, coyote is a more ambivalent figure than Wolf, who is viewed as being more benign and idealistic in its outlook. Wolf prefers an idealized world in which death is only temporary, childbirth is easy, and winter does not exist. In contrast, coyote thinks death should be permanent, childbirth should be difficult, and hardships and cold weather should be regular aspects of human experience (Lily Pete, in Smith, 1993: 3–4; Ramsey, 1977).

An example of this can be seen in this story told by Shoshone elder Corbin Harney:

> You see, the coyote and the wolf were talking long ago. Wolf was arguing that we should all look alike, the rocks should be the same, the sagebrush the same, the humans the same, and all the living things on this planet should be the same. We should think alike and act alike and so forth. But Coyote always said, "No, we should be all be different. We shouldn't look alike at all." And so today we look around us and nothing looks alike. Rocks are not alike. Humans are not alike. This is the root of why we don't believe in each other. It's just as Coyote said. There's no use believing in just one thing. Let's not believe it. Let's all

disagree, and everybody believes in different things. That's why I always say, it's easy to believe the bad things first, but the good thing is harder to believe and harder to come by. As Wolf said, "It's going to be really hard that way, because what you're saying is, let's not believe in each other."

So today, what Coyote said is what we've got. (Harney, 1995: 26).

The interesting aspect is that although children are taught to emulate wolf, through ceremonies like Kluckwalle, and view wolf is a much more sympathetic figure. Coyote actually presents the more realistic view of how the world truly functions, e.g., "if it weren't for coyote there would be too many people now" (Lily Pete, in Smith, 1993: 3). By functioning as such an entity, Coyote appears unsentimental, if conflicted in the way the world should function.

In many ways, this reflects the reality of the differences between wolves and coyotes as social and ecological entities. Wolves are very social and cooperative, and this ability to function within monogamous family groupings is a major reason that wolves easily integrate into human social groupings, which has reached its extreme form in some types of domestic dogs (Pierotti & Fogg, 2017). Coyotes are much less social, however, and largely forage alone on small prey, even though they do have monogamous pairing behavior. Coyote pups become independent when they are only a few months old, in contrast to wolves, who regularly remain as part of their family (pack) for several years, participating in cooperative hunting activities taking large prey, and helping to rear younger siblings. Tamed coyotes, even if friendly to humans, remain quite wild; they are more independent in their lifeways.

Two other large mammals associated with First Nations of the Pacific Northwest are bears, Grizzly (*Ursus horribilus*) and Black (*U. americanus*). North America's bears and humans have been intimately linked for over 20,000 years, coevolving together, although lacking the intimate link that wolves have with humans. For the Native Americans, who saw all animals as sacred and possessing individual spiritual powers, bears held a very special place in belief systems. Indigenous peoples saw bears as much more than simply powerful animals. They saw an animal with traits amazingly similar to themselves, including physical attributes such as the ability to walk upright for periods of time, human-like mannerisms, as well as the striking similarity of its skinned body to that of their own. In addition, existing on a diet of nuts, berries, roots, and meat, the bear was the only large omnivore other than human. Female bears were also recognized for their strong maternal devotion to their young (Marshall, 1995).

These traits, together with the animal's strength, made bears powerful totemic symbols, e.g., the act of hibernation gave rise to initiation ceremonies, and connected bears to the underground spirit realm. The digging of roots and following of the available seasonal food supply allowed bears to be considered "keepers of the plants," with vast knowledge concerning herbs and medicines (Pierotti, 2011: 59–61). Bear clans were very prominent and its members were greatly respected. Many rituals are well documented, revealing not only a deep connection to the animal but also a depth of understanding of its daily life. Bears are also of symbolic and spiritual importance to numerous peoples among Indigenous Americans (Rockwell, 1991). Several tribes, especially the Sioux of the Northern Plains and the tribes of the Pacific Northwest and Alaska, have had their ceremonies documented and preserved.

One of the great mysteries for scholars who study the stories and cultural traditions of Indigenous Americans is why are there so few stories about cougars (*Puma concolor,* a.k.a. mountain lion, puma, panther), and those few that exist seem to come from the Southwestern and Southeastern US. Cougars were found throughout the entire Pacific Northwest and California. For the last 10,000 years, cougars have been the only big cat in this range, outside of a few jaguars along the Mexican border. Indigenous Americans were very aware of the existence of cougars, but the only large strict carnivore in North America does not seem to have made a powerful impression on people, who have been careful to acknowledge wolves, coyotes, bears, badgers, foxes, otters, and even weasels in their stories and traditions. The irony is that cougars are by far the most dangerous predator on humans of any species in the USA. Over the last century, there have been more than 125 attacks and nearly 30 fatalities from cougar attacks. In contrast, if the frequency of wolf attacks on people is compared to that of other large carnivores, or wildlife in general, it is obvious that wolves are among the least dangerous species for their size and predatory potential, and that the risks of wolf attacks in Europe and North America are very low. Recent cases are rare, despite increasing numbers of wolves (Linnell et al., 2002: 5). There is little doubt that Indigenous Americans were aware of this difference as an overall pattern. This is similar to the situation in Africa, where lions (*Panthera leo*) are the species respected and honored by Indigenous cultures, whereas leopards (*P. pardus*), which kill orders of magnitude more humans, are treated as annoying pests (Marshall-Thomas, 1994). Predators who show no respect for humans appear to be shown little respect in return.

In contrast to predatory mammals, numerous other mammals were used as food sources including large, ungulates, rabbits and hares, and various rodents, e.g., beaver, marmots, and other ground squirrels. Among the ungulates, Bison (*Bison bison,* universally known as "buffalo" in colloquial English) are absent from the coastal habitats. They have occurred in the far eastern grassland areas of British Columbia, Washington, and Oregon, where they are found east of the Cascades (Sanderson et al., 2008). The next largest ungulate, Moose (*Alces alces*) are abundant in British Columbia and Alaska other than in the higher mountains, but are not found in Washington or Oregon, except in the northeastern areas of each state, and not at all in California (Jensen et al., 2018). The Roosevelt elk (*Cervus canadensis roosevelti),* also known as the Olympic elk and Roosevelt's wapiti, is the largest of the four surviving subspecies of elk (wapiti) in North America by body mass. This subspecies lives primarily on the western slopes of the Coastal and Cascade Ranges from northern California up to southern British Columbia. The smaller Mule Deer (*Odocoileus hemionus*) occurs throughout the area and most of western North America, with the smaller form, the Columbian Black-tailed Deer, being the form found along the coast from Big Sur in California to Prince Rupert. The Sitka Black-tailed deer occurs along the coast from there north into southeast Alaska.

Birds were locally important, especially ducks, geese, and grouse, which provide game large enough to be important aspects of the human diet. As usual, water was more productive than land for the coastal peoples. Tens of millions of ducks, geese, swans, cranes, alcids, and other waterfowl flocked to the coasts, lakes, and marshes. On land, several species of grouse, known in Northwest dialect as "chickens" or "partridges," were locally common, especially in forest openings and prairies.

In contrast to mammals, only a few birds, e.g., Ravens and Eagles, were culturally significant and clan moieties of many coastal First Nations. It could be argued that Raven is the most important cultural entity in the majority of cultures along the Pacific Northwest coast, as well as in eastern, coastal, Siberia (Goodchild, 1991). From a biological perspective, it seems like Ravens (*Corvus corax*) and their relatives in the Corvidae are the most intelligent group among birds (Heinrich, 1989). The subtitle of Heinrich's book *Mind of the Raven: Investigations and Adventures with Wolf-Birds* indicates something more important, i.e., that ravens are regularly associated with wolves and by extension, with humans (Pierotti, 2020). As we mentioned previously, Raven is regarded as the creator/trickster figure of the Northwest Coast, eastern Siberia, and the Arctic, and is linked to many important events, including the very creation of humans and of providing light to the world. The basis of this relationship may be found in the following account.

> The superb field biologist Bernd Heinrich studied Common Ravens, *Corvus corax*, feeding on carrion in the forests of New England (Heinrich, 1989). Heinrich was curious that in his study populations in Maine, ravens were shy near carcasses; in fact, they were so shy that he feared they 'might be almost paralytically afraid of dangerous ground predators' (Heinrich, 2000: 231). The idea that ravens feared mammalian predators was suggested by a prominent behavioral ecologist (Heinrich, 2000: 217). To test this idea, Heinrich traveled to Nova Scotia to observe a captive wolf pack and their interaction with ravens. He put down two piles of meat at the same time, knowing that based on their traditions, the wolves would feed as a group on only one of these piles. Local ravens thus had a choice, to feed with the wolves or to feed away from them. In every case the ravens chose not to avoid the predators, but to feed with them... Heinrich traveled to Yellowstone National Park after the reintroduction of wolves in 1994 and observed numerous carcasses, both attended by wolves and unattended. In all cases the carcasses attended by wolves also had ravens, magpies, and eagles feeding from the carcass. Not once did the ravens hesitate to go in and feed from a carcass or meat pile attended by wolves (Heinrich, 2000: 232–234). In fact, the ravens hung around the wolves even when there was no food present. This suggests that, rather than fearing predators, it is the absence of wolves that makes ravens shy and much more nervous around the body of an elk or deer. When wolves (or humans) are present, ravens and other birds are more confident and relaxed... The fear and shyness of ravens in Maine resulted not from possible threats from terrestrial predators, but from the absence of their fellow predators. For ravens there should not be carcasses around without their friends the wolves. In this context the Tsistsistas (Cheyenne and Blackfoot) tradition of 'calling' ravens, coyotes, and foxes to share in their kills... can be recognized as part of a tradition that may go back millions of years (Pierotti, 2011: 58).

Ravens are the ultimate omnivore, consuming virtually anything that could provide nutrition, including of course meat, fish, insects, molluscs, and fruit. Following this behavior, they are inveterate scavengers, which accounts for their attraction to wolves and humans, which are the two species that are the most frequent and consistent generators of carrion. Ravens are active participants in this process, being reported to guide Inuit hunters to caribou and musk ox (Heinrich, 2000), focusing on hunters carrying weapons, but ignoring others who are engaged in berry picking. Ravens have also been observed harassing sleeping bears to get them to hunt salmon, but also warning humans and wolves about bears that are approaching their dens or dwellings (Brad Josephs: https://www.youtube.com/watch?v=MEcrw7u_VZM&t=2421s: This video describes and shows ravens interacting with other species, including humans, wolves, and bears.)

In contrast, eagles on the Northwest Coast seem to be largely venerated for their appearance and power rather than for any social or ecological connections to humans. Despite their large size and impressive weaponry, ravens regularly bully and harass eagles, even stealing food from them. Bald Eagles, *Haliaeetus leucocephalus*, are common in the Pacific Northwest, where they function primarily as piscivorous scavengers around salmon streams. In the last century, as salmon runs have declined, Bald Eagles have been switching to feeding on seabirds and even newborn harbor seals as prey around the Olympic Peninsula and Gray's Harbor, Washington (Good et al., 2000). Bald Eagles are strongly associated with aquatic habitats, either rivers, lakes, or marine coasts, and are primarily piscivorous. They are less fierce predators than the Golden Eagle, *Aquila chrysaetos*, which is the sacred eagle of many tribes in the interior. In stories from the northwest, Bald Eagles are often portrayed as arrogant and foolish, so it is unclear why they are honored to the degree they seem to be.

Another important species ecologically is the Osprey, or fish hawk, *Pandion haliaetus*, which is found along the entire Pacific coast, from SE Alaska to Baja California. Ospreys are much more impressive fish hunters than are Bald Eagles, which is probably why they are often robbed of their catch by eagles. Osprey rarely takes salmon, at least in contemporary times, which may explain why they are ignored more than eagles. Ospreys are considered to be a type of eagle by some Native American tribes and are accorded the same respect as Bald and Golden eagles. Some groups describe ospreys as playing guardian roles in traditional legends; seeing one is sometimes considered to be a warning of danger. The Nez Perce considered Fish-Hawk a medicine bird; seeing an osprey in a dream or vision was a sign that a man had been granted spiritual power as a healer (http://www.native-languages.org/legends-osprey.htm).

Gulls of the genus *Larus* play a variety of roles in the folklore of different cultures. In some cases, seagulls are antagonists criticized for their noisy, aggressive, and greedy behavior. In others, they are noted for their endurance and perseverance. In some Northwest Coast tribes, Seagull was said to have powers over storms and weather. Gulls are also used as clan animals in some Native American cultures. Gulls are used as a clan crest in some Northwest Coast tribes (especially the Nuu-chah-nulth), and can sometimes be found carved on totem poles (http://www.native-languages.org/legends-seagull.htm).

In many Northwest stories, gulls are characterized as greedy and foolish foils for ravens. At least a dozen gull species migrate along the Pacific coast, but the only gull breeding along this coast north of the Columbia is the Glaucous-winged Gull (*L. glaucescens*), which is a specialist on the intertidal and mudflats, feeding largely on intertidal and shallow subtidal invertebrates. South of the Columbia, the only breeding species is the Western Gull, *L. occidentalis*, an offshore specialist that feeds predominantly on smaller schooling fishes and squid (Pierotti, 1981). These two species have a clinal hybrid zone that extends from the Oregon coast to the strait of Juan de Fuca, which seems to be a result of Bald Eagle predation on gulls nesting in exposed habitats, and has increased in strength during the twentieth century (Good et al., 2000).

Gulls are not a food source for humans; however, their eggs have been used as food in the past, and locally today. Tatoosh Island, which is part of the Makah Nation is a major breeding colony. In winter, several other species of gulls migrate south along the coast from the Arctic, but these are not producing eggs while they are present. Other seabirds that are more important as food sources, include Common Murres, *Uria aalge*, and Tufted Puffins, *Fratercula cirrhata*, but these large alcids produce only a single egg/breeding season. Cormorants (*Phalacrocorax* spp.), although relatively large, are not usually considered suitable food for humans, though there are records of Pacific Coast people eating them.

In winter there are several ducks, including the species of scoter, *Melanitta* spp., and Harlequin ducks *Histrionicus histrionicus*, are available along the coast. None of these species breed along the coast, but at streams, ponds, and lakes in the interior of Canada. A similar situation applies to White-Fronted Goose (*Anser albifrons*), Brant (*Branta benicla*), Snow (*Chen caerulescens*) and Ross's Geese (*C. rossii*), and Tundra (*Cygnus columbianus*), and Trumpeter Swans (*C. buccinator*), all of which winter along the coast but breed in Arctic tundra.

As for large terrestrial birds, Ruffed (*Bonasa umbellus*), Spruce (*Falcipennis canadensis*), and Blue Grouse (*Dendragapus obscurus*), can regularly be found in the forests along the coast. Three species of Ptarmigan (*Lagopus* spp.) can be found in high elevations along the NW coast and throughout Alaska. These are fine food species, but they are more available to cultures that live in the interior, and depend on terrestrial food sources when salmon are not running. They are not common.

As far as reptiles and amphibians are concerned, the only edible forms along the coast are frogs, which are considered sacred by many of the coastal people. Frogs symbolize wealth and abundance, when frogs are portrayed with tongues touching another individual this represents sharing either of knowledge or power. The Haida carved frogs on house posts to keep the posts from falling. (There are actually no frogs in Haida Gwaii, only toads, but the carvings do look like frogs rather than toads, raising interesting questions.) Frogs are able to inhabit both water and land and are associated with spring and renewal, signaling the end of winter and the beginning of hunting and fishing seasons.

Some animals were mythic: the double-headed serpent, the thunderbird, and others. The Thunderbird is clearly not an actual being, but the structure of the legend has some intriguing aspects, in particular its use of lightning to kill a whale. Lightning and thunder are actual meteorological phenomena, whose existence is undoubted, but how they work is not well understood, even by western science. Occasional thunderstorms that would be more likely during unusual oceanic conditions, e.g., El Nino events when oceanic waters are warmer than usual, might have resulted in whales being killed or stunned by lightning strikes. Given the way Indigenous cultures deal with unusual events, an explanation would be provided:

...in the realm of metaphysics, Indigenous people recognize the significance of unusual or unique events, which the European tradition would characterize as anecdotes rather than data... In the realm of "scientific" knowledge, this means that the European tradition uses the metaphor of a statistically "average" event to carry more weight than the single "unusual" event ... Unusual events are often attributed to the activities of "spirits" in Indigenous traditions, although this English term is inadequate to translate the actual meaning of the concept... to Indigenous knowledge traditions, such "spiritual" causes or happenings are not seen as "supernatural." This implies that they are not outside the realm of what is real, but part of the natural order of things and thus readily subject to interpretation and understanding (Pierotti, 2011: 8).

Thus, an unusual event, e.g., a thunderstorm at sea that led to the death of a whale, required an explanation from the world that was understood. Loud sounds and deadly strikes from the sky placed the events in the sky realm, i.e., the world of birds. A giant bird capable of launching a deadly serpent was as good an explanation as any that was available.

The most famous cryptid of the Pacific Northwest is the sasquatch—the word derives from Coast Salish languages. This entity has been examined by Wayne Suttles (1987: 73–99). Such beings may not exist. Wild "men" are regularly represented on totem poles and in masks of important ceremonies, including the Kluckwalle. The best book on the topic is *Raincoast Sasquatch* (Alley, 2003), which discusses a number of cases, including new cases previously unreported, and presents a solid ecological basis for the possible existence of such organisms. The sasquatch are local representatives of a worldwide archetype, represented also by Bigfoot, the Yeti, the Chinese *yeren* ("wild man"), the Maya *sinsim*, the Russian *almasti*, the old European 'savage' or "wodewose," and many others (Bartra, 1994). Such hairy, nonverbal, powerful, and solitary beings, rather than real tribal peoples, inspired the stereotypes of "savages" in the writings of Thomas Hobbes, John Locke, J.-J. Rousseau, and similar authors (Bartra, 1994).

Herbaceous Plants

Plant foods are concentrated in meadows, glades, and wetlands, often within forests. Many species of berries, roots, tubers, stems, shoots, leaves, and even flowers. Many trees, such as cedar, had edible inner cambium (bark), gathered when soft in spring and eaten either ground to meal, or in long strands like noodles (Turner, 2005). Some fungi and lichens were edible; one species commonly eaten throughout the interior was black tree moss (*Bryoria*), baked a long time in an earth oven to break down its carbohydrates and render it soft and edible. Found patchily at high altitudes was whitebark pine (*Pinus albicaulis*) with edible nuts, a favorite food of local grizzlies. In parts of the region, these plant and fungal foods did not supply many calories compared to fish or mammals, but were often essential and always valuable for nutrients. On the Plateau and in the southern parts of our region, plant foods could dominate subsistence.

At the far north end of the region, the Tlingit used many plant foods, some listed by Jacobs and Jacobs (1982). These include fritillary, a potato-like root (*k'unts'*), a carrot, and wild greens, Hudson Bay tea was used as described, sprouts of salmonberry and wild celery, hemlock sap, wild onions, and of course countless berries (see below).

In the interior habitats of Washington and Oregon, root foods were extremely important. The Sahaptins of the central Columbia River are well documented and known to harvest at least 25 species (Hunn & Selam, 1990). Many are species of *Lomatium*, wild parsley. Other genera include *Perideridia*, wild carrot; *Fritillaria*, a lily with many small bulbs sometimes called "rice roots"; *Brodiaea*, wild hyacinth; *Calochortus*, Mariposa lily; and *Claytonia lanceolata*, a potato-like root. All these genera are widespread in the northern interior, and are important, to varying degrees, over much of the Northwest. Higher elevations produce *Balsamorhiza*, balsamroot; *Erythronium*, avalanche lily or dogtooth violet, an extremely important bulb throughout the interior north; and several others.

Throughout almost all of the Northwest, a critical resource was camas, *Camassia quamash* and *C. leichlandii*. Early explorers reported yields up to a bushel a day. Hunn watched traditional harvesters digging up 3.5 kg per hour, producing about that same yield. He found that women harvested 1000–1300 kg per person per year for their families, representing about 30–50%of their food requirement. Nutrient assessments of starch, protein, and other nutrients showed these roots are quite high in calories and food value (Hunn & Selam, 1990: 170–177). Roots were steam-cooked in large earth ovens, using heated rocks. Camas could take 2–3 days to cook in this manner (John Hudson, quoted in Jacobs, 1945: 19).

South through Oregon into interior California, the climate becomes milder. Plant seeds and especially acorns became more abundant, being about as important as fish to northwestern California peoples, and steadily more important southward. Acorns were leached with water, usually after grinding to meal first. Oaks are common in Oregon and California, but very local in Washington and absent further north except for the southern tip of Vancouver Island. Seeds such as tarweed were important farther north, even to Alaska (e.g., John Hudson cited in Jacobs, 1945: 20).

Berries were perhaps the most important, and certainly the most culturally salient, plant foods in most of the area (Turner, 2005). Huckleberries and blueberries (*Vaccinium* spp.) are especially common and widespread, especially in higher elevations. These species tend to be fire-followers, which is important for the reasons discussed below. Highbush cranberries (*Viburnum*) are locally common. Blackberries and raspberries (*Rubus*) occur in many species. Particularly common and important in the coastal northwest is the salmonberry (*Rubus spectabilis*), a raspberry of indifferent flavor but extremely productive. It was often mixed with tallow. In spring, it also has edible shoots that are like bitter celery. Strawberries (*Fragaria*) abounded in drier wooded areas before logging and grazing eliminated them. Dogwood, wild rose, elderberry, salal berries, wild crabapples (*Malus oregona*), wild cherries (*Prunus* spp.), saskatoon berries (*Amelanchier alnifolia*), salal berries (*Gaultheria shallon*), and any other wild fruit not found to be deadly poisonous were all used. Dry, sour ones were eaten with salmon oil or animal fat to bring out sweetness and flavor. Many were made into cakes and dried, which brings out their sweetness.

A unique resource was the soapberry (*Shepherdia canadensis*). Found in the dry interior, it produces a small cranberry-like berry that contains a bitter saponin. Beaten up in water, it produces a thick froth, rather similar to cranberry-flavored whipped cream. It was widely traded for this luxurious use. Special carved and ornamented spoons were made to whip and serve it.

Humans compete with bears, and many species of birds, for berries. Many tales recount the problems thus occasioned. A hungry grizzly can supposedly eat 100,000 berries a day (information from rangers at Denali National Park, talking while ENA watched a grizzly pick up whole huckleberry bushes, one after another, and pass them through her teeth, pulling off berries, leaves and twigs and swallowing all together). Berries are a plant's way of getting birds and mammals to disperse (and fertilize) their seeds, and bears assume considerable responsibility for distributing berries around the landscape.

An astonishingly large number of sweets and sweeteners afforded themselves (Turner, 2020). In addition to berries, these include a complex sugar secreted by older Douglas firs. More common are sugars developed by baking roots in an earth oven. Long-chain sugars (such as inulin) and other carbohydrates are broken by slow, lengthy wet-cooking into simple, digestible sugars (see Deur & Turner, 2005). Minor resources included nectar in flowers, sap of several trees, and tender shoots of fireweed and other plants. Sap of bigleaf maple (*Acer macrophylla*) supported a maple syrup and sugar industry in the late nineteenth century, before eastern maple syrup became widely available.

Mushrooms were apparently not much valued in precolonial times, though many are now considered extremely valuable. The Northwest is one of the mushroom centers of the world, producing enormous quantities of such valuable species as chanterelles, morels, and matsutake. Far more important than their food value is their value as hidden nutrient-cyclers. The sheer volume of mycelia under a typical Northwest coast forest reaches astronomical figures. In northern hardwood forests, active microbial biomass (fungi plus bacteria) ranged from 38 to 103 mg/g of organic matter; active microbial biomass per unit area ranged from 40 to 800 mg/m^2 forest (Taylor et al., 1999). They form symbiotic networks with trees, trading nutrients and even carrying messages. Healthy forests depend on these mycorrhizal nets. A recent problem has been overlogging to the point of damaging or destroying critical mycelial and bacterial soil systems, leading to the failure of reforestation.

Plants also supplied fiber for basketry and cordage, bark for clothing (yellow cedar bark shreds into a comfortable water repellent fabric), and medicines, notably devil's club, *Oplopanax horrida*. Wild tobacco (*Nicotiana quadrivalvis* or locally *N. attenuata*) was widely smoked.

Critical in all the above is the extreme contrast between fantastically rich waters—shoreline and sea—and relatively poor land resources. On land, there were contrasts between rich but very local berry and root patches—mostly found in clearings, prairies, and plains—and the coniferous forests, which were very poor in foods that humans could exploit. Not only are they minimally productive of edible material; most of what they do produce is tens or hundreds of feet up, out of normal human reach.

Trees

The most common tree is Douglas fir (*Pseudotsuga menziesii*). More closely related to pines than to true firs, this species is the "fir" of west Canadian English, true firs being called "balsams." Its superior wood now makes it a staple of world trade, but it was of less use aboriginally, because the hardwood is difficult to work, but makes excellent firewood. A useful statistic is that a mature Douglas fir may have 70,000,000 needles. (For this, and for information about forest management conflicts, see Satterfield (2003). The figure took on some meaning when ENA lived in Seattle beneath an enormous Douglas fir, and had to sweep the patio regularly.) Douglas firs can reach over 300 feet in height, surpassed only by redwoods and one or two Australian eucalypts.

Since Douglas fir grows only in sunny conditions, and usually only in disturbed habitats, it is naturally a follower of fire, flood, and blowdown. It is thus the ideal tree for management for clearcut-and-replant practices, as a result, dismal cornfield-like stands of "Doug fir" cover much of the accessible Northwest today. Eventually, Douglas fir is replaced by shade-tolerant trees like red cedar and hemlock, but Douglas firs live so long that even very rare stand-replacement events prevent this succession from occurring in many areas. The cedar-hemlock-true fir forests of the coasts may take 500 or more years to establish. Now eliminated except in protected areas, they will not reappear soon, if ever.

Much more useful to Native people was the red cedar (*Thuja plicata*, technically a cypress, not a cedar). Its wood is softer and easy to work, but extremely durable, making it the preferred material for canoes, totem poles, and Northwestern Indigenous woodwork in general. One can carve, split, or bend it with convenience, and it is very tractable and cooperative—with few knots, splinters, checks, or splits. The tree grows to an enormous size. A single log can make a dugout big enough to carry a whole crew and several tons of freight. Dugouts are still made of cedar today, mostly for ceremonial voyages, but were still made for ordinary use well into the late twentieth century.

Cedars invaded the Northwest rather slowly after the glaciations, not becoming common in more northern areas till about 6000 years ago; cedar and spruce are still spreading northward. Cedars could be overharvested. On Anthony Island in Haida Gwaii, where the red cedar is at the very edge of its range, cedars declined dramatically after about 500 CE, when the island became the site of a large village with many canoes and totem poles (Lacourse et al., 2007). Cedar bark was so valuable to the Haida that it was called "every woman's elder sister" (Boelscher, 1988: 22). It was pounded and shredded for clothing and other textiles. The related yellow cedar replaces the red in cold and wet areas. Its wood is useful, and its shreddy, soft bark makes perfectly serviceable clothing.

Another useful tree was the red alder (*Alnus rubra*), much smaller but with hardwood that does not split easily and thus is ideal for tools such as mauls. The Pacific yew (*Taxus brevifolia*) provided the best bows (as did the European yew in old England). Many small trees with distinctive characteristics exist, such as the Oregon crabapple (*Malus fusca*), a true apple with extremely hard wood and small fruits that are sour but good (see long, detailed study by Reynolds & Dupres, 2018). Much commoner and more visible are the true firs, *Abies* species. Many species exist, some of which grow to 250'. Despite priority over the name, they have lost it to the Douglas fir in the Northwest; "fir"—without a qualifying adjective—always means Douglas fir. In western Canada, true firs are called "balsams." In the northwestern USA, they are "true firs."

Old-growth coastal forests in the Northwest are incredible sights Vaillant (2005) gives a stunning and romantic description, and a harrowing story. Many species of evergreens grow to 200' or more, some over 300'. The redwoods (*Sequoia sempervirens*) of northwestern California can reach almost 100 m; by far the tallest trees in the world. Douglas firs, and possibly other Northwest Coast trees (grand fir, sugar pine, and red cedar), also surpassed 80 m in the past. The sheer amount of wood is incredible. Wildlife is thin, and mostly far up in the canopy. Stands were often over 500 years old, since fire and other catastrophes are rare.

Soil fertility was maintained by nitrogen fixation by lichens and root-knot bacteria. Alders are symbiotic with *Frankia* bacteria, which fix nitrogen at an amazing level, turning alder leaf mold into a fertilizer. Alder is adapted to invade infertile locations, such as stony flood washes and recently burned (or, today, logged) forests; here it fixes N and prepares the land for the conifers that it protects with its shade. Eliminating this alder step in reforestation has been devastating to logged Northwest forests. Nitrogen-fixing lichens grow in the canopy, largely on older evergreen trees; they make a case for saving old-growth forests.

Carbon fixation was carried out actively by fungi as well as by plants. More exotically, bears catch some salmon, scavenge more, and the fish and bear-dung fertilize the land, sometimes as heavily as a modern farmer fertilizing a crop (Gende & Quinn, 2006). This kept alive many trees that needed much nitrogen and could not get it otherwise, such as Sitka spruce.

Interior forests are different: fires are much more frequent, and the dominant trees are the fire-following pines and Douglas firs. (On forest fire regimes, see Agee, 1993). These do not grow under their own shade and thus are slowly replaced by shade-tolerant hemlock, cedars, balsam firs, and larches if the fires do not return. Douglas firs can live hundreds of years. Replacement is just beginning in the 100-year-old forests around Seattle and much of the Olympic Peninsula. Soil has much to do with it. Replacement is far along on rich, deep soils, but barely starting on barren ridgetops. In the drier and more lightning-prone parts of the interior, fire intervals are much more common than this, so pine and Douglas fir remain dominant indefinitely. Douglas firs do not tolerate extreme cold, though they can endure a lot; they are replaced in high mountains and north of central British Columbia by spruces of various species. The interior north is dominated by white spruce, with black spruce on wet sites. Sitka spruce grows in areas where sea spray carries mineral nutrients it requires; it can only grow in the narrow belt where this occurs. Its long latitudinal range, from northern California to southern Alaska, contrasts strikingly with its exceedingly narrow confinement to a coastal strip. Other important conifers of the Northwest include two species of larch and various small junipers.

Middle mountain slopes and plateaus that frequently burned were occupied by lodgepole pine (*Pinus contorta*). Humanly usable resources are few in pine forests but include a wide range of fungi, as well as edible cambium and several herbs. Drier, sunnier interior habitats were also dominated by pines, especially ponderosa (*Pinus ponderosa*) in the hotter areas and lodgepole in the cooler ones. These are fire-following trees. Lodgepole pines, in particular, succumb to senility and insect predation if stand-replacement fires fail, and are now being rapidly reduced by bark beetles, especially *Dendroctonus* spp., which farms fungi in tunnels in the pines. Global warming lets more and more beetles survive, and fire suppression eliminates natural stand replacement, so bark-beetle-killed ghost forests are becoming more and more common. When they burn, their dry dead wood creates major firestorms.

The typical replacement situation throughout most of the Northwest involves alder or cottonwood coming in first. Alder species invade less fertile areas. Faster-growing but nitrogen-demanding cottonwood and willow take over in fertile, well-watered sites. Maples, wild cherries, and several other species of trees occur as minor components in relatively open forests. The bigleaf maple (*Acer macrophylla*) is a common understory tree in the more open conifer forests and can grow to enormous size there. Other maples are also common understory trees.

The extensive openings created by wetlands, prairies, high mountain meadows and tundra, coastal barrens, and the like are covered by a wide variety of grasses, composites, herbs, and bulbs. Most of these are useful to humans for food, medicine, basketry, cordage, bedding, and many other purposes. The variety of food plants is incredible, and the nutritional value of most of them is high (see Kuhnlein & Turner, 1991 for an extensive database).

The vast lava plateaus of interior Washington and Oregon were originally covered by grassland or sagebrush. These environments were rich in edible roots and bulbs, as well as other plant resources, plus deer and pronghorn. Root crops are low in calories and take a lot of work, therefore, fish remained the main resource in most of the region, but roots were (and locally still are) gathered in enormous quantities.

Vast areas of rugged, high mountains were almost worthless to early humans except for hunting marmots and mountain goats, and a few plant resources. Much of the forested area was thin on usable material, thus, fire to produce openings was often deliberately set.

At the other extreme of productivity were the mouths of the major rivers. In these high-energy zones, the nutrients of a large percentage of North American land were deposited. Tens of millions of salmon and countless lesser fish swarmed in the waters. The extensive mud banks were veritable factories of clams and oysters. Waterfowl in millions flocked there. The rich alluvial soil produced roots, berries, and herbs, as well as fast-growing trees. Bits and pieces of this riverine richness still exist along the lower Skeena, Fraser, Skagit, and Columbia, but so terribly hurt by mismanagement that we can form no real idea of what it was like 300 years ago.

References

Agee, J. K. (1993). *Fire ecology of Pacific northwest forests*. Island Press.
Alley, J. R. (2003). *Raincoast sasquatch: The bigfoot/sasquatch records of SE Alaska, coastal British Columbia, and Northwest Washington from Puget Sound to Yakutat*. Hancock House.
Bartra, R. (1994). *Wild men in the looking glass: The mythic origins of European otherness*. University of Michigan Press.
Boelscher, M. (1988). *The curtain within: Haida social and mythical discourse*. University of British Columbia Press.
Deur, D., & Turner, N. (Eds.). (2005). *Keeping it living: Traditions of plant use and cultivation on the northwest coast of North America*. University of Washington Press.
Fogg, B. R., Howe, N., & Pierotti, R. (2015). Relationships between indigenous American peoples and wolves 1: Wolves as teachers and guides. *Journal of Ethnobiology, 3*, 262–285.
Gende, S. M., & Quinn, T. P. (2006). The fish and the Forest. *Scientific American, 295*, 84–89.
Good, T. P., Ellis, J., Annett, C., & Pierotti, R. (2000). Bounded hybrid superiority: Effects of mate choice, habitat selection, and diet in an avian hybrid zone. *Evolution, 54*, 1774–1783.
Goodchild, P. (1991). *Raven tales: Traditional stories of native peoples*. Chicago Review Press.
Harney, C. (1995). *The way it is: One water . . . One air . . . One mother earth*. Blue Dolphin.
Heinrich, B. (1989). *Ravens in Winter*. Simon & Schuster.
Heinrich, B. (2000). *Mind of the raven: Investigations and adventures with wolf-birds*. HarperCollins.
Hunn, E., & Selam, J. (1990). *Nch'i-Wana, the big river*. University of Washington Press.
Jacobs, M. (1945). *Kalapuya texts* (p. 11). University of Washington. Publications in Anthropology.
Jacobs, M., Jr., & Jacobs, M., Sr. (1982). Southeast Alaska native foods. In A. Hope III (Ed.), *Raven's bones* (pp. 112–130). Sitka Community Association.
Jensen, W. F., Smith, J. R., Carstensen, M., Penner, C. E., Hosek, B. M., & Maskey, J. J. (2018). Expanding GIS analysis to monitor and assess north American moose distribution and density. *Alces: A Journal Devoted to the Biology and Management of Moose, 54*, 45–54.
Kuhnlein, H. V., & Turner, N. J. (1991). *Traditional plant foods of Canadian indigenous peoples: Nutrition, botany and use*. Gordon and Breach.
Lacourse, T., Mathewes, R. W., & Hebda, R. J. (2007). Paleoecological analyses of lake sediments reveal prehistoric human impact on forests at Anthony Island UNESCO world heritage site, Queen Charlotte Islands (Haida Gwaii), Canada. *Quaternary Research, 68*, 177–183.
Linnell, J. D. C., Andersen, R., Anderson, Z., Balciauskas, L., Blanco, J. C., Boitani, L., Brainerd, S., Breitenmoser, U., Kojola, I., Liberg, O., Loe, J., Okarma, H, Pedersen, H. C., Promberger, C., Sand, H., Solberg, E. J., Valdmann, H., & Wabakken, P. (2002). *The fear of wolves: A review of wolf attacks on humans*. Norsk Instituut for Naturforskning. Retrieved from https://mobil.wwf.de/fileadmin/fm-wwf/Publikationen-PDF/2002.Review.wolf.attacks.pdf
Marshall, J. (1995). *On behalf of the wolf and the first peoples*. University of New Mexico Press.
Marshall-Thomas, E. (1994). *The tribe of Tiger: Cats and their culture*. Simon and Schuster.
Pierotti, R. (1981). Male and female parental roles in the Western Gull under different environmental conditions. *Auk, 98*, 532–549.
Pierotti, R. (2011). *Indigenous knowledge, ecology, and evolutionary biology*. Routledge.
Pierotti, R. (2020). Learning about extraordinary beings: Native stories and real birds. *Ethnobiology Letters, 11*, 253–260.

References

Pierotti, R., & Fogg, B. R. (2017). *The first domestication: How wolves and humans coevolved.* Yale University Press.

Ramsey, J. (1977). *Coyote was going there: Indian literature of the Oregon country.* University of Washington Press.

Reynolds, N. D., & Dupres, C. (2018). The Pacific crabapple (*Malus fusca*) and Cowlitz cultural resurgence. *Journal of Northwest Anthropology, 52,* 36–62.

Rockwell, D. (1991). *Giving voice to bear: North American Indian rituals, myths, and images of the bear.* Roberts-Rinehart.

Sanderson, E. W., et al. (2008). The ecological future of the north American bison: Conceiving long-term, large-scale conservation of wildlife. *Conservation Biology, 22,* 252–266.

Satterfield, T. (2003). *Anatomy of a conflict: Identity, knowledge, and emotion in old-growth forests.* University of British Columbia.

Smith, A. M. (1993). *Shoshone tales.* University of Utah Press.

Suttles, W. (1987). *Coast Salish essays.* University of Washington Press.

Taylor, L. A., Arthur, M. A., & Yanai, R. D. (1999). Forest floor microbial biomass across a northern hardwood successional sequence. *Soil Biology and Biochemistry, 31,* 431–439.

Teit, J. (1919). Tahltan tales. *Journal of American Folk-Lore, 32,* 198–250.

Turner, N. J. (2005). *The Earth's blanket.* University of Washington Press.

Turner, N. J. (2020). 'That was our candy!' Sweet foods in indigenous peoples' traditional diets in northwestern North America. *Journal of Ethnobiology, 40,* 305–328.

Vaillant, J. (2005). *Golden spruce: A true story of myth, madness, and greed.* W. W. Norton & Company.

Chapter 4
Traditional Cultural Areas

Origins

Humans entered North America from Asia, probably more than 16,000 years ago. Current genetic evidence implies dispersal from a single Siberian population toward the Bering Land Bridge no earlier than about 30,000 years ago (and possibly after 22,000 years ago), then migration from Beringia to the Americas sometime before 16,500 years ago (Goebel et al., 2008; Raff, 2022). Earlier dates are suggested by equivocal. Interesting evidence from both North and South America, especially some human footprints in New Mexico dated at 21,000–22,000 years old (Raff, 2022).

The tool types of these early times are similar to recent finds in south-central Siberia at 24,000 and 17,000 years before present. Human skeletal finds from this last area have been genetically sequenced (Raghavan et al., 2015; Willerslev & Meltzer, 2021), and prove to be somewhat intermediate genetically between Europe and East Asia; showing many resemblances to earlier Chinese remains, and to European and west Asian populations. The split between east Siberians and Native Americans is about 20,000–23,000 years old; the Athabascans may have separated later, about 13,000 years ago (Goebel et al., 2008; Raghavan et al., 2015). There is no believable evidence of early contact from Europe to the Americas (despite some claims in the media), and indeed all Native Americans, living or long gone, are more closely related genetically to each other than to any Old World populations, and most closely related to East Asians among Old World groups. Inuit and their relatives, however, are somewhat more distant (Raff, 2022). There is also an ancient but minor difference between northern and southern North American populations, South America being close to the southern populations but with a strange and ancient connection with New Guinea and Australia, thanks to an even more ancient East Asian population (Raff, 2022). The Northwest Coast is largely northern by affiliation (Willerslev & Meltzer, 2021).

© The Author(s), under exclusive license to Springer Nature Switzerland AG 2022
E. N. Anderson, R. Pierotti, *Respect and Responsibility in Pacific Coast Indigenous Nations*, Studies in Human Ecology and Adaptation 13,
https://doi.org/10.1007/978-3-031-15586-4_4

There is no question that Native Americans entered the Americas via the Bering Strait within the last few tens of thousands of years. Native Americans often hold that "we have always been here," but this applies to the origins of their existing languages and societies, not to their ultimate genetic ancestors—as their origin stories show, since most relate migrations, and all tell of times before humans, when the nonhuman animals were the people (Pierotti, 2011, Chap. 5). Pierotti had an interesting experience in the early twenty-first century when he hosted Siberians and took Native American students from Haskell Indian Nations University to the Altai region of Siberia (Calhoon et al., 2003). During this exchange, many Native students recognized both physical and cultural similarities with Indigenous Siberians, and acknowledged that the Bering Strait idea had more substance than they had previously thought.

Archeological finds in the Northwest go back to about 12–16,000 years ago, but the area must have been settled before that. Humans were in South America by 15,000 years ago, and must have passed through the Northwest of North America to get there. An early site, On Your Knees Cave, on the Prince of Wales islands in Alaska, goes back to 10,300 years ago, has obsidian from Mt. Edziza, on the mainland some 200 km away (Moss, 2011; Raff, 2022; Turner, 2014: 1: 57; on Mt. Edziza, see Reimer, 2018). A wet site on Haida Gwaii has even preserved wooden tools that look like recent ones, but are over 10,000 years old (Moss, 2011: 90). Sea levels during the last glaciation were hundreds of feet lower than today, and many areas occupied during that time are now inaccessible to archeologists. Contemporary archeology has turned up a long record of gradual learning to use the land more and more efficiently. There are few, if any, sharp breaks in the record; apparently, people and societies have been developing rather smoothly (Moss, 2011), with no evidence of the vast migrations and sudden cultural intrusions that delight archeologists who work in Eurasia.

It is now widely agreed that the main early route into the Americas was along the Northwest Coast. The famous "Interior corridor" between the Cordilleran and eastern Canadian ice sheets, so beloved of "Clovis First" devotees (Raff, 2022) and other early scholars (Pierotti, 2011), probably did not open early enough to be a competitor, though evidently people came down it as soon as they could (Willerslev & Meltzer, 2021). The Pleistocene glacial maximum, which covered Canada with ice from Atlantic to Pacific and made travel impossible, may have reached maximum earlier than we used to think—more like 19,000 than 16,000 years ago (Clark et al., 2009). Thus, people could probably have come down the coast by 15,000 BC.

By 13,000 years ago, people were established in Haida Gwaii (the Queen Charlotte Islands) and elsewhere along the coast (https://www.sfu.ca/sfunews/stories/2019/09/evidence-shows-human-presence-haida-gwaii-2200-years-earlier-than-thought.html). There is complete cultural continuity in Haida Gwaii; apparently the ancestral Haida settled there and never moved. They were drying salmon in large villages 7000 years ago (Cannon & Yang, 2006), e.g., at Namu, where occupation goes back 11,000 years. Salmon and shellfish increase there from 6000 years ago (Moss, 2011; Turner, 2014: 1: 84). A full review of animals in archeology on the Northwest Coast is provided by Gregory Monks (2019), covering the entire prehistoric period thoroughly. An excellent and thorough review of the archeology of the Salish Sea islands (Hutchings & Williams, 2020) provides a good short introduction to coast prehistory; the islands currently represent a good microcosm of the whole central Northwest Coast.

The Archaic period lasted from before 10,000 BCE to about 4400, and was characterized by small, mobile groups wandering the landscape, hunting and fishing. After 4400, the Pacific period set in, lasting until contact with European settlers. Settled life rapidly advanced, with "storing large volumes of food; increased population density, sendentism, [increasing] household and community size; escalated warfare; development of canoe-based land-use patterns; more institutionalized social status differences," and more intensive management of resources (Sobel et al., 2013: 31). Pit houses appear in the interior by very early dates. They were common everywhere by 3000 years ago (Turner, 2014: 1: 85). By this time, high-elevation sites were also occupied, evidently for berrying and other resource extraction (Turner, 2014: 1: 97).

Development was fairly slow in the early millennia, though large pit houses appear quite early in the record. Storage and sophisticated management techniques allowed social and cultural complexity to take off through much of the region in the 2000–1000 BCE period (Monks, 2019). Very large, complex societies were established by 1000 BCE. In the Fraser River canyon, truly astonishing developments existed by 1800 years ago. At Bridge River, a tributary that joins the Fraser just north of Lillooet, by 1400–1300 BP, there was a huge village with dozens of house pits, some large enough to imply houses holding some 50 people (Prentiss et al., 2008, 2012, 2018). The modern Stl'átl'imx (Lillooet) maintain a complex, sophisticated environmental use and management system rooted in the past, and excellently surveyed in *A Complex Culture of the British Columbia Plateau* (Hayden, 1992).

Inequality rapidly arose. Chiefs and commoners—more accurately, high-ranked and lower-ranked lineages—appear very different indeed in the archeological record. This village, and other large villages on the middle Fraser, were abandoned around 1000 years ago. This appears to be the combined result of overhunting, overfishing, also presumably impacted by the Medieval Warm Period's negative effects on fish (the water may have grown too warm for healthy runs). People reverted to more egalitarian and nomadic lifestyles.

Prentiss and her collaborators review several possible reasons for the rise of inequality in a village like this. Social learning and imitation were involved. Self-aggrandizing individuals probably arose; "competitive signaling" by feasts and donations presumably occurred, as it did later. Some coercion was probably involved. There must have been rebels, surely. Inequality could emerge partly through competition between individuals—presumably their families would slowly evolve into ranked lineages. Prentiss tends to favor a more moderate situation, in which "house size evolves to solve problems to do with labor management, kin relations, and defense" (Prentiss et al., 2012: 544). One assumes some ideological and ceremonial reflection of this, but no obvious evidence appeared. Fishing, and control of good fishing spots, was apparently the basis of it all.

Further research showed diminishing returns to hunting and fishing as the site grew more populous. More and more mouths had to be fed from less and less available food. This would have meant competition, with the better providers winning out. Powerful lineages could have arisen as the more successful families consolidated their hold. The Prentiss group (2018) opts for Malthusian pressure as the driver of inequality, social complexity, and bigger, better-built villages. It would not be the first or only time in history when food problems drove hierarchy and dominance, but other causes might well have operated too: increased trade and the need to protect it, increased wealth, increased raiding, and any other increased stresses due to expanding economies. When the fishery declined sharply around 1000–900 CE, probably because of natural cycles in abundance (Pierotti, 2011: 65) the population dispersed and the village was abandoned. It was reoccupied just before settler contact, but abandoned for good in the 1850s, as gold rushes drove Indigenous people away from the Fraser.

Pit houses got larger as populations increased. On the coast pit houses began to give way to plank houses by 4000–5000 years ago; square house-floors a thousand years older suggest the existence of even earlier ones (Ames & Sobel, 2013: 131; Moss, 2011). One plank house survives, as an archeological ruin, from 800 BCE (Sobel et al., 2013: 31). Many were found at the Ozette site, where a landslide killed more than 100 people on the NW coast of the Olympic Peninsula in the sixteenth century. Archeology at Ozette has revealed that people lived there in this spot for around 2000 years before the landslide. (Good information on this site can be found at https://makahmuseum.com/about/ozette-archaeological-site) A single house could contain 35,000 board feet of planks, and a village could have a million board feet tied up in housing (Ames & Sobel, 2013: 138), to say nothing of the amount used for containers, tools, and fuel. Houses 400 and even 1200 feet long are claimed in early explorers' accounts, especially for the Lower Columbia and Lower Fraser areas, where resources were rich. These would have been "townhouse" units, housing whole villages.

The whole Northwest Coast shows remarkable continuity, with in situ development of specific cultures. The best-known areas are, unsurprisingly, those nearest the archeology faculties of the University of British Columbia and the University of Washington. In these areas, fairly complex cultures are already known by 3000 years ago, with modern cultures established well over 1000 years ago. Something of a cultural peak was reached in the Marpole phase along the lower Fraser River and in neighboring areas, 2000–3000 years ago (for Fraser culture history see Carlson et al., 2001). This may correlate with drier conditions that changed resource availability (Lepofsky et al., 2005). Major stone fortifications in this area show a great deal of warfare was present in the late prehistoric period (Schaeper, 2006). The global warm period (altithermal) around 9000 to 4000 years ago shifted on the Coast, as elsewhere, to a cold period about 3800–2600, which has been documented by the leading archeological expert Madonna Moss (2011: 138). Worldwide, there was a sharp temperature drop around 4200–4000 years ago.

The same correlation of dry periods with reduced salmon availability, leading to exploitation of a wider diversity of resources, and resultant rapid rise in cultural complexity has also been independently noted for the interior Plateau (Prentiss et al., 2005), where complexity of village life increased sharply around 2500 years ago when drier conditions set in. Climate reached almost modern levels about 1100 years ago at the start of the Medieval Warm Period, the relatively warm period from the 900 s to 1300 CE. Some increase took place thereafter, despite the Little Ice Age from 1300 to 1800. During the period when modern human beings were evolving over the last one hundred thousand years, there have been only two generally stable periods of climate (Pearce, 2007: 237). The first was when the ice sheets were largest and the world was coldest, and the second is the period in which we are living now. Ironically, the Western scientific tradition has treated this most recent period as if it were typical (Pierotti, 2011: 36–37).

A more recent site is Sunken Village, on Sauvie Island in the Columbia near Portland. Here, vast acorn processing and storage facilities were unearthed. Acorns and hazelnuts were staples. "The acorn-leaching technology and basketry of Sunken Village show marked similarities to those of much more ancient Jomon period acorn-processing sites throughout Japan" (Turner, 2014: 1: 101), with similar basket and bag techniques; this is independent invention rather than diffusion.

Another revealing, though tragic, find was of a young man who apparently fell and froze some 300 years ago at the edge of a glacier in British Columbia near the Alaska border. Recent melting revealed his body. He had eaten salmon, crab, and coastal vegetables, so he was evidently coming from the coast. Particularly interesting was a magnificent spruce-root hat he wore, a masterpiece of basket technology (Turner, 2014: 1: 105 shows a photograph). It appears that he not only has living relatives, but is probably an individual remembered in local oral tradition (Moss, 2011: 142–144). As an actual member of still-extant society, he was cremated according to Tlingit practice, and his ashes scattered near where he was found.

Languages

By the time the white voyagers and traders reached the area in the mid-eighteenth century, an incredible linguistic diversity existed in the northwest. This evidently dates back to the very early settlement. There was maximal opportunity for unrelated groups to establish themselves, and for related peoples to diverge to such a degree that their original relationships are now untraceable.

There were six very different language phyla along the coast: the Haida on their islands; the Na-Dene (Tlingit and Athapaskan) on the northern coasts and throughout the northern interior; the Tsimshian (including the Tsimshian, Southern Tsimshian, Nisga'a, and Gitksan languages) in the lower Skeena and Nass drainages and neighboring coast; the Wakashan (including Haisla, Heiltsuk, Kwakwala and Nuu-chah-nulth) along most of the rest of the coast; the Salish along the "Salish Sea" coasts and much of the interior; and the tiny Chemakuan language family on the Olympic Peninsula.

Several additional language groupings existed in the interior. One group includes Cayuse and Molalla in Oregon. Highly important in prehistory and history are the Sahaptian languages of the middle Columbia River in Washington and Oregon, including Sahaptin and Nez Perce. Possibly distantly related are the Chinookan languages of the lower Columbia River, important trade languages before and after Anglo contact.

Contacts and Trade

In historic times, the apex of cultural richness and complexity was in the north, among the Haida, Tsimshian, and their neighbors. In prehistoric times, however, the climax may well have been around the Salish Sea. This is the body of water whose Canadian portion is the Gulf of Georgia, whose American portion is largely Puget Sound, and whose shared portion is the Juan de Fuca Strait and the San Juan Islands, over which the US and England had a minor war, rarely reported in the history books (Mathews, n.d.). Since it is geographically and ecologically one congruent body of water, with its core area stretching from the Fraser River delta to central Puget Sound, it has recently acquired the much more apt name of Salish Sea. The international border between the US and Canada runs right through it, splitting a major ecological, linguistic, and cultural focus in two.

Contact across the Bering Strait or even directly across the North Pacific—at least via lost and wrecked boats from Asia—maintained cultural connections. During and after glaciation, the Bering Strait comprised a cultural highway, as Boas pointed out long ago (Fitzhugh & Crowell, 1988). One language, Yup'ik (related to Inuit and Inupiak), was and is spoken on both sides of the strait. Songs, tales, and other lore link the Siberian Chukchi and Koryak with the American Tlingit and others. (Chukchi, Koryak, and the Itelmen language of Kamchatka are related to each other but not to New World languages. On interactions, see Kerttula, 2000).

Metal came to the Northwest by trade and in the wrecks; the Northwest people also learned to work naturally occurring copper from the Copper River (Alaska) and elsewhere (Acheson, 2003). The earliest Boston and English traders brought more, up to tons per voyage (Acheson, 2003: 229). All this helped greatly with the woodworking, especially iron salvaged from wrecks and later from traders. This metal spread far inland, being often traded for interior products such as furs and *clamons*—elk hides prepared for armor (Cooper et al., 2015). These hides can be made so tough that they will stop arrows and even long-shot musket balls.

Kinship

The social structure of Northwest Coast peoples was based heavily on kinship—extended very widely, to include adopted kin, groups descended from a common but mythic ancestor, and kin related by wide-flung marriages; even animals, plants, and geographical features were reckoned as kin. Position within the social system was all-important. Social positions were loci of power, not only political but spiritual. "A traditional chief had a moral and religious obligation to transform chaotic cosmic energy into socially useful power by being a conduit for it down through his spine, ceremonial cane, or totem pole, which, above all, was the 'deed' to 'his' rank, name, house, and territory" (Miller, 2014: 96). The chief had the obligation of maintaining the land and its resources by his personal virtue, his faithfulness to rules and responsibilities, and his generosity and concern (Trosper, 2009).

Kinship was fundamental to all social organization and action, including resource maintenance (Anderson, 2014; Trosper, 2009). Northwest Coast society was made up of kinship groups, often including moieties: patrilineages or patrilineal kindreds in the south, matrilineages in the north and northern interior, and various intermediate accommodations—including something close to double-descent—in the middle, especially among Wakashan speakers.

The Na-Dene, Tsimshian and Haida peoples are all matrilineal; their combined territories make up largest matrilineally-dominated area on the planet. The Haida, Tsimshianic groups, and Tlingit had large and powerful matrilines, and are among the most strongly matrilineal societies in the world. (This disproves, among other things, the old idea—sometimes still heard—that matriliny is the result of hoe-based agriculture. On Tlingit matriliny see Kan, 1989, 2015). As late as the late twentieth century, Canada's strongly patrilineal and patriarchal legal system conflicted, often dramatically, with Indigenous law and practice. A person with a Native mother and Anglo-Canadian father was a full member of both societies, and could move easily in both worlds. Under Canadian (settler society) law, an individual with a non-Native mother and Native father was a rootless person, not really part of either world, because of Canada's attempt at termination of First nation identity, by making women who married non-Indians and their descendants nonstatus.

Fortunately, since those days, because of the need to create their own constitution, Canadian law has become less narrowly patriarchal, giving more recognition to First Nations people, largely through restoring the status of women, their children, and off reserve residents. Kinship really gives identity and personhood on the Northwest Coast; one is a member of one's family in a way incomprehensible, or at least alien, to many non-Indigenous people. The Canadian governmental attempt to remove First Nations identity/status was meant to function as cultural genocide. These matrilineages are further grouped into larger groupings: the Eagle and Raven moieties among the Haida, three or four comparable phratries among the Tlingit and Tsimshian. One must marry outside one's group. All have Eagle and Raven groups; a Haida Eagle cannot marry a Tsimshian Eagle.

The patrilineal societies were largely in the southern areas—Washington and Oregon, and along the Fraser River in British Columbia. They were not so extremely lineal-oriented as the matrilineal groups; tending toward bilateral reckoning. The lineage system was comparable, however.

A different matter pertains to the situation in the middle: the Nuu-chah-nulth, Kwakwaka'wakw, and some immediate neighbors. Here reckoning was equally through paternal and maternal lines. "The immediate family grouping (derived from four grandparents) is technically called a *kindred*, while the huge extended family, which was and is transnational or intertribal (through eight great-grandparents), is technically called an *intersept*" (Miller, 2014: 79). Names, titles, and privileges were taken up by any member of the kindred who wanted to potlatch for them, and could get enough support and resources to do so. Individuals thus often had, and still have, some titles inherited through the paternal line and others through the maternal line. A potlatch, of which much more below, was a major ceremonial event involving feasting, ceremonies, and dances, with major gift-giving by the host and some return from the guests. The word is Chinook jargon, derived from Nuu-chah-nulth *pach'itl*, "to give, especially at a potlatch."

Among the Nuu-chah-nulth this accompanied a kinship terminological system similar to that of the Hawaiians, with every relative in one's own generation being "brother" or "sister," and every one in the parent generation being "father" or "mother." This still persists for the parent generation, even in English, with the slight transformation that relatives from the previous generation, other than actual mothers and fathers become "uncle" or "auntie." Since virtually all Nuu-chah-nulth can trace at least some degree of relationship to all others, the terms can even be extended to all or most Nuu-chah-nulth in the generation above one's own. At the very least, they extend to older members of one's local descent group.

Specialization provided a quite different way of classifying people. Expert hunters, basketmakers, plant collectors, carvers, and other highly skilled people were recognized, and believed to have received power from spirit beings. One widespread powerful figure was the shaman (see below, p. xxx). The word *shaman* is originally from the Tungus languages of Siberia, and comes to us via early Russian explorer sources, but was known earlier in Chinese, because of the court "shamans" of the Tungus Jin dynasty (1115–1234; see Nowak & Durrant, 1977). The practice was universal among small-scale Siberian societies (e.g., Humphrey & Onon, 1996; Kenin-Lopsan, 1997), and is continuous across the Bering Strait, with Northwest Coast shamans supposedly acting very much like Siberian ones in that they achieve trance states, which provide access to information not available through ordinary knowledge and experience.

References

Acheson, S. (2003). The thin edge: Evidence for precontact use and working of metal on the northwest coast. In R. G. Matson, G. Coupland, & Q. Mackie (Eds.), *Emerging from the mist: Studies in northwest coast culture history* (pp. 213–229). University of British Columbia Press.

Ames, K. M., & Sobel, E. A. (2013). Houses and households. In R. T. Boyd, K. M. Ames, & T. A. Johnson (Eds.), *Chinookan peoples of the lower Columbia* (pp. 125–145). University of Washington Press.

Anderson, E. N. (2014). *Caring for place*. Left Coast Press.

Calhoon, J. A., Wildcat, D., Annett, C., Pierotti, R., & Griswold, W. (2003). Creating meaningful study abroad programs for American Indian postsecondary students. *Journal of American Indian Education, 42*, 46–57.

Cannon, A., & Yang, B. Y. (2006). Early storage and sendentism on the Pacific northwest coast: Ancient DNA analysis of Salmon remains from Namu, British Columbia. *American Antiquity, 71*, 123–140.

Carlson, K. T., et al. (2001). *A Stó: lō-coast Salish historical atlas*. Douglas & McIntyre.

Clark, P. U., Dyke, A. S., Shakun, J. D., Carlson, A. E., Clark, J., Wohlfarth, B., Mitrovica, J. X., Hostetler, S. W., & McCabe, A. M. (2009). The last glacial maximum. *Science, 325*, 710–714.

Cooper, H. K., Ames, K. M., & Davis, L. G. (2015). Metal and prestige in the greater lower Columbia River region, northwestern North America. *Journal of Northwest Anthropology, 49*, 143–166.

Fitzhugh, W. W., & Crowell, A. (1988). *Crossroads of continents: Cultures of Siberia and Alaska*. Smithsonian Institution Press.

Goebel, T., Water, M. R., & O'Rourke, D. H. (2008). The late Pleistocene dispersal of modern humans in the Americas. *Science, 319*, 1497–1502.

Hayden, B. (Ed.). (1992). *A complex culture of the British Columbia Plateau: Traditional Stl'átl'imx resource use*. University of British Columbia Press.

Humphrey, C., & Onon, U. (1996). *Shamans and elders: Experience, knowledge, and power among the Daur Mongols*. Oxford University Press.

Hutchings, R. M., & Williams, S. (2020). Salish Sea islands archaeology and precontact history. *Journal of Northwest Anthropology, 54*, 22–61.

Kan, S. (1989). *Symbolic immortality: The Tlingit potlatch of the nineteenth century*. Smithsonian Institution Press.

Kan, S. (Ed.). (2015). *Sharing our knowledge: The Tlingit and their coastal neighbors*. University of Nebraska Press.

Kenin-Lopsan, M. B. (1997). *Shamanic songs and myths of Tuva*. Akadémiai Kiadó.

Kerttula, A. M. (2000). *Antler on the sea: Yup'ik and Chukchi of the Russian Far East*. Cornell University Press.

Lepofsky, D., Lertzman, K., Hallett, D., & Mathewes, R. (2005). Climate change and culture change on the southern coast of British Columbia 2400-1200 Cal. B. P.: A hypothesis. *American Antiquity, 70*, 267–293.

Mathews, T. (n.d.). The Pig War of San Juan Island. *The Tablet*. Retrieved July 09, 2012, from www.wahmee.com. Archived from original on September 07, 2008.

Miller, J. (2014). *Rescues, rants, and researches: A review of Jay Miller's writings on northwest Indian cultures*. Northwest Anthropology, memoir 9.

Monks, G. G. (2019). Zooarchaeology on the northwest coast of North America. *Journal of Northwest Anthropology, 53*, 191–242.

Moss, M. L. (2011). *Northwest coast: Archaeology as deep history*. Society for American Archaeology.

Nowak, M., & Durrant, S. (1977). *The tale of the Nišan Shamaness: A Manchu folk epic*. University of Washington Press.

Pearce, F. (2007). *With speed and violence: Why scientists fear tipping points in climate change*. Beacon Press.

Pierotti, R. (2011). *Indigenous knowledge, ecology, and evolutionary biology*. Routledge.

Prentiss, W., Chatters, J. C., Lemert, M., Clarke, D. S., & O'Boyle, R. I. (2005). The archaeology of the plateau of northwestern North America during the late prehistoric period, 3500-200 BC. *Journal of World Prehistory, 19*, 47–118.

Prentiss, A. M., Cross, G., Foor, T. A., Hogan, M., Markle, D., & Clarke, D. S. (2008). Evolution of a late prehistoric Winter Village on the interior plateau of British Columbia: Geophysical investigations, radiocarbon dating, and spatial analysis of the Bridge River site. *American Antiquity, 73*, 59–82.

Prentiss, A. M., Foor, T. A., Cross, G., Harris, L. E., & Wanzenried, M. (2012). The cultural evolution of material wealth-based inequality at Bridge River, British Columbia. *American Antiquity, 77*, 542–564.

Prentiss, A. M., Foor, T. A., Hampton, A., Ryan, E., & Walsh, M. J. (2018). The evolution of material wealth-based inequality: The record of Housepit 54, Bridge River, British Columbia. *American Antiquity, 83*, 598–618.

Raff, J. (2022). *Origin: A genetic history of the Americas*. Hachette Book Group.

Raghavan, M., et al. (2015). Genomic evidence for the Pleistocene and recent population history of native Americans. *Science, 349*, 841.

Reimer, R. (2018). The social importance of volcanic peaks for the indigenous peoples of British Columbia. *Journal of Northwest Anthropology, 52*, 4–35.

Schaeper, D. M. (2006). Rock fortifications: Archaeological insights into pre-contact warfare and sociopolitical organization among the Stó:lō of the lower Fraser River canyon, BC. *American Antiquity, 71*, 671–705.

Sobel, E. A., Ames, K. A., & Losey, R. (2013). Environment and archaeology of the lower Columbia. In R. T. Boyd, K. M. Ames, & T. A. Johnson (Eds.), *Chinookan peoples of the lower Columbia* (pp. 23–41). University of Washington Press.

Trosper, R. L. (2009). *Resilience, reciprocity, and ecological economics: Northwest coast sustainability*. Routledge.

Turner, N. J. (2014). *Ancient pathways, ancestral knowledge: Ethnobotany and ecological wisdom of indigenous peoples of northwestern North America* (p. 2). McGill-Queen's University Press.

Willerslev, E., & Meltzer, D. J. (2021). Peopling of the Americas as inferred from ancient genomics. *Nature, 594*, 356–364.

Chapter 5
Social and Cultural-Ecological Dynamics

Chiefs and Houses

Local understandings of concepts like "land," "property," and "resources" are very different from those of Canadian and United States bureaucrats. These terms, and still more their equivalents in Indigenous languages, referred to things that were intensely, complexly, and personally related among cultural traditions and local communities. Myths, tales, historical narratives, and direct personal ties, including kinship with nonhuman persons, created or expressed these relationships (Cruikshank, 2005; Nadasdy, 2003; traditional European small-scale farmers were closer to First Nations' understandings than to the bureaucratic ones, which implies that situation as well as culture matters in these cases).

On the coast, especially north of the current national boundary, society consisted of lineages ranked in order of power or nobility. Within these were hereditary chiefs, ordinary people of status, and commoners. There were rather few members of the last category; this was a world where there were more elites than masses. This is, however, probably a post-European-settlement phenomenon, due to population decline and consequent shortage of people to fill the ranks of the titled (see Boelscher, 1988: 52 for debate over this idea). Below this were slaves, a distinctly separate class, made up largely of captives from other tribes and their descendants. Raiding for slavery was universal and important. Enslaved persons could be killed, and sometimes were (Frederica De Laguna, 1972 gives accounts, scattered through a huge book, and hard to find; see also Boelscher, 1988: 59–62). The large coastal groups did most of the raiding, with the powerful Haida, Lekwiltok, and some others most feared. Southerly groups such as those of Washington and Oregon were more often captured than captors.

> Viola Garfield's description of the Tsimshian can stand for the whole northern coast: "Chiefs delegated work to their immediate relatives, wives, children and slaves and set a goal of the quantities to be collected. They supervised men's work while their senior wives supervised the labor of younger women and female slaves. Feasts, potlatches and major undertakings

© The Author(s), under exclusive license to Springer Nature Switzerland AG 2022
E. N. Anderson, R. Pierotti, *Respect and Responsibility in Pacific Coast Indigenous Nations*, Studies in Human Ecology and Adaptation 13,
https://doi.org/10.1007/978-3-031-15586-4_5

like the building of a new house, were planned several years in advance; and surpluses were accumulated in accordance with such long range plans...all resource properties belonged to lineages or their titular heads. However, the privilege of using areas belonging to a man were extended to his sons during his lifetime.... Chiefs should be able leaders, good speakers, haughty and proud before strangers, and humble and generous toward tribesmen. The ideal leader was an able organizer and speaker, and a model of good taste and conduct." (Garfield et al., 1950: 15, 17; see also Trosper, 2009).

The use-right of sons does not conflict with the normal rules of matrilineal inheritance because it was limited to a man's lifetime. It existed at the discretion of the chiefs. Senior branches of lineages supplied the chiefs; cadet branches could flourish only by supporting them. Note the present tense; the descriptions still apply, except that "haughty" behavior is now largely confined to interaction with hostile government bureaucrats (ENA, personal research).

Marianne Boelscher (1988); later Marianne Boelscher Ignace; Ignace and Ignace (2020) drew a contrast between structural and ideal-type accounts of Northwest Coast society compared with accounts based on observation of ongoing practice, in the spirit of Pierre Bourdieu (1977, 1990). Bourdieu described the process in which ongoing processes of social action, usually negotiation but also conflict and separation, lead to structures and systems. Somewhat related is Anthony Giddens' concept of "structuration" (Giddens, 1984). The present book is heavily influenced by Boelscher Ignace, Bourdieu, and Giddens. A corollary is that structures are not static, crystallized things, but constantly shifting benchmarks, renegotiated almost daily.

This is indeed the reality of Northwest Coast society, where potlatches, feasts, visits, co-work, and other events, rise from an ongoing process of creative and dynamic interaction (Ignace & Ignace, 2020) build on the same insight. The problem with older ethnographies is that they usually focused on memory culture, which was easily idealized and abstracted by both consultants and ethnographers. Boelscher's ethnographies are particularly detailed, excellent studies of the ways that practice constructs systems. For parallels one must usually go to Indigenous sources (such as Alfred, 2004; Reyes, 2002, 2006; Sewid, 1969), though settler ethnographers with long-standing ties to particular groups have contributed (e.g., Ridington, 1988; Ross, 2011).

There has been debate on whether the Northwest Coast was a "class society." Enslaved persons clearly comprised a distinct class. There remains a question of whether there was a major distinction between chiefs, nobles, and commoners, or the chiefs were merely *primi inter pares* (firsts among equals). Some of the great chiefs of the mid-nineteenth century were rulers of villages with large bands of followers, but this situation was to some extent a product of the settler society's fur and whale oil trade. Scholars who have worked with more egalitarian groups like the Gitksan tend to think pre-fur-trade chiefs merely gave symbolic leadership to families (Daly, 2005: 196). Scholars working with the more powerful groups of the outer coast, such as the Haida and Tsimshian, are more open to the term "class" (Garfield et al., 1950: 28; Roth, 2008). Noble families were acutely aware of their status, as all ethnographies and Indigenous writers about truly coastal groups agree. "Class" is an ambiguous concept at best. Northwest Coast people's more powerful groups were in process of change and differentiation, making their societies "boundary phenomena."

Chiefs came from noble families. In most documented groups, members of noble families considerably outnumbered commoner or slave ones. Wage work and trade gave commoners a chance (Drucker & Heizer, 1967). Even in earlier times, nobles were numerous, commoners rather rare (Boelscher, 1988; Suttles, 1987). It seems that commoners were people down on their luck, morally dubious, or the descendants of such, rather than a stalwart ordinary class (see e.g., Boelscher, 1988: 51–52 for people unable to assemble resources for potlatching).

Societies were also organized into "houses." The actual houses were usually large and had distinctive names. The house as a named institution, however, was often coterminous with the lineage, or with the group of close bilineal-reckoned relatives typical of central coast First Nations. Houses thus became comparable to the "houses" of Europe, as in the house of Windsor, house of Tudor, or house of Gotha—though obviously on a smaller scale. This comparison was first taken seriously by Claude Lévi-Strauss (1982) and has been much discussed since, especially by Jay Miller (2014, passim).

A particularly interesting institution was the personal name (Miller, 2014). A child would have a personal name, but anyone of substance eventually succeeded to a hereditary title within his or her lineage, and new names could be created. Names were honored titles, comparable to those of feudal Europe. As Macbeth was first addressed as Cawdor, later as Glamis, and finally as Scotland, so a Northwest Coast person would be addressed by the latest, highest name he or she had acquired. The difference was that Macbeth took over the estates with those names, whereas Northwest Coast names are strictly personal titles; they imply decision-making power over lineage estates or other property, but they are not the actual names of properties. This is well illustrated in the story of the *Death Feast at Dimlahamid* (Glavin, 1998, 2000), a detailed, illustrative narrative that deals with the hierarchical structures of the Gitksan and Wet'suwet'en First Nations as they fought for control over their traditional lands, against the Province of British Columbia and the Canadian timber industry. Their victory, at great personal and national, cost, forever changed Aboriginal law in Canada through what has become known as the Delgamuukw decision, named after the hereditary chief who served as lead plaintiff in the case (Borrows, 1999).

Unlike many aspects of Northwest Coast culture, name-taking continues today. One of the most obvious immediate reasons for a potlatch is taking a new name. This could have interesting kinship ramifications; among the matrilineal Gitksan, the father and his family had the work of producing and assembling spiritual and material wealth for a child's naming ceremonies (Adams, 1973: 38–39). Thus, normally, one expects potlatching to serve social statics (the successor gets his or her ancestor's name), but potlatching to take up a vacant title is a standard way of upward mobility for those who accumulate wealth (Boelscher, 1988: 67). The title of Glavin's book (1998 above) comes from a death feast (or funerary potlatch), which shares many features with a potlatch, for an elder who has passed, and the celebration of their name, and title, being taken by a younger member of the Nation.

Many chiefly names that are mistakenly considered to be personal names in earlier literature are actually, in effect, titles. Examples include Shakes in Tlingit Alaska, Legaix among the Tsimshian near Prince Rupert, and Wi Seeks in Gitksan in Interior BC (Glavin, 1998). One could be, for instance, the Capilano of North Vancouver. Capilano—Salish *keypilanoq*—was the title of the head chief of the village near which Capilano Community College now stands.

The hereditary chiefly name Maquinna, passed on for centuries on northern Vancouver Island, inspired an excellent discussion of this whole concept by the current Maquinna, Chief Earl George of Ahousaht (George, 2003). A highly educated anthropologist, he explains in anthropological terms the cluster of rights and responsibilities that the name actually labels. As Maquinna, he has major responsibilities spread over the central west coast of Vancouver Island. Another autobiography with enormous and fascinating detail on names and rank is that of the wonderful and impressive Kwakwaka'wakw noblewoman Agnes Alfred (2004).

Among outsider accounts, Roth's (2008) account of Tsimshian naming, and Boelscher's of Haida (1988: 151–166) are particularly thorough and excellent. A very detailed section of W. W. Elmendorf's ethnography of the Twana of Puget Sound (1960) describes naming in the more southerly reaches. Roth points out that a name has a reality of its own, and the bearer of a name comes in a real sense to *be* the name—complete with all its duties and responsibilities. Again, readers of Shakespeare's *Macbeth* should understand. Macbeth did not just get the letters "Cawdor" and "Glamis" after his name, let alone the word "King" in front; he became a very different social entity as he took over those titles. Similarly, Popes assume new names when they assume the title, e.g., in 1978 Karol Józef Wojtyła, became Pope John Paul II. Inheritance, marriages, and house traditions are similar to those of old Europe. "Today, Northwest natives are well aware of such parallels between the dynastic marriages of European and of Tsimshian noble houses" (Miller, 2014: 97). Some assimilation to European norms has probably taken place since white settlement, but the parallels existed long before contact.

Like kingship in medieval Scotland, Northwest Coast names are often spiritual and religious in meaning. God appoints kings, and kingship is a divine duty. Ancestors are not "gods" or "worshiped," but are involved in the appointment of a new Legaix, and some names are so spirit-linked that they convey direct spirit power. Names are unique; only one person at a time can have a specific name, which confined to their descent groups. There are male and female names, chief and nonchief names, and names that go with particular functions.

Names are not all formal titles. Ordinary names include casual personal monikers; even these usually have stories attached to them. Family histories describe the origins of names identified with particular descent lines. These follow from tales that often involve meeting supernatural beings and getting spirit powers, but may merely commemorate secular events. Such events can be minor or even silly, but once they become grounds for a name, they become canonical stories for the group (see Roth, 2008, esp.: 40 ff.; once again there are parallels in medieval Europe, such as nicknames that conveyed respect even when intrinsically disrespectful, like "Charles the Fat").

An individual may take many names during life: a youthful personal name, an adult name, and then any number of titles. As an example, Pierotti's child name was Tsakabaru (*Tsakkopah* In Newa (Shoshoni\e) pronounced Chikoba, which means "breaks things off," he also has a sacred name not revealed to others). Taking a new name of any status, as well as any other change of status such as a marriage (at least between nobles), must be recognized with a potlatch. This—plus a laudable desire for hospitality and fun—keeps the potlatch alive and flourishing today. Even groups whose native language has died out often maintain names.

Thus, naming is a major political act of extreme importance. Anyone of consequence would potlatch for a new name, or be given a new name at a potlatch. Boelscher's long essay on Haida naming (Boelscher, 1988: 151–166) is a particularly thorough and insightful account of the way that structure and practice interacted to produce a complex, fluid, responsive politics of name-taking and name-giving.

Names were often taboo when the wearer died. All the Puget Sound languages have one name for the mallard duck, except for one group, the Twana, that calls them "red feet," because a chief within historic time had had a name sounding like the usual "mallard" word (Elmendorf, 1960: 393). Puget Sound Salish has special terms for various types of naming taboos.

The naming and house system reached a peak of development among the Tsimshian peoples (Adams, 1973; Beynon, 2000; Menzies, 2016 [Southern Tsimshian]; Miller & Eastman, 1984). These groups—the Coast Tsimshian, Southern Tsimshian, Nisga'a (Nishga), and Gitksan—were divided into four phratries, probably created by dividing into two the moieties basic to neighboring societies. The key concept is *halait:* supernatural, spiritually derived power, or anyone manifesting that power, as in a dance, ritual, or chiefly display of authority. *Halait* were spiritual groundings of chiefship and power. They are validated and explained by *ada'ox*, "the family-owned histories of the origin, migrations, and eventual settlement...of the Houses...they indicate how the latter acquired their crests and powers" (Anderson & Halpin, 2000: 15). These last include the *naxnox*, hereditary privileges in the form of masks, totem pole motifs, dances, songs, myths, and ritual regalia.

These are displayed in actual ceremonial contexts. Initiation and name-taking was necessary to assume *naxnox*. A chief would have to go to heaven (or have a vision of it). This made him a *semoiget*, "real person." On return, he would often be able to display a terrifying power, such as cannibalism (biting people or pretending to) or eating dogs. At least among these Tsimshianic groups, dogs were and considered inedible and even poisonous, so tremendous power was necessary to eat one and survive. In practice, the devouring was faked (at least in reported cases), but—just as wearing a mask makes one into the being represented by the mask—the fakery had a transcendent reality.

The full displays in the winter ceremonies included potlatches, feasts, dances, storytelling, and other performances. Here and elsewhere on the Coast, they were carefully planned and truly spectacular. All this had everything to do with ecological management and resource ownership. The Gitksan leader and spokesman Neil Sterritt explains it as follows:

> For centuries, successive Gitksan and Wet'suwet'en leaders have defended the boundaries of their territories. Here, a complex system of ownership and jurisdiction has evolved, where the chiefs continually validate their rights and responsibilities to their people, their lands, and the resources contained within them. The Gitksan and Wet'suwet'en express their ownership and jurisdiction in many ways, but the most formal forum is the feast, which is sometimes referred to as a potlatch (quoted in Anderson & Halpin, 2000: 31).

Residence on most of the coast was in very large plank houses that contained whole lineages or large families. In front of these, among the northern and central groups, were the famous "totem poles," which displayed family crests and portrayed historically interesting family stories or other events. Most of these events were mythical. Sometimes, however, they were recent, or nonlocal, e.g., Abraham Lincoln was carved on one pole. Further south, these were replaced by large carved house posts among the Nuu-chah-nulth and Salish. In mid-coastal Washington and further south, there was little or no large-scale carving.

Inland, away from the great fishing stations, societies rapidly became less complex and rich. In the north, the deep interior consisted largely of hunters of large mammals (usually ungulates). They had exactly the opposite dynamic from the rich, settled coastal tribes, because they had to be sparse on the ground and highly mobile, with few possessions. Most were Dene (Athapaskans), who lived in small matrilineal groups, and even those groups broke up into families for hunting and foraging during summer and fall. Societies were strongly egalitarian. Individuals sought spiritual power for hunting, healing, and other skills, but in these groups, there was no formal chieftainship, or supernatural powers comparable to the *naxnox* of the Tsimshian, except in zones of mixing and mutual influence with the big coastal groups.

Farther south, in the Plateau region, root foods became more important, and were managed intensively. Salmon were important there too, if less so than on the coast. Societies intermediate in size and complexity arose. Intermediate complexity also characterized the first tier of Athapaskan groups in Canada and Alaska.

These more southerly interior groups, speakers of Salishan or Sahaptin languages, lived in medium-sized, semisubterranean houses. To build one, a large circular pit was dug down a few feet into the ground, with level floor and often a bench ring (dug out less deeply) along the side. Four large pillars supported a log frame roof that was then covered with lighter materials and then with earth. Such a house was warm in winter, cool in summer, and resistant to all weathers. This style is very ancient in the region. Similar houses were once constructed all over western North America and northeastern Asia. (Versions of them often survive as ceremonial structures, such as the kivas of the Pueblos and the ceremonial houses of the Navaho and the central California Nations.) Semisubterranean houses were replaced by lighter, more portable, or easily-built dwellings in most regions in the last few millennia.

Among all interior groups, coastal ideas like ranked lineages and potlatching faded rapidly as one went inland. Societies were less hierarchical, the more so as one went east and south. The traditional explanation for this is that the groups "borrowed" ideas such as potlatching from the Coast. However, current archaeology suggests that the complexity developed slowly throughout the region. The Athapaskans and Plateau peoples were not latecomers; they were always part of the process. The more concentrated the subsistence base, the more complicated life and communities became.

Throughout the Pacific Northwest, the relative importance of fishing at set places for anadromous fish is a good predictor of social complexity and village size. This pattern was established by at least 3000 years ago. The groups most dependent on rich, concentrated fisheries were usually the most complex and socially ranked. A conspicuous exception is the Haida, who maintained one of the most socially complex and highly structured societies with largely oceanic resources.

In the southern part of the coastal region, the fish were as common as in the north, but the societies were generally simpler; the reason appears to be that the rivers (and therefore the runs) were smaller and often more restricted seasonally, while the landward resources—seeds, nuts, game—were richer. Control was less centralized, and control of specific resource points and the labor required to make full use of them was less vital. "The exceptions prove the rule": more centralized control occurs on the lower Columbia and in northwest California. The Columbia and Klamath River systems are large systems, and once had huge fish runs, in an otherwise rather barren area; and, as the model predicts, the groups along it had highly complex societies and large towns. Elsewhere, people could and did scatter out in small villages near sites that varied with regard to productivity.

Environment and Naming Social Units

An example of naming from the environment is provided by the use of orcas as related and eponymous animals. There have been some suggestions that some whaling activities were targeted at this species, but this seems unlikely, considering how important this species is in the stories of the Indigenous peoples:

> Orcas have been a symbol of the West Coast for many thousands of years. They are an important part of the culture of many Indigenous peoples, belief systems, symbolism, art and storytelling. The orca is a symbol often centered around luck, compassion and family. Orcas are known to some Indigenous communities as the guardians of the sea. To some people, orcas represent the strength of love and the bonds of family because of their strong group behaviour (https://georgiastrait.org/work/species-at-risk/orca-protection/killer-whales-pacific-northwest/ accessed 5 January 2022).

As an example, orcas are often represented on totem poles as powerful entities. In contrast, when salmon are represented on totem poles it is usually as prey, with the salmon being held by bears, eagles, or occasionally humans. In contrast, the Orca, which is a particularly strong feature of Haida art and myth, is a popular crest. All Raven lineages use forms of the Killer Whale as a crest (https://www.historymuseum.ca/cmc/exhibitions/aborig/haida/hapmc01e.html, accessed 5 January 2022). The general form of the stories behind this symbolism is as follows:

The killer whale (also known as Orca or Blackfish) is an important animal to the Native American peoples of the Northwest Coast. Killer whales are considered a particular symbol of power and strength, and catching sight of one is considered a momentous omen. Some tribes, such as the Tlingit, view the killer whale as a special protector of humankind *and never hunted killer whales* (although they were accomplished whale hunters of other species.) The Kwakiutl tribes believed that *the souls of marine hunters turned into killer whales upon their death, just as the souls of forest hunters turned into wolves*. For this reason, there were a number of special rituals regarding the killing of a killer whale, so that its spirit could be reborn as a human once again (http://www.native-languages.org/killer-whale.htm, accessed 5 January 2022, emphasis added).

A more specific example comes from the Nuu-chah-nulth:

human observations of predatory behaviour by orcas (or killer whales) led to these animals also being perceived as non-human whalers from which chiefly prerogatives could be obtained. Wolves, the main figures in Nuu-chah-nulth ceremonial life, had the power to transform into orcas, explaining their frequent presence in the art with thunderbirds and whales (MacMillan, 1999).

These statements basically illustrate that, given the belief systems we are addressing in this book, it seems very unlikely that a species that is regarded in this fashion would be the target of attempts to take their lives for food. In such cultures, there are clear divisions between species that are acceptable as food, and fellow predators who are worthy of special respect (Pierotti, 2011). As an example, Gispwudwada is the name for the Killer whale "clan" (phratry) in Tsimshian. This clan is considered analogous or identical to the Gisgahaast (variously spelled; also Gisk'aast) clan in British Columbia's Gitxsan nation and the Gisk̲'ahaast/Gisk̲'aast Tribe of the Nisga'a.

Some species of mammals, i.e., predators, were not taken as food, but had important totemic and spiritual roles, such as wolves, bears, and orcas. As an example, wolves (*Canis lupus*) serve as a totem of important clans or phratries of First Nations of the Pacific Northwest. The *Laxgibuu* or *Laxgyibuu* (variously spelled) is the name for the Wolf "clan" in Tsimshian. It is considered analogous or identical to identically named clans among neighboring Gitxsan and Nisga'a nations. The name *Laxgibuu* derives from *gibuu*, which means wolf in the Gitxsan and Nisga'a languages. In Tsimshian the word is *gibaaw (gyibaaw or gyibaw)*, but Tsimshian still use the word *Laxgibuu* for Wolf clan. The chief crest of the *Laxgibuu* is the Wolf. Other crests used by some matrilineal house-groups of the *Laxgibuu* include black bear. Some *Laxgibuu* house-groups are related to Wolf clan groups among the Tahltan and Tlingit First Nations to the north. According to their ancient creation story, the Quileute Nation of the Olympic Peninsula were changed from wolves by a wandering Transformer. This story was incorporated (and greatly distorted) in making the series of Twilight films set on the Olympic Peninsula (https://quileutenation.org/history/).

The Nez Perce (Ni Mii Puu) nation have a story in which wolves and grizzly bears form the constellations, which humans use to navigate (Pinkham, 2006). The interior Northwest is full of stories in which various mammals, especially wolf and coyote, work to create various aspects of the contemporary world, including the story about badger and coyote being companions (Ramsey, 1977), which has now been shown to be supported by actual studies in behavioral ecology, where these two species have been shown to be regular companions, hunting cooperatively (Minta et al., 1992; Pierotti, 2011).

Wolves are crucial components of the ceremonial and spiritual traditions of coastal nations, such as the Makah, Quilleutes, and Nuu-chal-nuulth (Ernst, 1962). The *Kluckwalle* (*Qua-ech'*) is a multi-day ceremony that seems to concern the fact that these cultures considered wolves to be the dominant species in the forests because of their courage and bravery. As one informant states, "The wolf is the bravest of any animal in the woods. They are the killers. They don't fear anything, which is why they can run the country undisturbed. That is why the wolf is chosen.... The spirit of Kluckwalle is something separate that comes to each person..." (Ernst, 1962: 48).

Warfare

The Northwest Coast people were generally a highly warlike set of societies. The Haida and the Lekwiltok Kwakwaka'wakw enjoyed reputations as particularly fierce and terrifying warriors, at least in the nineteenth century. Tsimshian groups conquered along fur trade routes, developing powerful chiefs. By contrast, many of the southern groups—from southwest British Columbia south along Puget Sound and the coast—seem to have been relatively peaceful, but perhaps only relatively.

Fairly typical for most of the region is the following paragraph about the people of the Fraser River area in southern British Columbia:

> Relations [of the Lillooet] with the Chilcotin were generally hostile, and stories are common of their raiding the Lillooet for salmon and attacking isolated hunting parties and wandering children for slaves.... The Halkomelem on the Lower Faser River... engaged in warfare with the Lower Lillooet over elk hunting in the Pitt and Harrison River areas, and hunters of both tribes were massacred. In the distant past, the Lillooet made war on the coastal people and took many slaves. Sometime earlier, the Shuswap waged war on the Lower Lillooet, driving them from their lands and fisheries between Anderson Lake and Birkenhead River.... The lower Lillooet in the early 1800s were subjected to frequent raids by the Nlaka'pamux (Thompson). Allied forces of Nlaka'pamux, Shuswap, and Okanagan attacked the Lake Lillooet but remained friendly with the Fraser River Band of Lillooet... (Kennedy & Bouchard, 1998: 176).

All this action and violence took place in an area of only about 10,000 square miles. Within that area—and, within most of the Northwest Coast—all groups were at least occasionally hostile to each other. All extensive collections of tales, gathered from early anthropological research, include many tales of such wars.

Especially on the northern coast, peace could be maintained by scattering eagle or swan down over participants in a ceremony. This ensured peace at the ceremony, and if done properly, or during a treaty ceremony, would ensure peace between fighting elements, potentially for all eternity (though reality interfered with that vision).

Potlatches often involved a peace-treaty or truce component, with eagles or swans down being featured as symbolic of peace. Thus, warlike tendencies were not uncontrolled or uncontrollable, and peace—necessary to maintain trade and management—was guaranteed most of the time for most communities.

War, strong chieftainship, and highly localized resources went together. If a group was focused on a single river with a huge run of fish, they desperately needed to protect it, and it greatly helped to have slave labor to help process the fish during the run. Any and all labor was valuable. This combination of factors tended to make these peoples warlike: there was a need to make themselves secure, and then to raid outward for slaves. Other groups had to be warlike to defend themselves, and so the process created continuous feedback loops.

Despite this situation, in the more evenly productive landscape of Puget Sound, warfare and strong chieftainship seem rather subdued by comparison. Groups could spread widely over the countryside and maintain large—and thus more secure—populations. This may be part of the explanation for the individualism and personal freedom noted for these societies (Angelbeck, 2016; Angelbeck & Grier, 2012; Miller, 2014). Angelbeck notes that the highly individualized quest for personal guardian spirits was a key part of this, as it may have been for interior and Great Plains peoples. Each boy and many, if not all, girls went on a vision quest at puberty or in early teens, seeking guardian powers. These were not usually spoken of openly. They allowed a person to avoid irksome social duties by saying "my spirit doesn't like/permit that" (Angelbeck, 2016). Others did not really know what one's spirit was, or what it caveats it had placed on the relationship, so such statements were hard to counter. No one was ever allowed to eat, or even harm, physical manifestations of their spirit guides. This is why prey animals are rarely, if ever, found as clan symbols or spirit guides.

Such situations are sometimes reciprocal. As an example, grizzly bears are considered to be close relatives of the Bear clan of the Tlingit people, which at some levels, appears to be more than a metaphor. A story is told of a Tlingit elder from the Bear clan who died. As his body was being carried by truck to the village of Angoon, "a half dozen or so bears materialized out of the woods and lined the road from the dock to the village... As the truck [carrying the body] rolled by, some [bears] actually stood up...and whenever bears came to close to the village, the old folks normally went out and talked to them, asking them to stay away because they frightened the little children" (Bernice Hansen quoted in C. Martin, 1999: 37, also cited in Pierotti, 2011: 60). Although to a Western-trained scientist this story may be hard to accept, this story was confirmed by a Tlingit student at Northwest Indian College, who assured Pierotti that the elder involved had been his relative. The anthropologist Robin Ridington tells the story of Japasa (Asah), of the Duune-za First Nation, a man whose spirit guide was Fox. As Asah lay dying, preparing to "follow his dreams towards Yagatunne, The Trail to Heaven, every night fox tracks were found around his tent until he passed. The foxes were said to be guiding his spirit towards Yagatunne" (Ridington, 1988). It is also said that a man of Asah's village who had moose as a spirit guide always had moose tracks around his grave, like those around a salt lick, showing heavy activity and continual presence.

Angelbeck and Grier (2012) even employ classic anarchist theory (as in Kropotkin, 2004) to explain Puget Sound Salish and nearby dynamics, as Robert Bettinger (2015) did for California, and as James C. Scott (2009) did for northern southeast Asia. The result seems a bit overdone—most hunting-gathering societies are individualist and free; it is the northern Northwest Coast societies that are the anomaly. One need not invoke political theory. Still, the point is worth making, and the high level of individualism in the Northwest is worth remembering.

Feasts and Potlatches

The social requirements of chieftainship in a warlike world are great. Chiefs had to attract and keep warriors through generosity. A pattern of competitive feasting and wealth giveaways is sure to develop. Feasts come in many forms. The Tsimshian word *liligit*, "calling people together," includes occasions memorably listed by Joan Ryan, who held the Sm'ooygit Hannamauxw title of the Gitksan people:

> [S]ettlement feasts (which occur after funerals); totem pole- or gravestone-raising feasts; welcome feasts (to celebrate totem pole-raising events); smoke feasts; retirement feasts; divorce feasts; wedding feasts; restitution feasts; shame feasts; reinstatement feasts (pertaining to Gitksan citizens who have disobeyed Gitksan laws [but repented]); first game feasts; welcome feasts (to celebrate births); graduation feasts (to celebrate recent achievements, either academic or spiritual); cleansing feasts (to restore spirits after serious accidents); and coming out feasts (to mark the transition from teen years to adult years). The feast system is a vehicle by which the Gitksan Nation carries out activities and transactions that affect the daily lives of the House members (Ryan, 2000: ix).

Rules and etiquette for these Gitksan feasts are documented by the People of 'Ksan (1980: 112–120). Feasts are occasions for showing solidarity, support, and good will. They are, of course, generous sharing of food. They are occasions for fun and enjoyment, and solemn and serious play such as the masked dances. Similar occasions entail feasts elsewhere on the coast. Almost all such occasions involve marking some significant change in status of an individual, said change requiring announcement, recognition, and validation by the whole community.

These rounds of feasts had the interesting side effect of producing a genuine gourmet cuisine—a self-conscious high-tone cookery designed to impress with its quality and sophistication. This may be unique among nonagricultural peoples worldwide. Anyone who has participated in a full-scale Northwest Coast feast is familiar with the countless ways to cook salmon, black cod (sablefish), and dozens of other local foods. The cuisine brings out the fine qualities of local foods, including some of the best sea food in the world, with techniques and mastery that rival French dining in taste if not in complexity.

Franz Boas famously provided the world with over 300 pages of Kwakwaka'wakw recipes (Boas, 1921: "preservation of food," 223–304; "recipes," 304–601), supplied to him by George Hunt, a man of Tlingit and English ancestry who had been adopted into and married into the Kwakwaka'wakw. Hunt in turn collected them from his wife (whose death he was mourning at the time) and sisters, and doubtless other women of his family (as pointed out by Martine Reid in Alfred, 2004: 236)—giving women a rare and unfortunately poorly acknowledged presence in early ethnography. Men do much of the preparation and cooking also, however, so Hunt added his own details. Most of these recipes are fairly simple, but some are complex cuisine. Boas and Hunt included methods of catching the foods (from octopus to duck) and preparing them, and some of the proper etiquette and ritual behavior connected with them and with the feasting, so these are far more than mere directions for cooking. For instance, detailed rituals for treating "the first dog-salmon of the season" (302–304), serving mountain goat fat (452–456), and dealing with a beached whale, an incredibly lucky find (464–468) are included. They were memorably analyzed, along with feasting in general, by Stanley Walens in *Feasting with Cannibals* (1981).

The first 52 pages of recipes concern salmon—every part of the fish from head to tail, as well as eggs and "guts"—with appropriate attention to different species and different conditions. Salmon may be subjected to as many as three processes in gourmet cooking, such as successively smoking, roasting, and barbecuing or boiling. Plant foods get their due attention, with recipes for avalanche lilies, berries, fern roots, inner pine bark, seaweeds, and much more, showing that early writers recognized the importance of these foods. Later in the book can be found accounts of feasts, including proper division of animals, serving of plant foods, and etiquette (Boas, 1921: 750–776). Typically, one specific food was the focus of the feast, being eaten with varying accompaniments of oil or the like. One specific type, a "feast of salmon backbones," sounds less than stellar, until one learns that filleting—as was done for smoking and drying—leaves much of the best meat attached to the backbone. It is significant that rituals are not separated from preparation methods and serving etiquette. All are part of the same conceptual unity. A rare ethnographic awareness of taste is Marian Smith's note on the Puyallup-Nisqually: "There was considerable interest in delicacies and epicurean tidbits" (Smith, 1940: 228). These included both soups and solid foods, and both animal—mostly fish—and plant items. Smith's whole chapter on Puyallup-Nisqually foods is a distant, but worthy, second to Boas' recipes (Smith, 1940: 228–251).

Generosity was such a major value to these peoples, and so useful in society, that T. F. McIlwraith recorded: "A native store-keeper once ruefully commented on the fact that he would gain no advantage from the goods on his shelves, since he hoped merely to sell them, not to give them away" (quoted Kramer, 2013: 738). This certainly reverses most Anglo-American attitudes and provides a viable alternative to greed-driven capitalist concepts. A recent major article by Grant Arndt (2022) notes a widespread view of settler society as selfish and narrow; Arndt discusses this in a proper Boasian (and Kantian) way, seeing it as stemming from direct experience, abstracted and culturally constructed—becoming "ontological" only at a high level of Indigenous abstraction, but still retaining a basis in empirical reality.

The high point of feasting is the familiar potlatch (from Nuu-chah-nulth *p'ačiƛ*, "to give," via Chinook Wawa) of the Northwest Coast. The greatest feasts were connected with potlatches, which also involved giving away vast amounts of

Feasts and Potlatches

property and goods. The most generous chiefs gave away fortunes, including vast amounts of wealth goods and foods assembled by their followers, clients, and enslaved dependents. The magnitude of this giveaways so frightened Europeans, including clergy, that for almost a century there were constant efforts to ban these ceremonies in British Columbia:

> Integral to the meaning of the potlatch today, especially among the Kwakwaka'wakw and other Coastal First Nations, is the Canadian governments banning of the ceremony through legal means. Potlatching was made illegal in 1885, and the prohibition was not lifted until 1951 (Cole & Chaikin, 1990). Such attempts at suppression were not new. Missionaries and federal officials had been trying to ban the custom since they first arrived in British Columbia. The lobbying of the federal government to legislate the ban, can be seen as evidence of just how ineffective their initial attempts at suppression were (Ibid: 14). The purpose of the ban was explicit. It was intended to stamp out aboriginal people and their culture. Coastal First Nations were persecuted, chiefs and noblewomen were jailed for practicing their culture, masks were confiscated, Big Houses were torn down, and ceremonial objects were burned (quoted from The Bill Reid Centre: https://www.sfu.ca/brc/online_exhibits/masks-2-0/the-potlatch-ban.html; Accessed 29 July 2021; see also Trosper, 2009).

Guests coming to a potlatch contributed small amounts on occasion. Chiefs who attended were expected to reciprocate with further—hopefully larger—potlatches at some point.

The direct purpose of many potlatches is in effect to get an actual vote on the change in a person's status. If it requires community acceptance, having everyone show up for the feast shows that the change is indeed accepted. If few or no helpers will contribute to the feast, or if few show up, or if they show up but insult the host, the change is obviously not approved. Few indeed would give a potlatch if this were a risk, but it did happen in the past and could occasion fighting or war.

Funeral potlatches have been the most important and widespread potlatches in recent times, but in the past nametaking was apparently the most important. The Gitksan Death Feast for Waigyet (Elsie Morrison) carefully and respectfully described by Glavin (1998, Chap. 5), seems to be both a celebration of the old woman's life and a way of identifying her successor as Waigyet. Tsimshian chief William Beynon provided an account of a series of potlatches by the Gitksan in 1945 (Beynon, 2000). Potlatching was illegal at the time, but the Gitksan country on the Skeena River was remote, and Beynon was hardly about to be excluded. His account remained unpublished until 2000, probably because of the ban described above. Beynon's father was Welsh, his mother Tsimshian, which made him one of the lucky ones: White by Canadian law (patrilineal in those days), Tsimshian by matrilineal Tsimshian law.

Buried in a footnote to Agnes Alfred's life story is a good short summary by Martine Reid of explanations for the potlatch (Alfred, 2004: 244–245). It discusses several less direct or "latent" reasons for the potlatch. One hypothesis is that of Wayne Suttles and Stuart Piddocke (Piddocke, 1965; Suttles, 1987) that the potlatch arose to even out resources. A group would give a potlatch in good times, and thus in bad times would be able to call on reciprocity from more fortunate people. However, this hypothesis did not hold up well (Adams, 1973; John Douglass, interview with ENA, 1984). It turned out that potlatches would not serve this function, because they must be planned years in advance, before scarcity or abundance would be known. Ordinary reciprocity works instead for the purpose (as Adams and Douglass both emphasize). On the other hand, Suttles' data show some evidence for redistribution and evening-out in his major research area, Puget Sound and the Salish Sea (Wayne Suttles, 1987, and interview with ENA, 1984).

A more recent argument is that the potlatch structure functioned to maintain resilience through reciprocal exchanges over millennia through an ecological economics approach (Trosper, 2009). Trosper, a Salish-Kootenay tribal member, provides a detailed economic analysis that suggests that human systems can link environmental ethics as developed by NW First Nations with sustainable ecological practices through institutions such as the potlatch. He further suggests these governance and exchange principles might serve as a model for the creation of resilient societies in the contemporary world.

Potlatches may locally have served to equalize resources, but this has never been proved. What they clearly do is reinforce and maintain reciprocity, sharing, mutual concern, and local harmony (Trosper, 2009). Suttles (1987: 23–25, 60–63) has discussed the whole controversy, with pros and cons. He noted that the potlatch has a great deal to do with social competition and status, social reciprocity, spiritual life, and group solidarity, but continued to see it having a beneficial impact on production and exchange, allowing more resources to be produced and distributed. He pointed out: "A man with a temporary abundance of food had three choices: (1) he could share it with his fellow villagers, if they could consume it... (2) he could preserve it, if it was preservable and he had the labor force and time...; or (3) he could take it to his in-laws in another village...and receive in return a gift of wealth.... If he got more wealth than he gave, he could always potlatch and convert the wealth into glory..." (Suttles, 1987: 60).

Since he wrote, the simple equalization theory has been largely abandoned by writers familiar with the region (Daly, 2005: 59–60). One problem with this way of thinking is that the potlatching group all gets together to finance the enterprise, and relatively poor individuals must often give up quite a bit; even guests may have to give more than ethnographies usually imply. The stronger but more general theory of potlatching as maintaining regional reciprocity continues to flourish (Trosper, 2009).

Other cultures around the world also mark such transitions with spectacular giveaways. Genghis Khan's Mongols, for instance, shared out loot at great feasts (Buell & Anderson, 2010). The giveaways, demonstration of powers and possessions, and feasts powerfully increase and reinforce the status of the potlatchers. This is certainly one reason for them. John Adams (1973), discussing this, shows that they also can serve indirectly to redistribute population, notably from larger but troubled groups to smaller and cohesive ones, because of shifting after the potlatch.

In early times there was another compelling reason to potlatch. Philip Drucker and Robert Heizer (1967) argued that one root purpose was organization and mobilization for conflict, including—focally—competition for warriors or for chief-status allies. A successful chief would gain loot by fighting. This could be donated to warriors directly, or translated into wealth for the whole group, thus organizing and leading his own people. Either way, he can thus attract more warriors to his side or form alliances with neighboring groups. A less successful chief must either defer or attempt to do better. If he cannot be generous enough. If his warriors defect, he is as good as dead. Potlatches can thus be extended to substitute for actual fighting, with the most generous person assumed to have power. Competitive gift-giving then replaces physical conflict; "fighting with property," as Helen Codere (1950; see also Rosman & Rubel, 1971) famously called it. Her long and detailed study of the economics of potlatching showed how it became a peaceful alternative to war; the potlatcher could get his rivals in his debt, overawe them, outdo them, show he had many followers, and show how powerful he was, thus forestalling threat. This does not mean that potlatches are basically competitive, a nineteenth-century claim that does not hold up well to modern studies. Competition there certainly was (and is), but the marking of status change remains the core function (as was explained many times to ENA during research), and maintaining intergroup relations, including reciprocity, a major effect (Trosper, 2009). Codere also pointed out that the spectacular potlatches of the early twentieth century were responses to colonial rule, including outlawing of them, and to the heightened competition for titles occasioned by tragic decline of population.

Similar patterns of competitive feasting and generosity are well known from West Africa, highland Southeast Asia, early-day Afghanistan (Robertson, 1896), and many other areas of the world. Often, professional poets—the bards of Ireland, scops of Anglo-Saxon society, griots of Africa, song leaders on the Northwest Coast—sang the leader's praises at such feasts.

These singers and tale-tellers graced potlatches. We know little of their professional status, but the major songs, dances, and masks were lineage property. They were usually given to the lineage ancestors by supernatural beings or in non-natural experiences. We also are somewhat unclear on the range of chiefly power; the great chiefdoms in which one man extended dominion over several villages seem likely to be products of the fur trade (as among the Tsimshian) in the early post-contact period (Matson et al., 2003).

Chiefs could be powerful, and there was an ideology of this; the Nuu-chah-nulth connected chiefs and whales (Coté, 2010; Harkin, 2007; Sapir, 2004). As Nuu-chah-nulth anthropologist Ḳi-ḳe-in points out: "[T]he Nuuchaanulth people historically had policing, a judiciary system, and a most sophisticated and inclusive system of ownership and title...the entire sea and land territory and everything in it are vested in the Taayii Ha'wilth," the head chief (Ḳi-ḳe-in, 2013: 29; see also Ḳi-ḳe-in, 2000). He did not have European-style autocratic control, but acted as leader of his wide, kin-linked community. The appropriateness of the term "chiefdom" has been challenged (see Miller, 2014: 94), but the chiefs had enough power over enough area to make the term useful, even though the Northwest Coast chiefdoms were small in extent and low in population compared to the great chiefdoms of old Polynesia or Mesoamerica.

The survival of potlatching for 200 years after the pacification of the northwest shows that it served purposes other than mere war mobilization. Name-taking remains the direct motive. Sometimes it was defense of a name, as when a potlatch was given to wipe away shame because a holder of a high title had—for example—

stumbled at a feast, or been insulted. Potlatches could, by the same token, shame an opponent.

Potlatches are about giving away vast amounts of food and other useful goods, but they are also about prominently displaying what can rarely be given: names, crests, dances, sacred possessions (Roth, 2008: 102ff). A potlatcher may point out that the food shared out is the product of the potlatcher's titled endowment property (Daly, 2005: 59, citing Philip Drucker as well as Daly's own research). Once again, this institution recalls medieval Ireland and England. The king could and did feast his followers and give away gold and coin, but he could not give away the crown jewels, the ritual sword and mace, or the intangible names and titles. As on the Northwest Coast, many of the material possessions had supernatural qualities. A Tsimshian inherited copper is not just a piece of copper sheeting, just as Excalibur was not just a sword.

A particularly common and widespread type of potlatch was the memorial potlatch, held first when a high-born person died (the death feast), then again when his or her totem pole—today, headstone—was raised (see notable discussion in Fiske & Patrick, 2000). This was the occasion to validate the succession: the name-taker had to host the potlatch, feeding and giving lavish gifts to those who came. Chiefs and others of status had to give potlatches for marriage, to call everyone to witness and support the couple. Assumption of any new name or title provided other occasions for name-taking. A girl reaching puberty, thus the age when arrangements for marriage can begin, has a "coming-out" or "coming-of-age" potlatch.

Potlatching is influenced by simple ambition, or lack thereof. A friend of ENA was in line to be a hereditary chief in one of the Nuu-chah-nulth groups, but did not want the responsibility at the time, so let his younger brother potlatch for the position; later the younger brother got tired of it, so the elder potlatched and took over the job.

The demographic collapse on the Northwest Coast after settler-society contact led to a pile-up of titles: there were too few people competing for too many names. This, plus the availability of trade goods, caused potlatching to spin into heights of lavish expenditure that opened the door for criticism by missionaries. They managed to get the potlatch outlawed in Canada from 1884 to 1951. Potlatching continued anyway but typically went underground in a somewhat distorted form. This was a major psychological blow to the people (Atleo, 2011). All the missionaries accomplished was giving a couple of generations of Native people a contempt for the overt racism in settlers' law.

Potlatching became extremely competitive among the Kwakwaka'wakw (Kwakiutlan) peoples. This was partly because their complex system of descent does not produce clear lineages, so many people have a chance at inheriting a name and set of privileges. It was also partly because the central location of the Kwakwaka'wakw made them a crossroads for wealth, opportunity, and communication. Their potlatches were spectacular, involving vast destructions of property, such as huge fires burning whale and seal oil (highly precious commodities). North and south of them, potlatches and feasts were calmer, more routine, and more concerned with actual orderly succession. This did not, however, prevent politicking and gossip (cf. Boelscher, 1988 for Haida, Beynon, 2000 and Roth, 2008 for the Tsimshian).

Boelscher (1988) pointed out that in potlatching, as in other matters involving rhetoric, oratory, and public presence, there is a tremendous amount of individual jockeying for position, often by denying one's own position with exaggerated modesty. Rhetoric is highly formulaic and stereotyped, but people manage to assert individuality through practice. As so often, autobiographies are particularly useful sources (e.g., Alfred, 2004; Blackman, 1982; Sewid, 1969).

The gift-and-obligation side of potlatching was famously studied by Mauss in his excellent analysis of gifts (1990; French original 1925) and has been debated ever since. Complicated and highly variable rules of reciprocity, with or without interest, were critically important to theory and practice (see Daly, 2005; Roth, 2008; Trosper, 2009). Potlatching may involve war mobilization, social glue, name and title validation, status-change validation, validation of land ownership, and much else. Gift transactions, however defined and managed, represent more means than ends. They are necessary and obligatory, however, what they communicate is what reveals the ultimate purposes of potlatches.

Potlatches in the nineteenth century became spectacular, with great destruction of property, consumption of food, and supposedly, killing of enslaved dependents (a doubtful charge). This gave it a reputation for wasteful "savagery" that was pushed very hard by missionaries intent on stamping out Native religion. Unfortunately, Ruth Benedict, in her famous book *Patterns of Culture* (1934), emphasized this aspect of the potlatch, and it is repeated by modern anthropologists who should know better. The gaudy, vivid colonial stereotype of the potlatch is unfortunately revived by Graeber and Wengrow (2021: 182–185), who should have known better; they apparently did not read any Indigenous accounts of the institution, let alone attend a potlatch, most of which are quiet affairs. The spectacular, destructive potlatches occurred in a specific social context, i.e., competition for titles in the late nineteenth century, when trade goods were abundant and shortly after disease had killed most Indigenous people, leaving too many titles for too few competitors (Codere, 1950). The vast majority of potlatches, including all modern ones, are mild affairs. ENA's experience is that they involve a quiet feast, with good local food given out and small amounts of cash or minor gift items, all in a calm, familial setting. The new status is assumed or the late relative is commemorated, and everyone goes home.

From the point of view of resource management, potlatches are important not only because of the enormous amount of resources mobilized and consumed, or given, but also for the statement of group rights and privileges. Rights to lands and waters are among the rights displayed and ritualized, allowing no one to forget them (Reid, 2015). This presumably had considerable importance in maintaining ownership in former times when conflict was more open.

Ownership: A Key to Management

Ownership on the Northwest Coast was and is important and complex. Lineages are involved in resource ownership. Nancy Turner and colleagues (2005; Turner, 2014, vol. 2, passim) provide a review, but the full complexity defies summary and must be found in the ethnographies, especially in Richard Daly's ethnography of the Gitksan and Witsu'weten that are involved in the Delgamuukw case (Daly, 2005; Glavin, 1998). The latter was based on courtroom-certified evidence of ownership; thus providing an extremely full, detailed, accurate account. It applies only to the two groups mentioned, however, broadly similar ownership patterns exist throughout. They are generally even more precise and self-consciously maintained along the coast itself, least precise and tightly enforced in the deep interior, although the Delgamuukw case involves interior First Nations.

Along the coast and up the major rivers, every fishing spot, productive cedar grove, and berry patch was tightly owned. Good hunting grounds for localized species like marmots and mountain goats also had ownership; however, these species needed to be treated with respect, e.g., the ancient community of Dimlahamid was reportedly destroyed in a landslide caused by vengeful mountain goat spirits after disrespectful hunting practices (Glavin, 1998, and below). Widely necessary goods that were locally abundant but otherwise nonexistent, notably mineral resources like salt and obsidian, might be open to all (Daly, 2005: 263). Complex arrangements for sharing, loaning, and even renting have been widely reported, especially for fishing sites (Reid, 2015). If a run failed or a slide blocked a stream, a group would be in desperate straits, and neighboring groups would then allow fishing access to controlled areas, which supports the reciprocity concept. Daly (2005) reports boundary markers, special passes, such as braided straps, used by people borrowing access, and other highly specific ownership applications. Male ethnographers underappreciated the importance of female ownership of plant resources, which was incredibly complex and detailed (Turner, 2014: 2: 90–91), in a manner comparable to land rights of intensive farmers in other cultures. Rights were loaned, land was opened, usufruct was differentiated from outright ownership, and sharing was subject to complex rules and norms. Generosity was always idealized, but ownership strictly enforced.

Among the Haida and Makah, lineage "members share rights to ancestral villages, to fishing, hunting, and gathering locations, including:

- major salmon-spawning rivers
- halibut and cod banks offshore
- important berry patches (especially cranberry and crabapple)
- bird nesting sites (particularly species of sea gull, Cassin's auklet, and ancient murrelet)
- rights to stranded whales on the coastline near lineage-owned lands
- trap and deadfall sites for land mammals along rivers
- access to house sites in ancestral villages" (Boelscher, 1988: 35; Reid, 2015).

Tlingit land and resource ownership—again by matrilineage—startled early explorers; French and Spanish in the eighteenth century found they could not take (at least without a fight) firewood, wild game, fish, berries, and other resources that they assumed would have been free for the taking (De Laguna, 1972: vol. 1: 119). Descent groups owned crests and other cultural privileges in the same way (De Laguna, 1972: vol. 1). Early explorers found that the Nuu-chah-nulth would not allow them to collect shells on the beach without proper compensation (Inglis & Haggarty, 2000: 96). Poaching on someone else's land, even to take small items, was cause for combat.

Among the Nuu-chah-nulth, the concept *topati* (*tupaati*) covers things owned by families. Topati includes "various kinds of ceremonially recognized property (names, songs, dances, ceremonies, hunting territory, specified parts of a captured whale) whose use is restricted to a given family and is subject to certain principles of ownership, inheritance, and transfer" (Sapir & Swadesh, 1939: 222). Often these were the property of an *ʔuuštaqimł*, or "descent group." Larger social units owned larger territories, whole whales, and many ritual prerogatives. The lands in question were *ḥaḥuułi*, descent group homelands, which had to be maintained and cared for, not just exploited (Coté, 2022: 29). The all-important Wolf Ritual (Tlo:kwa:na; Sapir et al., 2007), still a vitally important part of Nuu-chah-nulth life, is owned more widely, but components of it may be locally owned.

Especially in the northern coast, lineages (matrilineages in the north) had crests: natural or supernatural entities that were symbols and that provided some supernatural power for them. Usually, these were impressive animals, all predatory species, including killer whales, grizzly bears, sharks (including dogfish), wolves, Steller and California sea lions, Bald Eagles, frogs and dragonflies (these being spiritually powerful beings), and ravens. Some plants served as crests, as did weather (Haida crests included cumulus and cirrus clouds), the moon, and various spirit beings. People could add crests when interesting events involving crest-type beings happened to their families. These crests were typically shown on totem poles, blankets, and even on the skin of individuals; a Haida man would have his lineage crest tattooed on his body (Boelscher, 1988: 142). Similarity to feudal European usage led to these being called "crests" from the beginning of Anglo settlement.

Among the Kwakwaka'wakw, the *numaym* bilateral descent group owned "masks, headdresses, paintings, sculptures, ceremonial dishes…landed estates made up of hunting and gathering territories, streams, fishing sites, and the locations of weird…These territorial rights were fiercely defended; their legitimate owners did not hesitate to kill trespassers" (Lévi-Strauss, 1982: 168). Note that masks and other regalia involve not only the physical entities but the rights to make those particular artistic forms and designs—in other words, Indigenous copyright obtains for artistic representations. Other societies vested ownership in the "house," not just the actual houses, but the social form compared by Lévi-Strauss to the "great houses" of Europe; he speaks of "house" societies (Lévi-Strauss, 1982: 168–187), characterized by such units and often by hereditary titles whose succession he found interestingly similar between the Northwest Coast and old Europe and Japan.

Almost equally important, especially in the coastal societies, were lineage rights to particular songs, dances, designs, and other intellectual property. Lineages and phratries even had their own origin myths, very different from other lineages' stories. In fact, Indigenous rules regarding intellectual property were, and still remain, complex and sophisticated, rather like modern copyright legislation. A song could be common property and available for everyone to sing, or owned by the moiety or phratry, or by the lineage, or by a family or even by an individual. Whoever owned it could allow someone else to sing it, but a substantial payment or exchange was usually required. This could be indirect; lineage A might generously allow someone from lineage B to sing a lineage A song, but of course, lineage B was expected to do something equivalently generous for lineage A at some point. Showing off one's lineage dances, songs, and arts was—and in many areas still is—an integral part of potlatching and feasting. Often it involved competitive display; at other times it seems closer to an arts festival or powwow.

A particularly fine discussion of Northwest Coast song—specifically Makah song—by Linda Goodman and Makah singer Helma Swan provides notable discussion of ownership (Goodman & Swan, 2002; see esp. pp. 41–43) and singing contexts. The concept of ownership was similar to the modern idea of copyright. Thomas McIlwraith, in his early days as ethnographer of the Nuxalk, inadvertently sang a song belonging to one family, whereupon an elder of that family had to adopt him on the spot to spare everybody serious shame and difficulty (quoted in Kramer, 2013: 737).

Individuals owned the products of their own labor, especially things like bows and arrows, clothing, food, and personal articles, but also songs they composed, and ordinary art objects they created. An individual might pass songs and crests on to his or her children; a man could pass them to his children even among the matrilineal Haida (Boelscher, 1988: 38), though that could be a touchy proposition. (The desire of a man to pass on privileges to his biological children, and the opposition of his wife's brothers to this, is a well-known feature of matrilineal societies in many parts of the world.) Normally, privileges including songs and crests were the property of matrilineages in the northern coast, passed on accordingly.

Among the matrilineal societies, women had great authority and owned property. Chiefs were generally (but not always) male, being the brothers of the women who were the hereditary heads of the lineages; but women had real power. There is a famous story of a Haida chief taking refuge on a fur trader's boat because he had dared to sell some furs belonging to his wife. She and her brothers were out after him. He had to stay on the boat, under the protection of the trader, for months before he could work out an arrangement to go home.

Like other matrilineal societies worldwide, these were not "matriarchies," but did give more power to women than patrilineal societies generally do. On the other hand, among the patrilineal societies of the southern Northwest Coast, women were, and remain, by no means meek or oppressed; they held their own and had considerable real power, partly through controlling access to and management of resources, including many plants and shellfish, and other subsistence and political resources.

These societies were organized in moieties; among the Tsimshian, the moieties had split into two parts, giving a four-phratry system, but the original moiety structure was still identifiable (Miller, 1982). Crest privileges and their artistic expression were taken so seriously that upper-class artists "worked in great secrecy, punishing with death any unqualified person unfortunate enough to stumble on their workshops hidden in the forest" (Miller, 1982; true to Tsimshian form, however, the threat was enough, and killings were very rarely carried out).

In short, ownership among the Northwest Coast peoples was extremely highly developed and complex—as far from the old stereotype of "primitive communism" as one could possibly get. This was related to the need for control over, and management of, resources (see detailed account of plant resource ownership patterns in Turner et al., 2005). Even the ownership of songs and art motifs could be seen as related to maintaining group solidarity and identity for defense of hard goods—although it was very much more than that, being integral to all aspects of social life. The interior groups were far less property-conscious, but even the much less possessive Plateau groups had sharply defined rights to fishing spots, berry patches, and the like. Only among the migrant hunters of the interior north did ownership dwindle to personal possessions (including things like songs and spirit powers) and some specific rights to traplines and berry patches.

References

Adams, J. W. (1973). *The Gitksan Potlatch: Population flux, resource ownership and reciprocity.* Holt, Rinehart & Winston.
Alfred, A. (2004). *Paddling to where I stand: Agnes Alfred, Qwiqwasutinuxw noblewoman.* University of Washington Press.
Anderson, M., & Halpin, M. (2000). Introduction. In W. Beynon (Ed.), *Potlatch at Gitsegukla: William Beynon's 1945 field notebooks* (pp. 1–31). University of British Columbia Press.
Angelbeck, B. (2016). Localized rituals and individual Spirit powers: Discovering regional autonomy through religious practices in the coast Salish past. *Journal of Northwest Anthropology, 50,* 27–52.
Angelbeck, B., & Grier, C. (2012). Anarchism and the archaeology of anarchic societies: Resistance to centralization in the coast Salish region of the Pacific northwest coast. *Current Anthropology, 53,* 547–587.
Arndt, G. (2022). The Indian's white man: Indigenous knowledge, mutual understanding, and the politics of indigenous reason. *Current Anthropology, 63,* 10–30.
Atleo, E. R. (2011). *Principles of Tsawalk: An indigenous approach to global crisis.* University of British Columbia Press.
Benedict, R. F. (1934). *Patterns of culture.* Routledge.
Bettinger, R. L. (2015). *Orderly anarchy: Sociopolitical evolution in aboriginal California.* University of California Press.
Beynon, W. (2000). *Potlatch at Gitsegukla: William Beynon's 1945 field notebooks.* University of British Columbia Press.
Blackman, M. (1982). *During my time: Florence Edenshaw Davidson, a Haida woman.* University of Washington Press.
Boas, F. (1921). *Ethnology of the Kwakiutl.* 2 vols. United States Government, Bureau of American Ethnology, Annual Report for 1913-1914.

Boelscher, M. (1988). *The curtain within: Haida social and mythical discourse*. Unversity of British Columbia Press.
Borrows, J. (1999). Sovereignty's alchemy: An analysis of Delgamuukw v. British Columbia. *Osgoode Hall Law Journal, 37*, 537–596.
Bourdieu, P. (1977). *Outline of a theory of practice*. Cambridge University Press.
Bourdieu, P. (1990). *The logic of practice*. Stanford University Press.
Buell, P. D., & Anderson, E. N. (2010). *A soup for the Qan*. Brill.
Codere, H. (1950). *Fighting with property*. J. J. Augustin.
Cole, D., & Chaikin, I. (1990). *An iron hand upon the people: The law against the potlatch on the northwest coast*. Douglas & McIntyre.
Coté, C. (2010). *Spirits of our whaling ancestors: Revitalizing Makah and Nuu-chah-nulth traditions*. University of Washington Press.
Coté, C. (2022). *A drum in one hand, a Sockeye in the other: Stories of indigenous food sovereignty from the northwest coast*. University of Washington Press.
Cruikshank, J. (2005). *Do glaciers listen? Local knowledge, colonial encounters, and social imagination*. University of British Columbia Press.
Daly, R. (2005). *Our box was full: An ethnography for the Delgamuukw plaintiffs*. University of British Columbia Press.
De Laguna, F. (1972). *Under Mount Saint Elias: The history and culture of the Yakutat Tlingit*. Smithsonian Institution, Smithsonian Contributions to Anthropology 7.
Drucker, P., & Heizer, R. (1967). *To make my name good: A reexamination of the southern Kwakiutl potlatch*. University of California Press.
Elmendorf, W. W. (1960). *The structure of Twana culture, with comparative notes on the structure of Yurok culture [by] A. L. Kroeber*. Research Studies, Washington State University, Monographic Supplement 2, part 1.
Ernst, A. H. (1962). *The wolf ritual on the northwest coast*. University of Oregon Press.
Fiske, J., & Patrick, B. (2000). *Cis Dideen Kat: When the plumes rise*. University of British Columbia Press.
Garfield, V. E., Wingert, P. S., & Barbeau, M. (1950). *The Tsimshian: Their arts and music*. Publications of the American Ethnological Society XVIII. J. J. Augustin.
George, E. (2003). *Living on the edge: Nuu-Chah-Nulth history from an Ahousaht Chief's perspective*. Sono Nis Press.
Giddens, A. (1984). *The constitution of society*. University of California Press.
Glavin, T. (1998). *A death feast in Dimlahamid*. New Star Books.
Glavin, T. (2000). *The last great sea: A voyage through the human and natural history of the North Pacific Ocean*. Douglas & McIntyre.
Goodman, L. J., & Swan, H. (2002). *Singing the songs of my ancestors: The life and music of Helma Swan, Makah elder*. University of Oklahoma Press.
Graeber, D., & Wengrow, D. (2021). *The dawn of everything: A new history of humanity*. Farrar, Straus & Giroux.
Harkin, M. E. (2007). Swallowing wealth: Northwest coast beliefs and ecological practices. In M. E. Harkin & D. R. Lewis (Eds.), *Native Americans and the environment: Perspectives on the ecological Indian* (pp. 211–232). University of Nebraska Press.
Ignace, M., & Ignace, R. (2020). A place called Pípsell: An indigenous cultural keystone place, mining, and Secwépemc law. In N. J. Turner (Ed.), *Plants, people and places: The roles of ethnobotany and ethnoecology in indigenous peoples' land rights in Canada and beyond* (pp. 131–150). McGill-Queen's University Press.
Inglis, R., & Haggarty, J. C. (2000). Cook to Jewitt: Three decades of change in Nootka sound. In A. L. Hoover (Ed.), *Nuu-chah-nulth voices, histories, objects, and journeys* (pp. 92–106). Royal British Columbia Museum.
Kennedy, D., & Bouchard, R. (1998). Northern Okanagan, Lakes, and Colville. In D. E. Walker (Ed.), *Handbook of North American Indians: Plateau* (Vol. 12, pp. 238–252). Smithsonian Institution.

References

Ḵi-ḵe-in. (2000). Responses, in "A conversation with Ḵi-ḵe-in" by Charlotte Townsend-Gault. In A. L. Hoover (Ed.), *Nuu-Chah-Nulth voices, histories, objects, and journeys* (pp. 203–229). Royal British Columbia Museum.

Ḵi-ḵe-in (2013). "Hilth Hiitinkis—From the beach." In C. Townsend-Gault, J.R Kramer, & Ḵi-ḵe-in (eds.), Native art of the northwest coast: A history of changing idea (pp. 26–30),. : University of British Columbia Press.

Kramer, J. (2013). 'Fighting with property': The double-edged character of ownership. In C. Townsend-Gault, J. Kramer, & Ḵi-ḵe-in (Eds.), *Native art of the northwest coast: A history of changing ideas* (pp. 720–756). University of British Columbia Press.

Kropotkin, P. (2004). *Mutual aid, a factor of evolution*. Wm. Heinemann.

Lévi-Strauss, C. (1982). *The way of the masks*. Trans. by S. Modelski (Fr. orig. 1979). University of Washington Press.

MacMillan, A. D. (1999). *Since the time of the transformers: The ancient heritage of the Nuu-chah-nulth, Ditidaht, and Makah*. University of British Columbia Press.

Martin, C. (1999). *The way of the human being*. Yale University Press.

Matson, R. G., Coupland, G., & Mackie, Q. (Eds.). (2003). *Emerging from the mist: Studies in northwest coast culture history*. University of British Columbia Press.

Mauss, M. (1990). *The gift*. Trans. by W. D. Halls. (Fr. orig. 1925.). Routledge.

Menzies, C. R. (2016). *People of the saltwater: An ethnography of Git lax m'oon*. University of Nebraska Press.

Miller, J. (1982). Tsimshian moieties and other clarifications. *Northwest Anthropological Research Notes, 16*, 148–164.

Miller, J. (2014). *Rescues, rants, and researches: A review of Jay Miller's writings on northwest Indian cultures*. Northwest Anthropology, Memoir 9.

Miller, J., & Eastman, C. (Eds.). (1984). *The Tsimshian and their neighbors of the North Pacific coast*. University of Washington Press.

Minta, S. C., Minta, K. A., & Lott, D. F. (1992). Hunting associations between badgers and coyotes. *Journal of Mammalogy, 73*, 814–820.

Nadasdy, P. (2003). *Hunters and bureaucrats*. University of British Columbia Press.

People of 'Ksan. (1980). *Gathering what the great nature provided*. University of Washington Press.

Piddocke, S. (1965). The potlatch system of the southern Kwakiutl: A new perspective. *Southwestern Journal of Anthropology, 21*, 244–264.

Pierotti, R. (2011). *Indigenous knowledge, ecology, and evolutionary biology*. Routledge.

Pinkham, A. V. (2006). We Yao o Yet Soyapo. In A. Josephy (Ed.), *Lewis and Clark through Indian eyes* (pp. 137–162). Alfred Knopf.

Ramsey, J. (1977). *Coyote was going there: Indian literature of the Oregon country*. University of Washington Press.

Reid, J. L. (2015). *The sea is my country: The maritime world of the Makah*. Yale University Press.

Reyes, L. L. (2002). *White Grizzly Bear's legacy: Learning to be Indian*. University of Washington Press.

Reyes, L. L. (2006). *Bernie Whitebear: An urban Indian's quest for justice*. University of Arizona Press.

Ridington, R. (1988). *Trail to heaven: Knowledge and narrative in a northern native community*. University of Iowa Press.

Robertson, G. S. (1896). *The Kafirs of the Hindu-Kush*. Lawrence and Bullen.

Rosman, A., & Rubel, P. G. (1971). *Feasting with mine enemy: Rank and exchange among northwest coast societies*. Waveland Press.

Ross, J. A. (2011). *The Spokan Indians*. Michael J. Ross.

Roth, C. F. (2008). *Becoming Tsimshian: The social life of names*. University of Washington Press.

Ryan, J. (2000). Preface. In M. Anderson & M. Halpin (Eds.), *Potlatch at Gitskegukla: William Beynon's 1945 field notebooks* (pp. ix–x). University of British Columbia Press.

Sapir, E. (2004). *The whaling Indians: West coast legends and stories, legendary hunters*. Canadian Museum of Civilization.

Sapir, E., & Swadesh, M. (1939). *Nootka texts: Tales and ethnological narratives with grammatical notes and lexical materials*. Linguistic Society of America and University of Pennsylvania.

Sapir, E., Sayachapis, T., Totisim, Swadesh, M., Thomas, A., Thomas, J., & Williams, F. (2007). *The origin of the wolf ritual*. Canadian Museum of Civilization.

Scott, J. C. (2009). *The art of not being governed: An anarchist history of upland Southeast Asia*. Yale University Press.

Sewid, J. (1969). *Guests never leave hungry: The autobiography of James Sewid, a Kwakiutl Indian*. Yale University Press.

Smith, M. W. (1940). *The Puyallup-Nisqually*. Columbia University Press.

Suttles, W. (1987). *Coast Salish essays*. University of Washington Press.

Trosper, R. L. (2009). *Resilience, reciprocity, and ecological economics: Northwest coast sustainability*. Routledge.

Turner, N. J. (2005). *The Earth's blanket*. University of Washington Press.

Turner, N. J. (2014). *Ancient pathways, ancestral knowledge: Ethnobotany and ecological wisdom of indigenous peoples of northwestern North America* (p. 2). McGill-Queen's University Press.

Turner, N., Smith, R., & Jones, J. T. (2005). 'A fine line between two nations': Ownership patterns for plant resources among Northwest Coast indigenous peoples. In D. Deur & N. Turner (Eds.), *Keeping it living: Traditions of plant use and cultivation on the Northwest Coast of North America* (pp. 151–179). University of Washington Press.

Walens, S. (1981). *Feasting with cannibals*. Princeton University Press.

Chapter 6
Traditional Resource Management

Complexity and Challenges

The Northwest Coast has always seemed anthropologically anomalous. Despite being "hunters and gatherers," the Native peoples of this region maintained high populations, lived in large villages, and had a complex social organization (Coupland et al., 2009; Matson et al., 2003; Reid, 2015; Taylor & Grabert, 1984).

This region seems much less anomalous now that we have better understanding of nonagricultural populations elsewhere in the world. Complex nonagricultural populations with large settlements and complicated ownership rules extended all along the Pacific Coast, even into Baja California, and were also characteristic of southern Florida and several other parts of the New World, as well as of northern Australia and Mesolithic Europe (Graeber & Wengrow, 2021). All these societies depended to a great extent on marine and riverine foods, often anadromous fish.

Northwest Coast knowledge of the environment was as thorough, detailed and comprehensive as one would expect from over 16,000 years of continuous occupation of and dependence on the land. Ethnobiology of most of these groups has been documented in detail. The most comprehensive work is the corpus of Nancy Turner's classic ethnobotanical studies, many coauthored with First Nations people, including her magnum opus *Ancient Pathways, Ancestral Knowledge* (2014; see esp. vol. 2: 198–244). Turner has worked with, and collected knowledge from, essentially everyone who has done ethnographic or ethnobiological research on the Northwest Coast in the last 40 years (see Turner [ed.] 2020, and the dozens of acknowledgments in her article "That Was Our Candy!" 2020). The most comprehensive work on animals and places has been done by Eugene Hunn, especially in *N'Chi Wana* (1990) coauthored with Yakama Elder James Selam. His students such as Thomas Thornton (2008) go beyond ethnobiology to discuss major ontological and epistemological issues.

Knowledge can be as simple as knowing bird songs that announce plant seasons. In spring, when salmonberries flower and set fruit, the woods are suddenly full of Swainson's Thrushes returning from their winter homes. Their exquisitely beautiful songs sound everywhere: *woodily woodily woodily woodily*, rising up the scale. Almost everywhere this bird is common, the local groups hear its song as saying their words for "ripen"; some think it actually ripens the berries. For instance, the Saanich (W̱sáneć) of Vancouver Island hear this as "wew̱elew̱elew̱elew̱eś"—"ripen, ripen, ripen" (Turner & Hebda, 2012: 103). The Kwakwaka'wakw also have this belief (Boas & Hunt, 1905: 298–299, and Boas, 1910: 65 with song transcription). Such playful wording of bird songs is widespread in North America, even among ornithologists. The Sahaptin of what is now Washington state see the Bullock's oriole as the harbinger of spring, and disrespect towards this species leads to "failure of golden currant and serviceberry crops" (Hunn & Selam, 1990: 147). These berries are, at one stage of their growth, orange like the oriole.

Northwest Coast peoples managed the environment extremely well (Alberta Society of Professional Biologists, 1986; Deur & Turner, 2005; Hunn & Selam, 1990; Kirk, 1986; Menzies, 2006; Miller, 1999; Turner, 2005). The basic principle was to value, and when appropriate to use, everything alive, which resulted in total-landscape management. The whole region was managed to varying degrees, with productive areas being intensively managed. Because so much was useful, the first rule of management was to keep everything thriving. Since management was by village or extended family, which was the level at which enforcement of respectful magement was possible, it was also necessary to manage for long-term, wide-flung payoffs. Focusing on maximizing short-term payoffs could deplete resources to benefit one or a very few people, though rapid and flexible adaptation for things like large potlatches were always necessary. That made short-term exploitation a selfish crime against society, and it was correspondingly condemned with often draconian judgments.

These cultures were the first to admit that they were far from perfect. Many had tales of overharvesting followed by famine, or of selfish or disrespectful behavior resulting in negative consequences. These tales were retold with cautionary import. The Tsimshianic groups were particularly rich in conservation lore, well documented by Tsimshian anthropologist Charles Menzies (2012, esp.: 165ff). There was a learning curve, and everyone knew it. Many individuals, moreover, tell of their own overharvesting, and of resulting social sanctions that were applied in no uncertain terms (e.g., Atleo, 2011: 50). Such stories are far more revealing of how management really developed than are the romantic stereotypes of the "Indian at one with nature." Competition and cooperation over resources drove protection of one's tribal resource base with awareness of the consequences of failure (Johnsen, 2009; Pierotti, 2011: 37–38).

The Northwest Coast people were, above all, exploiters of the aquatic world. Fish runs are highly concentrated in time and space, forcing people that depend on them to concentrate in large social agglomerations on a regular basis. This in turn requires some sort of comprehensive social organization, with strong authority structures provided by kinship groupings or chieftainship or—typically—both. The best fishing stations are well worth fighting for, and the resulting conflict forces even more tight military organization on the people. Last, the enormous but brief salmon runs of the Northwest require huge bursts of labor, since people need to develop a virtual assembly line of filleting and drying the catch for preservation through the rest of the year. These periodic abundances required careful conservation practices (see discussion of First Salmon ceremonies below). Variable abundance has some negative consequences, because slavery becomes a likely economic consequence under such circumstances. Raiding for persons to enslave adds to fighting for fishing sites as an economically important activity.

However, humans do not live by fish alone. A diet of fish would be too high in protein and vitamin D (Deur & Turner, 2005: 33). We now realize that the Northwest Coast groups practiced extensive horticulture. (Those that did not get much plant foods had to eat a great deal of fat, like the Inuit; this was one reason for the importance of the fat-rich ooligan.) They managed, increased, transplanted, and selectively harvested plants, and had at least one or two domesticated species. Tobacco was the only widespread domestic crop—which says something about human preferences, even though in these cultures tobacco was used almost exclusively for ceremonial purposes—but "wild" plants like camas were managed to the point of quasi-domestication. Some authorities now refer to the Northwest Coast people as horticulturalists, rather than hunter-gatherers.

Complexity has traditionally been explained by the "rich" or "abundant" resources of the Northwest, but this is a false premise. The resources consist largely of fish, which are not always easy to catch, and depend upon year classes which can be strong or weak. The great salmon runs of the major rivers are often incredible, but one must average out the productivity of the rivers over the thousands of square miles of barren mountains and nutrient-poor forests that make up 99% of the area of the Northwest Coast. (The exception is the Plateau, largely sagebrush or grass steppe.) In the southern areas—from central Washington and extreme southern British Columbia southward—root crops, berries, and nuts become increasingly abundant, but not common enough to be major staples until south of the Columbia River. Moreover, salmon runs sometimes fail, especially in the north (Pierotti, 2011: 65). When a run failed, all the people could do was move away, hopefully finding some relatives to visit, or switching to hunting.

For example, maintaining the population of eight to ten thousand on Haida Gwaii was no easy undertaking. The land was rather poor in berries and roots (though not as poor as it is now). There were few large game animals, only a tiny herd of caribou. The survival of these caribou into recent times shows the success of conservation practice in Haida Gwaii; at any point, a few good hunters could have exterminated them. There were not even any large salmon streams. The people had to carry out dangerous open-sea fishing for halibut and hunting for sea mammals. Even more difficult was the need to cross the wide and treacherous Hecate Strait to trade for ooligan oil (necessary for nutrition) and other goods with the often less-than-friendly Tsimshian. The Haida needed formidable toughness and intelligence, which is reflected in their stunning art and oral literature. Similar, if less dramatic, cases could be made for other groups along the coast and into the deep interior.

Under the circumstances, it is not surprising that all possible resources were possible for use, which provided, and still provides, strong incentive for managing the entire landscape sustainably. Worldwide, other things being equal, the higher the percentage of natural resources that are used, the greater the need to maintain the ecosystem (ENA, research in progress). Conversely, when people depend on a handful of non-native foods, as in the modern USA, there is much less incentive to maintain the natural environment, and, indeed, every incentive to destroy it. The wide spectrum of foods and industrial resources used by Indigenous peoples in the Northwest contrasts strikingly with the extreme biological poverty of crops grown on any significant scale by white settlers.

Recall that not only were salmon and sea fish important, but river fish, which are now neglected, such as suckers and lampreys, were major resources. Royle (2021) provides a thorough review of use of suckers (Catostomatidae) in northwest North America, providing data on every cultural group known to use them. Some groups disliked them for their numerous intermuscular bones, and even cautioned children against eating them, but other groups considered suckers as important and worthy of care as salmon. Some had First Sucker rites similar to First Salmon ceremonies. Suckers have dropped out of use today, and are even considered "trash fish" by most settler-society fishers, again because of the annoying bones. Indigenous users sometimes processed them in ways that made bones soft and edible. Lampreys remain favorite foods, though now very hard to get (Miller, 2012). Dams and pollution kill them. They are perceived as parasitic on fish considered more valuable, such as salmon and trout, which discourages protection. Shellfish, all edible plants, fungi, and even lichens were drawn on for food.

Wet sites that preserve many types of remains for archeologists have given us new insights into the Northwest. At Ozette (an abandoned Makah town) in Washington state, whaling was far more important than had previously been realized, even though the Makah have long been known to hunt gray whales. In addition, wooden arrows without stone points were far more common than stone-tipped ones (Croes, 2003), showing that hunting for elk and black-tailed deer was considerably more important than the relative rarity of classic "arrow point" discoveries had suggested. (Extremely hard wood that will take a sharp point, such as that of *Holodiscus discolor* and crabapple, makes wood-tipped arrows deadly.) Baskets manufactured there and elsewhere show cultural continuity and adaptability over long time periods. The importance of basketry reveals the need for transporting and storing food (Bernick, 2003; Croes, 2003).

Tight grounding of conservation in spiritual beliefs and rituals, including origin myths, is found everywhere in the region. This goes a long way to disprove the occasional claim that conservation is actually a new idea, learned from the Whites. This claim has been advanced by some American anthropologists, such as Shepherd Krech (1999). Many early ethnographers segregated "religion" and "subsistence" in separate parts of their ethnographies, which allowed conservation, which to indigenous peoples is a spiritual matter and the sources of most important ceremonies and rituals (Pierotti, 2011: 17–23) to become lost somewhere between these concepts. David Arnold (2008), for instance, saw "religion" as inadequate for conservation, and "property rights" as needed—apparently without realizing that the Tlingit he was discussing saw property rights as part of religion. More accurate would be to take the religion-environment-subsistence interaction as the basic unity, with both "religion" and "subsistence" being artificial categories imposed from outside, upon cultures where subsistence and survival were strongly linked to the way these peoples conceived of dealing with nonhuman entities (Anderson, 2014; Pierotti, 2011: 37–39).

Calories and Food

Deur and Turner (2005: 33) note that the famous fish dependence of the Northwest Coast peoples must have been overstated, because living on fish would cause unhealthy overconsumption of protein and perhaps of vitamin D. The peoples in question had to have been eating a lot of carbohydrates, or fats (from sea mammals and ooligans). Records show that diet breadth was always great; these people worked to maintain diversity in all matters and were fully aware of nutritional issues prior to invasion and colonization.

> An account of a typical meal was provided by William Clark around 1805, in his famously creative spelling. He "was invited to a lodge by a young Chief" of the Clatsop in Oregon, and received "great Politeness, we had new mats to Set on, and himself and wife produced for us to eate, fish, Lickorish [shore lupine, *Lupinus littoralis*], & black roots [edible thistle, *Cirsium edule*], on neet Small mats, and Cramberries [*Vaccinium oxycoccos*] & Sackacomey beris [kinnikinnik, *Arctostaphylos uva-ursi*], in bowls made of horn, Supe made of a kind of bread made of berries [cf. salal, *Gaultheria shallon*] common to this Countrey which they gave me in a neet wooden trancher, with a Cockle Shell to eate it with." (Quoted in Gahr, 2013: 65, from Lewis & Clark, 1990: 118; her notes in brackets).

The dominance of fish and berries is notable, and the few plant foods mentioned were roots.

On the other hand, it should never be forgotten that roots, berries, and shellfish are all low to extremely low in calories, even if they are often high in crucial nutrients. Northwest Coast people could not have gotten a high percentage of their calories from such foods. What has been missed is not so much the plant foods, but the importance of oil, both from ooligan (eulachon) fish and from mammals (Steller sea lions, gray whales, harbor seals, deer, elk, and beaver). Early accounts stress the enormous importance of oils in trade, feasting, and food. The Makah used to compete to see who could drink the most whale oil at feasts (Colson, 1953). People were desperate for oils. Recall that "ooligan" is derived from a Tsimshian word meaning "savior," a word now used for Jesus. Watertight boxes of oil from the ooligan, a fish that is mostly fat by dry weight, were traded all up and down the coast. The biggest ooligan run was on the Columbia River, where the commercial catch in the peak year, 1945, was over 5.5 million pounds. The noncommercial (including Indigenous) catch was at least as high (Reynolds & Romano, 2013). The fish is almost extinct in the Columbia drainage now, thanks to dams and pollution, but is still abundant in Gray's Harbor and a few undammed rivers.

Oil provided necessary calories. In the Northwest, clothing is not much barrier against the constant rain, which soaks or trickles through anything. Thus, oil was used externally as "clothing," and consumed internally to keep metabolism up. Anyone who has lived and worked with Native people on the outer coast is aware of their frequent use of very light clothing for work in rain, their incredible caloric requirements entailed by working in that climate, and the incredible amounts of food needed to maintain body warmth while carrying out any task. Early descriptions of people paddling for four continuous days in cold rain (Drucker, 1951; Reid, 2015; Sproat, 1987 [1868]) may be hard to match now, but one can easily witness many feats of sustained effort in cold rain that at least make the old stories thoroughly credible. They are possible only if the workers are eating heavily. Today, the old oil sources are largely gone, and drinking oil straight is no longer in vogue, so recent theorists generally miss the importance of this in the past.

Finally, the region had few food taboos. Animals that ate carrion—including those famous tricksters, ravens and coyotes—or were otherwise more or less disgusting were avoided as food, though not as spiritual icons. The Haida did not eat killer whales, because humans (especially males, most especially those who die by drowning) are assumed to be reincarnated as killer whales. Killer whales are easily identifiable as individuals by their patterns of black and white. As of the 1980s, Haida would discuss which local killer whale was which recently deceased Haida man (Boelscher, 1988: 183 and personal observation by ENA). This practice can still be seen in the case of the Orca called Luna. the Canadian Department of Fisheries and Oceans (DFO) authorized an effort in June 2004 to rescue Luna and return him to his pod. The plan was opposed by the Mowachaht/Muchalaht First Nations, who believed Luna was the reincarnation of a former chief.

The Haida word for "killer whale," *sgaana*, also means "spiritual power." (their word *sgaaga*, means one who has power; one high power spirit figure was *Sins Sgaanaa*, "power of the shining heavens"; Boelscher, 1988: 172–173.) The Haida also regarded some foods as high status, some as low; the former were foods that came from large animals—whales being particularly high—or that were taken by collective effort, like whales, fur seals, salmon and berries. Low-status foods were those that involved individual and unobtrusive collecting: such as shellfish and similar beach foods, especially, though well-liked ones such as chiton were higher on the scale (Boelscher, 1988; ENA, personal observation). There seems to be little comparable information on other groups, except the general observation that salmon and large land mammals are highly regarded everywhere. Whales were certainly special among the whaling peoples.

Fish: The Key Resource

David Arnold (2008: 24) estimates that the Tlingit consumed about 150,000,000 fish per year, weighing some 6 million pounds, in aboriginal times, given the very conservative estimate of 500 pounds of fish per person. In fact, they could well have required twice that. Five hundred pounds per year would represent only about 1200 calories of fat salmon per day, and only half as many of lean fish. The Tlingit had little but fish and other sea animals to eat.

Indeed, at the other end of the region, the average Spokan ate almost 1000 lb. of salmon in a year (Ross, 2011: 359; the figure may better apply to all fish). This would be about 1,400,000 pounds per year for the whole tribe, representing well over 100,000 fish (Ross, 2011: 359, citing Allan Scholtz). The Spokan, unlike the Tlingit, got half their calories from plant foods. The nearby Sahaptin got 30–40% of food from salmon, and much of the rest from other fish. This appears to be about typical for inland groups, though the far-upriver groups got only 2–5% from salmon, while the central Fraser River tribes got 40–60% (Hunn & Selam, 1990: 148–150).

The Tlingit got their salmon at river mouths, where they were fat, while the Spokan caught fish far upriver, where they had used most of their fat reserves fighting their way up the Columbia. A reasonable guess for the Northwest Coast as a whole would be two pounds of fat fish or three of lean per day. These fish were often herring, halibut, rockfish, or others, as well as salmon.

Most fish are low-calorie fare. Three hundred calories per pound is typical for nonmigratory fish like halibut and rockfish. Salmon fatten up at sea for their runs up the rivers, and a fat salmon entering a river could be four times that rich in calories, but at sea or far upriver even salmon are not high in calories. Herring and oolichans (and their and other fish eggs) are high to very high in fat, but are not available except during brief runs. However, they are typically stored to increase the range of availability. Life outdoors in a cold, wet climate made caloric consumption a high priority. Bears and other predators took a huge number of salmon; their pressure on the resource base was considerable. Fish were abundant enough, however, that it took commercial fishing to destroy the resource because of overfishing which indigenous peoples would never have permitted.

Social groups had depended on the fish for millennia (Carothers et al., 2021). Accumulated wisdom specified what people could take and how to maximize returns. They used a variety of gears, traps, weirs, and other devices, including stone weirs to catch fish on receding tides (Menzies, 2016). Conservation was generally accomplished by limiting days of fishing, permitting escapements, especially of the largest, most fecund individuals, and improving breeding conditions, rather than by limiting gear or daily takes. Improving spawning success included things such as returning fish wastes to water for nutrient recycling, and protecting spawning streams. Cultures upriver in spawning areas often cleared redds—the gravel spawning beds that can be ruined by tree falls and other accumulated mass. Weirs would be opened at regular times, for escapement. Boas recorded an early and ongoing conflict over weir escapement that was resolved by marriage (Boas, 2006: 105).

Fish management was notably careful (Johnsen, 2009). Not only were spears and hooks universal, but very effective netting was made from grasses. Trawl nets and set reef nets were used in the Salish Sea, and seines, weirs, scoop nets, and other nets were used everywhere. Nettles might be grown especially for making nets, or tough grass traded in from other areas (Miller, 2014: 83). Given these sophisticated methods and the concentrated nature of runs, individuals or small groups could easily "rob a creek": take all the fish, leaving none for reproduction; thus ending fishing there for the foreseeable future (as was done by many settler-society fisheries). Indian agents' letters suggest it was ordinary confrontations and administrative decisions over fishing spaces, gear, closed seasons, and licenses, rather than the official policies of the Department of Indian Affairs, that redefined native fishing in accordance with settler interests. By extending so-called privileges to native fishers, Indian agents worked to conserve the resource for a settler society, and imposed forced assimilation of native fishers into state management practices (Schreiber, 2008).

Strict limits were set on fishing by the First Nations, which was usually religiously represented through powerful ceremonies in which all took place. Strong religious rules and ceremonies governed fishing. This is well documented all the way from the Tlingit in the north (Arnold, 2008) to the Yurok in the far south (Swezey & Heizer, 1977), and for most groups in between, such as the groups along the Skeena (Morrell, 1985). The first fish was subject to ceremonies, and often had to be taken by a religious leader (Hunn & Selam, 1990: 153; Mishler & Simeone, 2004: 126). This tradition of the First Salmon ceremonies of tribes in the Pacific Northwest and southeastern Alaska (Gunther, 1926) illustrates the importance of fish as prey to Indigenous Americans can be seen:

> When a salmon run began, no one was allowed to fish until the first catch was ceremonially welcomed. The fish was bathed, painted with red ochre, and sprinkled with bird down. The fish was ceremonially carried to the place where it would be prepared, laid on a bed of ferns, and dabbed with ochre. Women butchered the salmon, carefully removing the flesh from the spine in one piece so it could be roasted over a firepit. Prayers were offered so that the fish would look kindly on the people and return in great numbers... All of the flesh was eaten, sometimes by the entire community, sometimes only by elders and children... The bones, head, and entrails were carefully collected in a basket or mat and reverentially deposited in the river or the sea where the fish had been caught. This was intended to convince the salmon that it should come to life again and lead others of its kind to the fishing sites (cited in Pierotti, 2011: 64; this generic description seems based on Kwakiutl and Puget Sound Salish practice.)

People waited until the season was thus opened. This allowed leaders to provide for escapement, for spawning or for the tribes upriver.

In a classic paper Swezey and Heizer (1977), reported for the Klamath River Nations that the conservation for other peoples was a motive; because not allowing enough fish to get through the lower river weirs was cause for the upriver tribes to make war on those downstream. At the other end of the long Northwest Coast region, the Tsimshian report a war involving competitive destruction of each others' fishtraps (Barbeau & Beynon, 1987, vol. 2: 32). First Salmon Ceremonies not only allowed escapement for breeding fish, but also kept the peace. On the Chehalis River in Washington, the same story held: the upriver groups had to negotiate with the downriver ones to open the major weir and allow escapement (Hill-Tout, 1978c: 97–98). Other groups paid a royalty for the privilege of fishing the Chehalis (p. 101).

Most areas had a fish leader to start the season. The Lakes (Sin-Aikst, Syilx) and Colville Salishans fished at Kettle Falls. The Lakes scholar Lawney Reyes, recalling his childhood, reports:

> During the salmon harvest, the head authority at the falls was the appointed Salmon Chief. Before any nets or traps were set or fishermen were positioned, he stood near the lower falls, facing downriver, and prayed. The Salmon Chief welcomed the arrival of the chinook. He apologized and thanked those salmon that would be taken in the traps. He assured the chinook, that most would be allowed to go upriver to spawn and bring forth their young.... It was the responsibility of the Salmon Chief to see that the closely related tribes...shared equally in all salmon that were caught. Other tribes that came to fish at the falls were also treated fairly.... According to Sin-Aikst tradition, the first Salmon Chief was chosen by Coyote.... [who] made Beaver the Salmon Chief [and told him]...'You must never be greedy and you must see to it that no one else is greedy' (Reyes, 2002: 46–47).

The Lakes and their neighbors had, in fact, such a supervisor for all hunting and fishing activities—there was a hunting chief, as well as a salmon chief, and other fishing leaders. Such leaders of hunting and fishing had a special title, *xa'tús*, from a word for "first" (Kennedy & Bouchard, 1998a, 1998b: 241; Miller, 1998: 255). The salmon chief was locally called in English "salmon *tyee*" (Miller, 1998: 255), *tyee* being Chinook Jargon for "chief."

> Among the Lower Lillooet, resource stewards...directed the use of specific hunting grounds and some fisheries. These were hereditary positions, at least in the historic period, but required special knowledge. Spiritual qualities were not requisite for this position, but such qualities were required for the specialized hunters (Kennedy & Bouchard, 1998a: 182).

All other groups in the northern interior had these stewards also. The Spokan, for instance, had a fish leader or fish chief to oversee fishing, and a fish shaman to make sure that all the rules of proper resource management were kept; there was also a fish trap keeper (Ross, 2011). Rules that were pragmatic (no blood or waste in the water; it repels salmon) merged into purely ritual rules, all of which were believed equally necessary to keep the fish coming.

The Secwépemc (Shuswap), and some other groups, had recognized resource stewards, which was a named office (Ignace & Ignace, 2017: 193–194). They watched over fish, game, and plant stocks, organized and timed burning to maximize berries and other resources, maintained trails, and oversaw conservation. They appear to be a development from the "fish chiefs" and "game bosses" reported generally over western North America (see above).

The Nuu-chah-nulth and Tlingit, at least, and probably other groups, stocked eggs and fry in salmon streams (Deur & Turner, 2005: 19; George, 2003: 74). Gilbert Sproat describes this practice from the early nineteenth century (Sproat, 1987 [1868]: 148). Since Sproat was the first European to visit some of the areas he described, this was evidently a pre-contact practice.

A Saanich origin myth of salmon, and of the use of wild parsley (*Lomatium*) as "medicine" to lure it, tells of the creator hero-twins stocking salmon as they went around Vancouver Island (Turner & Hebda, 2012: 132). Earl Maquinna George describes the process in detail, also pointing out that the Nuu-chah-nulth would clear choking vegetation out of streams, leaving enough to keep the stream healthy. Salmon need some logs for shelter and protection for fry. The Nuu-chah-nulth also understood that bears were people too, and needed and deserved their fair share of the catch, for all to cooperate and consent (Atleo, 2011: 100–101). This probably indicates understanding such benefits as the fertilizing effect of bears on streamside vegetation.

The European settlers found fish incredibly abundant. Europeans were fortunate to have invaded during a time of unusually high runs, which they then assumed were typical, a situation the two cultures have been in conflict over ever since (Pierotti, 2011). An example can be seen in Michael Harkin's argument about the superabundance of resources in the Pacific Northwest and Alaska that:

> ...reciprocity "implies the *need for a total balance* in the exchange of substance between human and nonhuman beings," whereas predation presumably involves "*aggressive exploitation* of nonhuman beings as are useful" (Harkin, 2007: 217; emphasis added). Harkin goes on to argue that humans "never repay their debts to animals and thus never achieve genuine reciprocity... humans appear almost exclusively as predators rather than as equal partners" (218). This line of argument seems wrong at least two levels... Harkin seems to completely ignore the idea that ceremonies and rituals designed to placate the animals taken are as close to full reciprocity as any predator can get...Following Harkin's logic, no predator exists in a co-evolutionary or egalitarian relationship with prey species, which, although it ignores evolutionary ecology is a position regularly occupied by scholars who argue that Indigenous people do not really understand how to maintain a sustainable relationship with nature" (Pierotti, 2011: 64).

It is worth noting that Harkin also ignores the clearing of streams and redds, by these peoples, which clearly benefitted salmon in their efforts at reproduction. We address Michael Harkin's work because it is particularly thorough, valuable, and insightful, rather than because it is inadequate. Its thoughtfulness makes it worth unpacking. However, his argument fails to recognize that:

> Surviving in a fluctuating environment (A philosophical requirement for Indigenous lifeways) is a very different reality than living in "harmony" with the "balance of nature," as Indigenous life is often caricatured... The assumption of a "balance of nature" whereby ecosystems, ecological communities, and animal populations are presumed to exist under equilibrium conditions to which they are inclined to return any time they experience a perturbation, is a Western concept, (with) roots in the attempts of European philosophers to link science and religion through economic metaphors during the seventeenth century. "The creator had designed an integrated order in nature that functioned like a single, universal, well-oiled machine" (Worster, 1994: 39; Pierotti, 2011: 34).

Some of this abundance may be due to population recovery after Indian fishing was reduced in the contact period (Ames, 2005: 84), but an extremely important and little-remarked fact is, even the tiniest and most insignificant streams had runs at contact. Ames (2005: 83–85) remarks that there is no archeological evidence of depletion of the fish resource, and that people continued to rely on a wide range of sea and land foods throughout the millennia, although this seems to be white imagination. They always describe resources, especially fish and wildlife populations as boundless or limitless, just before they succeed in destroy them. (Evidence of local and temporary depletion has turned up since 2005 (e.g., Rick & Erlandson, 2008, 2009).

Normal natural processes might have exterminated some of these small-stream runs. Such streams have few fish and short run-times. Since a large percentage of salmon run eventually into creeks small enough to lose runs by chance or by individuals "robbing creeks," overfishing, especially as exemplified by western fishery 'management' can quickly wipe out even a huge river's salmon stocks (see Myers & Worm, 2004). Indigenous fishing was heavy. Only stocking? combined with careful conservation could have maintained the runs.

Such behaviors extend to Siberia. Indigenous elders on Kamchatka counsel leaving some salmon and opening weirs (Kasten, 2012: 79). Yukaghir also have this rule (Jochelson, 1926: 150). Kasten tells of one Kamchatkan elder who remembered asking, as a child, why some fish were spared, and being told those were females that would reproduce the fish. (Presumably somebody pointed out that males were also needed.) Today, the runaway market in Russia for caviar has doomed the female fish, which are often simply stripped of caviar and left to rot, to the horror of traditional persons. Kasten recommends reviving traditional ways of conserving before it is too late. Comparable accounts from Siberia make it clear that conservation was general there among Indigenous peoples (e.g., Koester, 2012; Wilson, 2012). In Alaska, the Alutiiq conserved and managed fish until settler societies forced overfishing (Carothers, 2012).

The evil of monopolizing fish was often written directly into the mythic charter of society. A widespread story involves Coyote forcing greedy downriver people to open their weirs so the salmon would run upriver. Usually, the greedy ones are two sisters, whom Coyote outsmarts, destroying their weir. Many versions are recorded. In some Coyote goes on to cause salmon to run up rivers where people gave him wives (Boas, 1917: 67–72, from Teit's Okanagan recordings). Sometimes Coyote turns the selfish sisters into sandpipers (Boas, 1917: 68, for the Okanagan; p. 101–102 for the Sanpoil; p. 121 for the Coeur d'Alene, p. 121; Sahaptin, 140–144; see also Teit, 1909: 630). The story, without the sandpipers, is also noted by James Teit for the Nlakapamux (Teit, 1898: 27, 1912: 301–304 and references there), in Louis Simpson's Wishram stories told to Edward Sapir (Sapir, 1990: 29–33, 290–293; the first was poetically redone by Dell Hymes in Sapir, 1990: 343–354), and in many other records (e.g., Hill-Tout, 1978a: 143–144 for the Salish; Jacobs, 1945: 236–237 for the Kalapuya of Oregon; Archie Phinney, 1934: 26–29, 380–381 and Walker and Matthews, 1994: 43–52 for the Nez Perce; Ramsey, 1977: 47–49, from Edward Curtis; Thompson & Egesdal, 2008: 177–192 for the Moses-Columbia). A similar story was the charter for conservation and rejection of overtake and monopoly among the Klamath River tribes, as noted above (Swezey & Heizer, 1977).

Some of the locations mentioned in James Teit's accounts are notorious for stream-blocking landslides. In Seattle, on the Duwamish, a natural volcanic dyke that once blocked the river is reported to be the ruins of the two women's dam. It can still be seen today (Cummings, 2020). An earthquake may have been the real "Coyote" that destroyed this blockade.

Another conservation measure was returning inedible parts of the fish to the water. Salmon die on spawning because the young depend on the nutrients from their parents' bodies; this enormous import of nutrients from the sea to the river headwaters allows the fantastic productivity of Northwest Coast streams, which would otherwise be extremely nutrient-poor. (Coniferous forests make oligotrophic streams.) The Native people know this, and some at least see it as equivalent to human parents feeding their children (Gottesfeld, 1994b).

Returning the bones to the water was widespread along the acidic, oligotrophic streams of the coast, but on the Plateau, the bones were often ground up for human consumption (Ross, 2011: 429), and in western Washington and Oregon and among Tsimshian groups the bones were burned, while the Nuxalk threw salmon remains in the forest. The streams in those areas are richer in nutrients and do not need the addition. The Gitksan returned beaver bones to the beaver ponds in hopes that the beaver would repopulate the pond, presumably being reincarnated (People of 'Ksan, 1980: 45).

The Tlingit of Alaska stocked or restocked streams and took out beaver dams that were blocking these streams. They also excavated shallow pools (*ish*) in small spawning streams, to prevent drying in summer, freezing in winter, and excess predation. They devised a complex trap system to divert salmon fry while trapping Dolly Varden trout, which eat salmon eggs. They not only prevented overfishing, but were careful to take about three males for every one female from the upriver runs, to maximize spawning. This was done with the usual recognition of salmon as people deserving and requiring respect. Steven Langdon calls this an "existencescape of willful intentional beings." Following Eduardo Viveiros de Castro (2015), he speaks of "relational ontology" and "abductive reflexivity" (Langdon, 2016), a rather abstract way of saying that salmon are people and must be treated as such if they are to survive into the future.

Sea Mammals

The Nuu-chah-nulth and closely related Ditidat and Makah of the outer coasts of Vancouver Island and the Olympic Peninsula depended to varying degrees on taking whales and northern fur and harbor seals. The Makah may have gotten 80–90% of their food from whales and seals, and some Nuu-chah-nulth groups were not far behind (Coté, 2010; MacMillan, 1999; Reid, 2015). They could stay out in canoes for days, keeping small fires on flat stones or other fireproof platforms in the canoes (Reid, 2015: 143–145). Earlier authors usually thought that Indigenous groups could not have caught many whales, but excavations at the Ozette site in Washington proved the contrary. Further historical scholarship by Joshua Reid and others has shown that the Native groups took not only the docile grays but also humpback, sperm, and other whales, supposedly (but dubiously) even blues and orcas (Reid,

2015: 144–145). Ozette was a village covered by a landslide on the Olympic Peninsula coast about 300 years ago. Archeological work, done in cooperation with the Makah, revealed a great deal of fine art, like that historically known, as well as large houses, and a complex use of plant materials, including wood from all the common trees (see Turner, 2014: 1: 108–110 for lists).

Whaling from an open canoe is very much like the whaling Herman Melville so graphically described in *Moby Dick*, but without the support of a large sailing ship. The difficulty of striking a swimming whale with an enormously heavy harpoon from an open boat is made clear by a famous photograph. An early photographer, Asahel Curtis, managed to immortalize the moment when Makah whaler Charles White struck a whale, and the picture has been reproduced countless times (see e.g., Reid, 2015: 175; a huge blowup of the picture decorates the Makah Cultural and Research Center in Neah Bay. This photograph is also said to be of Makah whaler Shobid Hunter, by Renker & Gunther, 1990: 424–425). It is certainly as dramatic a photograph as world fisheries afford. It is not surprising that whalers resorted to magical practices that could involve months of ritual preparation (Drucker, 1951; Jonaitis, 1999; Reid, 2015: 150). Whaling demanded months of retreat and purification for "magical" power, which was basically achieving the proper state of mind and physical strength.

On a more practical note, these groups had to be prepared for finding themselves far out at sea. A harpooned whale could run seaward, towing the harpooner's canoe on what Melville's whalers called a "Nantucket sleighride," perhaps better described as the first indigenous power boat. They could take days to paddle back to shore, especially if they were towing a whale carcass (Drucker, 1951). If the night was clear, they could "steer by the pole star," but fog was typical: "Combinations of regular swell patterns and winds enabled them to fix their approximate location, even in the regular fogs that conceal the coast. Experienced Makah mariners also used the water's appearance and the set of the riptide to approximate their location when out of sight of land," as well as the cries of gulls as they neared Tatoosh Island, the most northwestern point in the continental USA, and a major seabird colony (Good et al., 2000). They were also experienced at predicting the weather (Reid, 2015: 142). They thus had a navigational science similar to that of the Micronesians and other deep-ocean voyagers.

The Makah also developed a commercial fur sealing industry in the late nineteenth century, even purchasing and running some long-distance-voyaging schooners (Reid, 2015: 177–196). Unfortunately, overfishing by settlers' industrial craft virtually wiped out the whales and seals by the early twentieth century. There has since been some recovery, by all the exploited species, except Steller sea lions; these underwent a decline in the 1990s, probably due to food depletion by commercial overfishing, or to long-term ecological patterns, such as those that impact salmon abundance on a multi-decadal scale. Gray whales, which are specialists on benthic invertebrate prey, have recovered to such a degree they have been removed from the Endangered Species list, which has led to a new controversy as Makah requested reinstatement of the tribal treaty guaranteed whaling rights. A well-balanced account of this controversy can be found in Sullivan (2000; see also Reid, 2015).

Whale and seal products were traded widely, and these groups often had to depend on trade relationships for necessary food items; they were, to a significant extent, specialized whalers. Such dependence on trade for staple food is unusual, to say the least, among hunter-gatherers. The west coast pattern developed from about 3000 to 2000 years ago, but dependence on whaling probably increased after the latter date. The Nuu-chah-nulth naturally yearned to possess the safer living of salmon streams, and fought many a war to acquire such.

On the other hand, sea mammals are not only bigger than fish, they average much more weight devoted to insulation and their oil is of high quality. A seal or whale obviously equals a lot of salmon, and produces an enormous amount of oil as well as meat. It is quite possible that some Northwest Coast groups besides the Makah got most of their calories from sea mammals, not fish. This was certainly the case for Aleutian and other island Alaskan Native groups (e.g., Jolles, 2002).

One indicator of the Makah concern for sea mammal populations is that the Makah recognized that the gray whale population was in trouble early in the twentieth Century and ceased taking whales around 1920 when they feared the population was in dire danger, even at some cost economically and ecologically to the tribe. Ironically, when the gray whale was removed from the Endangered Species list in 1993, and the Makah wanted to resume whaling, this was opposed by settler "conservationists" who claimed they would be driving the population towards extinction (Sullivan, 2000). This issue reveals the difference between the true conservation efforts made by northwest peoples, and the false conservation concerns of settler colonists.

As far as taking female fur seals was concerned, there was more of an issue with conservation and management. The Makah engaged in the take of fur seals as a commercial enterprise starting in the late nineteenth century, apparently as the abundance of gray whales started to decline. They expanded their efforts and used larger commercially built boats (Reid, 2015: 177–196), which allowed them to expand their efforts south to California and North to the Bering Sea. What had been a sustainable hunt when restricted to local waters became destructive to such an extent that the US Government stepped in to regulate not only the Makah, but also the Canadians, Japanese, and Russians, all of whom were also engaged in pelagic fur-sealing that was starting to threaten to threaten Fur Seal breeding colonies on the Pribilof Islands which were owned by the USA as part of the Alaska Territory (Busch, 1987, Chap. 5). The Makah succeeded economically, but brought down management that blamed them for the problem. In consequence, the North Pacific Fur Sealing Commission which was a treaty signed by the four nation states (but not conspicuously by the Makah Nation; Reid, 2015). This set up a harvest program on the Pribilofs but limited pelagic sealing, which had been the Makah specialty. The sealing operation on the Pribilofs was conducted by Aleuts who had been forcibly removed from their ancestral islands and transported to the Pribilofs. Despite ignoring the needs and concerns of Indigenous peoples, this program worked reasonably well until the 1980s when fur seal numbers began to collapse again, possibly in response to climate change and El Nino events.

Birds too were carefully managed, with seabirds and their eggs being taken in a self-consciously sustainable way (Hunn et al., 2003), as is archeologically confirmed by continued abundance of the birds on even very vulnerable islands (Moss, 2007). In seabirds, removal of eggs does little harm to populations, because the adults are long lived (20–40 years), however killing adults can place such populations in peril (Pierotti, 2010).

Managing Shellfish

Shellfish were also carefully managed by developing breeding grounds and enforcing limits. Charles Menzies (2010, 2016: 119–130) has described abalone management by the Tsimshian. Abalone populations collapsed after settler contact and overharvest, however this can be misleading, because abalone are often hard to catch if there is natural predation, because they seek refuge in cracks and under ledges, which renders them difficult to exploit commercially. This means that 'easily taken' abalone disappear, even though the population itself can be quite healthy. This settler overharvesting led to ethnographic erasure of the importance of abalone in precontact times. Menzies' archeological work turned up massive amounts of abalone shells, leading to new realization of its former importance. The same could be said of many other resources and foods that we now are beginning to realize were once important. But now: "In the face of aggressive overfishing of *bilhaa* (abalone) by non-Indigenous commercial fishermen, the Canadian Department of Fisheries and Oceans closed all forms of harvesting of bilhaa" (Menzies, 2010) in spite of the fact that the Gitxaala (Tsimshian) and probably other groups had been harvesting these shellfish with extreme care for thousands of years, yet another example of European assumption of a limitless resource that they managed to destroy within a short time period. The new regulation caused some hardship for the Gitxaala. Menzies documents their caring regime, and advocates a return to a controlled fishery.

Some groups, including the Kwakwaka'wakw and nearby Salishan-speaking groups protected clam beaches by creating low rock walls along the seaward side to prevent beach erosion (Williams, 2006). These "clam gardens" are now the subject of major research (Augustine et al., 2016; Jackley et al., 2016; Lepofsky et al., 2015; Moss, 2011; Toniello et al., 2019). With the help of First Nations advisors and Randy Bouchard and Dorothy Kennedy, experts on the ethnography of the local Salishan groups, Lepofsky and her group have found large numbers of rock walls separating, marking off, shoring up, and preserving large areas of clam beaches. Large stones were moved and arranged to create beach terraces where sand accumulated. Clams could be seeded into these. Large areas not naturally sandy enough for clams were thus turned into highly productive clam beds. Local people painstakingly cleared the beaches of large boulders, cobbles, and logs for countless centuries. Clam digging—which was rigorously managed for sustainability—broke up the substrate, which otherwise became too hard for easy clam establishment. The settler research group tried these various techniques on sample beaches, finding the number of clams increased severalfold, suggesting that these techniques should be introduced to modern beach management.

What Lepofsky and her group found is really clam farming—as intensive and demanding as the oyster and mussel farming now practiced widely by Europeans in Washington state. It is at least as labor-intensive and landscape-changing as such modern shellfish farming, and quite comparable in both labor needs and results to the horticulture of many Native American groups in eastern North America. Along with the emerging knowledge of plant management, such efforts change our idea of the Northwest Coast peoples.

It is very easy to overharvest shellfish, and indeed many Native American groups did, as early as several thousand years ago (Rick & Erlandson, 2008, 2009). Huge shell middens stand up as islands or bars in marshy areas, and cover hundreds of acres along rivers and coasts in the Northwest; some were several meters deep. As with fish, only diligent conservation and management could keep the shellfish resource flourishing in near-settlement locations.

Game Management and Conservation

Game management is less easy to evaluate. Overall, there are few declines demonstrated in the archeological record—in fact, none was found in a major review (Butler & Campbell, 2004; Butler, 2005). They did not even find a shift in exploitation from big game to lower ranked resources. Elk were as common 200 years ago as 5000 years ago, suggesting successful management. Locally, the situation gets more interesting, with patterns of local decline or scarcity contrasting with considerable abundance (Kay & Simmons, 2002, see below). The fact remains that the Northwest Coast was rich in game when first contacted by White settlers, and unlike the situation hypothesized in California by William Preston (1996), this first contact came *before* introduced disease reduced the human population, presumably allowing game numbers to rebound.

An interesting recent finding is that distinct populations of grizzly bears correlate in distribution with language families in the central British Columbia coast (Henson et al., 2021). A northern population is almost exactly coextensive with the Tsimshianic family; south of that is a population roughly coterminous with Wakashan; to the east is a small population occupying the lands of the Nuxalk. The reasons for this correlation are obscure, probably involving human influences as well as local geography impacting distribution of fish and forest resources.

Game was conserved by conscious choice to shift hunting grounds when game became depleted, like rotating grazing pressure in cattle, and by the whole complex of ideas associated with respect for animals (see below): not to take too many, to kill with a shot (rather than wound and allow escape), to use all parts of an animal, and other restrictions. These rules are pan-North American; and all remain well known, for instance, in the Maya towns of Mexico (Anderson & Medina Tzuc, 2005; Pierotti, 2011).

Explicit conservation was common. Among the Puyallup-Nisqually, for example,

> [t]he hunter was said to have known the habits of the particular animals which were within his range. He knew the number of beaver dams, the location of the woodchuck holes. He knew the herds of deer and the runways they used. He is said to have kept track of the birth rate and the natural mortality so that he could so gauge how much he might kill each year without constituting a drain upon the supply" (Smith, 1940: 25).

Such knowledge is not unusual for Indigenous hunters, anywhere in the world, but rarely is it so explicit and general as on the Northwest Coast. "Any waste of the products of the land, especially of food, incurred the displeasure of the dwarfs"— small spirit-beings that enforced morality (Smith, 1940: 132). Only prime resources were taken, with express provisions to leave young for later and to allow escapement for fish (Smith, 1940: 272–273).

Among the Tutchone of the Yukon-Alaska border country, a Tutchone woman recalled group leaders in hunting season telling people to take mountain sheep while they could, but not too many (McClellan, 1975: 98). The neighboring Tagish trapped various animals for food and fur, but, as one person said: "We don't trap them all out, we leave a few for seeds" (McClellan, 1975: 101)—a delightful way of describing sensible hunting. The Nuxalk "are usually careful not to kill more than they need; and knowledge of the supernatural power possessed by animals sometimes leads men to treat them with great consideration" (McIlwraith, 1948: I: 74).

One of the best examples of this was presented from the Kluane people in the western Yukon in relation to the meaning of Traditional Ecological Knowledge:

> ...two different forms of knowledge are being discussed here: 1) patterns that have been shown to work over many generations which are the equivalents of premises or postulates in Western science (e.g., trophic dynamics), and 2) specific observations that are incorporated into these patterns (e.g., notation that one species preys upon or is preyed upon by another at certain times of the year). This can be illustrated by ... an elder and superior hunter named Moose Jackson. 'Moose said he can tell by looking at moose tracks how old they are... where the sun was at the time the moose was there and the direction the wind was blowing. This information, *combined with his knowledge of moose behavior*, is enough for him to know where the moose was going and what it was doing ... (and) is enough to tell him where the moose is... he does not actually have to follow the moose tracks, he merely goes to where he knows the moose will be" (Nadasdy, 2003: 107, emphasis added)... This is very detailed and local knowledge, but it is based upon careful empirical observation... Jackson has previous knowledge of moose behavior. I can readily imagine an experienced field biologist from the Western tradition operating at a similar level of sophistication... The main difference would be the biologist would write down information about wind direction, depth and symmetry of tracks, sun angle, etc. whereas Moose Jackson has this information carefully stored away. There would also be a difference in the way this material is passed on. The biologist would publish the most substantial of their findings, give talks at meetings, and perhaps train student assistants, but they would not include every detail of the activities that they employ in the field. In contrast, Moose Jackson would take younger hunters out and show them directly, and would tell them stories of previous moose hunts. The two approaches are quite similar in practice, but the biologist is paid a salary and buys his "meat." Therefore, this is not seen as a way of life in the sense that "the single most important aspect of Kluane people's way of life is hunting (Nadasdy, 2003: 63).

Among the Spokan of eastern Washington: "Land mammals were never stalked or taken at springs, as this was disrespectful and, if violated, was believed to be the main reason for game leaving an area" (Ross, 1998: 273)—this is important to recognize, as waterhole hunting is the surest way to wipe out local game. Equally intriguing was the same group's way of dealing with bears; "the successful hunter...sang the bear's death song" and observed "three days of strict behavioral and dietary taboos to avoid dreaming of the bear or being burned later by fire or struck by lightning" (Ross, 1998: 273, 2011: 314). This shows both respect for the bear and awareness of interaction and interdependence in ecology—though bears and lightning would not be associated by most peoples. The Spokan were competent enough hunters to wipe out the game, using tricks like taking scent glands from animals and luring them with the scents (Ross, 2011: 307–308). The Spokan lived on fish and plant materials (mostly berries and roots), about 50% each; thus, game was a relatively rare treat (Ross, 2011: 311).

The culturally related but linguistically different Sahaptins of the Columbia had similar values:

> ...wasting game was an offense against the animals themselves, against the moral order of nature, and was punished by bad luck in hunting or by sickness.... That these same conservation values are upheld by contemporary Indians as well, is indicated by their careful regulation of hunting by their own tribal members on reservation lands. On both Yakima and Warm Springs reservations considerable land has been set aside by tribal law for game preserves..." but "there are committed environmentalists among both Indians and whites and selfish poachers in both camps as well (Hunn & Selam, 1990: 141).

Native American "poachers" are to some extent a result of new laws, combined with the availability of firearms. This concept emerged because Western wildlife management restricted hunting to specific times of year, which contradicts the tradition in indigenous hunting, where it is recognized that does and fawns are often the best feeding, whereas Western approaches assume that only adult males should be taken, presumably for trophies, rather than for feeding families (Pierotti, 2011). Thus Indigenous hunters following traditional methods violated Western "laws" and became identified as "poachers." In former times, hunting often required cooperation, which allowed enforcement of group rules and ethical standards. Another form of conservation was the Sahaptin taboo against killing hoary marmots, high-mountain "groundhogs" that give piercing whistles that sound supernatural (Hunn & Selam, 1990: 142). Sahaptin autobiographies indicate surviving attitudes of respect for life, with rocks, trees, fire and water seen as alive and conservation of living resources stressed, even when traditional religion has been heavily influenced by Christianity (e.g., Beavert, 2017: 97; and below, Chaps. 7 and 8).

Perhaps the most important form of conservation was recognition that animals (and plants) had bosses or "keepers of the game," who could punish by removing their species if humans offended them. Stories of "keepers" are based upon actual experience with extraordinarily successful or experienced individual animals—if individual animals that would have contributed a preponderance of recruits to succeeding cohorts are killed, or even harassed to such an extent that they are driven away, then a local population that depended on their output could go "functionally extinct," which is equivalent to making their kind unavailable to humans, or of taking away luck in hunting or trapping (Pierotti, 2010, 2011).

For the Shuswap, "'Stinginess' [in distributing fish and game...invited bad luck for the future... Most of all, resource management was carried out through a value system that enforced the use of all parts of killed animals and sanctioned individuals who were wasteful" (Ignace, 1998: 208). A major origin myth of the Coeur d'Alene instructs hunters to take no more than two deer per hunt (Frey, 2001: 4). The Kaska "do not kill animals needlessly," realizing that "animals are best saved for times when they are most useful" (Nelson, 1973: 155). Indigenous peoples are strongly averse to using poison because animals eating the poisoned animals will die in turn, thus killing needlessly (Nelson, 1973: 244), anort "have a well-developed conservation ethic" (Nelson, 1973: 311). The Sahaptin learned from Coyote not to leave game foetuses in the field, and to sweat before hunting (Hunn & Selam, 1990: 233–234).

The Native peoples knew overhunting all too well, and knew exactly what its effects were and how much hunting would be suicidal (see Kay & Simmons, 2002; Pierotti, 2011). These teaching stories make no sense, unless people had overhunted, had realized the consequences to their cost, consciously learned better, and had even more self-consciously developed a stern morality of conservation which was incorporated through rituals and ceremonies. The fact that this morality was taught so diligently, with so many stories, proves that it met a felt need. Historians routinely use the existence of a law as proof that the outlawed activity was common enough to be a problem. The same can be said here, and this is incorporated in the two basic principles of Indigenous philosophy: *All things are Connected* and *All things are Related*, which are both scientific fact and philosophical basis for a system of ethics (Pierotti, 2011).

A strange Kathlamet story that personifies Hunger as an evil woman was told by Charles Cultee to Franz Boas in 1894 (Boas, 1901: 207–215). A man who had gained power from this woman took her mat, in which were the bones of the game animals and fish. He and his people returned the bones to the water, and so revived the game. The story does not specifically mention overhunting, but parallels with the rest of the region indicate that the power of Hunger Woman was punishment for that sin. However, supplying the Anglo-American settlers made this a negotiable issue, and one hunter supposedly took 20 to 30 deer per day in one early-nineteenth-century winter (Gahr, 2013: 66; several hunters were probably involved; in any case the toll was unsustainable). Trade in deerskins, fur, and other commodities led increasingly to depletion.

Martin and Szuter (1999, 2002) found that "no man's lands" between warring tribes were rich in game, while village areas were not. (Similar findings are reported from South America [Alvard, 1995] to Vietnam. ENA has observed it in spaces between villages in Mexico. There, the villages are peaceful, but people do not often hunt near village boundaries, because of remoteness and vigilance about trespass). Martin and Szuter argue that the game was killed out near settlements, though bothering and disturbing may have had more to do with driving them away. The game animals are smart enough to move to safe places.

Contra the old stereotypes, Native Northwest Coast people are not wantonly wasteful killers (*pace* Shepard Krech, 1999; see Appendix 3 below), not natural conservationists (*pace* Hughes, 1983), and most certainly not mere animals with no impact on nature, as so often alleged by settlers, and even some modern historians (Pryce, 1999: 85–90). A notable problem in recent politics has involved naïve Whites trying to hold Native peoples to an imagined conservationist standard alleged to have been part of their past, and to criticize them for not being the angelic children of nature postulated by some writers (Pierotti, 2011, Chap. 8; Ranco, 2007). Julia Cruikshank describes this attitude, with quotes quite painful to read, in *The Social Life of Stories* (1998: 58–63). Krech (1999: a major purveyor of this perspective) and Nadasdy (2003, 2007) also comment on it. Probably all who research this topic have encountered it.

Northwest Coast conservation stands in marked contrast with some attitudes of groups outside the Northwest Coast. For example, June Helm's careful description of the ecology of the northeastern Athapaskans and the quotes from earlier sources that she appends (Helm, 2000) show they not only shot and trapped what meat and fur animals they could; opposed government conservation ideas explicitly and at length, with the result that game was drastically depleted around settlements. They lived in an area far from the Northwest Coast, and formerly so thinly peopled that they had never previously overhunted it enough to learn conservation the hard way. If game was depleted, people simply moved elsewhere. Firearms and population growth created a problem that eastern Dene culture apparently could not solve; needs for cash led to overtrapping, there being no alternative source of income (Helm, 2000: 66–68, 79–89). Other studies of groups in the area confirm this. Some groups in eastern Canada also show less conservation, almost certainly as a response to the fur trade and to the impacts of cultural damage resulting from colonization (Martin, 1978; Tanner, 1979).

The contrast with the Northwest is similar to the well-documented contrast in South America between sparsely-populated groups like the nonconserving Matsigenka and Ache' (Alvard, 1995) and densely-populated groups like the Bari (Beckerman et al., 2002) and Tukano (Reichel-Dolmatoff, 1996). The latter are more consciously conservationist. In general, other things being equal, the more crowded the group is, the more carefully they manage what they have. The conservation ethos of Native America is not due to "simple oneness with nature" but to shrewd and calculated common sense and the development of a spiritual philosophy that emphasizes reciprocal relationships and dependence on other species (Beckerman et al., 2002; Pierotti, 2011; Trosper, 2009).

Domestication and Taming

The only true domesticate was the dog. In some groups, it was subject to serious selective breeding—most unusual for small-scale societies. A wool-bearing variety of dog was developed, probably by the Salish, and a woman's wealth could be reckoned by how many dogs she had; their fur, combined with geese or duck down and fireweed fluff were twisted into wool for weaving (Crockford & Frederick, 2011; Eels, 1985: 122). The painter Paul Kane managed to see and draw one; it looked rather like a lamb, or like the spitz dogs of Siberia. Archeology shows that it was very common in the Coast Salish, Makah, and Nuu-chah-nulth areas, and quite different from the larger ordinary village dogs (McKechnie et al., 2020). The Kwakwaka'wakw also had it, and a thorough account of it, including the ways its wool was spun, was provided by George Hunt (Boas, 1921: 1317–1318). It sadly became extinct in the mid-nineteenth century, but Russell Barsh (pers. comm. to ENA, 2007) has tried to breed it back. Another endangered special variety is the Tahltan bear dog, developed for bear hunting by the Tahltan people of northern British Columbia. It somewhat resembled the Karelian bear dog of Finland. Other groups apparently also had special hunting breeds in the old days.

Dogs were ambiguous creatures—nonhuman, yet in the human realm. Pamela Amoss (1984) memorably called them (quoting Hamlet) "'a little more than kin, and less than kind': the ambiguous Northwest Coast dog." Among the Tlingit, dogs could be killed for acting too human, as by walking on their hind legs (Kan, 1989: 320). The Upper Tanana of Alaska loved dogs enough to give small potlatches for favorite dogs that died, had birthdays, or recovered; some of these potlatches were children-oriented (Guédon, 1974: 207). The Tla'amin (Sliammon, Northern Coast Salish) love and cherish their dogs, and find them indispensable in hunting; they know dogs are close to wolves and share hunting power. They use false hellebore (*Veratrum viride*) in various ways to sharpen the dogs' sight and scent; this is part of a magical pattern found widely in North America (Anderson & Medina Tzuc, 2005; Anza-Burgess et al., 2020, esp.: 440–441). Everywhere, dogs were of considerable cultural interest.

A charming story—actual history, not myth—recorded by William Beynon concerns a conflict between Kisgegas village (a Gitksan town) and the Tsetsaut. This led to a few Gitksan men traveling—for the first time—to a newly established white settler fort, in 1825. They struck out lightly and cheerfully through hundreds of miles of trackless wilderness. Reaching the fort, they were impressed by the dog owned by Mr. Ross, the head of the establishment; for one thing, the dog had flop ears, not prick ears like Indigenous dogs. "Here they observed the white man, his possessions, and his strange ways, for the first time, and considered their adventure in the light of a supernatural experience. They marveled in particular at the white man's dog, the palisade or fortification around his houses, and the broad wagon road." (Beynon's actual telling of the story does not mention the supernatural aspect, and Barbeau may have invented that.) One, Wa-Iget, took this dog as a crest, *Ansem Midaw* "white man's dog," and had it carved on his totem pole at Kisgagas, under the name of *'Auge-maeselos* ("dog of Mr. Ross," *r* being pronounced *l*; Barbeau, 1929: 103). Another took the palisade, and built a small palisade around his totem pole (Barbeau, 1950: 10; Barbeau & Beynon, 1987: vol. 2: 162–163).

A creation story related to this theme is the origin of the Tsleil-Waututh Nation of British Columbia, whose First Ancestors were transformed from a wolf and created from the sediments of Burrard Inlet. God, or the Great Spirit, transformed a young wolf into a man, giving the family the linkage with wolves (Morin, 2015). This story has been altered and popularized into the *The She Wolf of Tsla-a-Wat* (Simeon, 1977), which tells of a female wolf who adopts a human child who is the last survivor of an epidemic, which decimated his people. This is one of numerous Indigenous accounts of how their people originated from a dynamic interaction between wolves and humans (Pierotti & Fogg, 2017).

Various wild mammals were probably tamed on more than a few occasions. Early accounts speak of Salish women trapping mountain goats in spring, when they were shedding their winter coats (Gustafson, 1980). The women pulled the loose fur off the goats and let them go. The wool was then spun with the dog wool. It was then woven on sophisticated native looms to make the beautiful traditional blankets. Mountain goats are somewhat tameable. Until discouraged by rangers, they were pests in Glacier National Park, begging for food from tourists (personal observation by ENA). It is fairly obvious that the Salish tamed their goat-antelopes. Goat hunting did go on, possibly as an unobtrusive cropping of semi-managed herds. Mountain goats are, however, less easily tamed than mountain sheep, and reproduce less rapidly, so they require careful management (Festa-Bianchet & Côté, 2008). Stories of overhunting them are common on the Northwest Coast, and used to teach respectful care (e.g., Glavin, 1998).

The rifle has changed game hunting in America; bow hunting required a close approach. One had to get near enough to make a killing shot, hopefully without losing a hard-to-replace stone point. The sensible thing to do with mountain sheep, mountain goats, and deer was to keep the herds as tame and unsuspecting as possible, taking laggards and easily-shot animals but never taking enough to alarm the group (Pierotti, 2010). Hunters were masters of stalking, ambush, and disguise, sometimes wearing animal skins and assuming the identity of either the prey or its predator (Willerslev, 2007). This is why archeology reveals a "poor management" practice of taking many females and young rather than concentrating on the conspicuous, wary adult males, as a modern rifle-hunter would. (On hunting in western North America in general, see Frison, 2004; Kay & Simmons, 2002.) Today, game animals outside of parks and towns stay out of rifle range, but learn with amazing speed where they are safe, and in such places, they will mix with humans quite comfortably. Mountain sheep and elk today know they are safer in towns than in huntable wilds, and thus become small-town residents, to the annoyance of local people who must drive around them (as shown by the authors' personal experience in many towns across the USA and Canada).

Plant Resource Management

The Northwest Coast peoples, like other western American hunter-gatherers, managed plant resources by cultivating, pruning, transplanting, fertilizing, and indeed all gardening activities short of actual domestication. Among the Salish, "all traditional territories were carefully tended by their occupants" (Miller, 1999: 17), and the same could be said of all other groups. Land was owned, and permission had to be asked for using it, but sharing was general and permission usually granted. Many authors have surveyed these practices for the Northwest (e.g., Blukis Onat, 2002; Deur & Turner, 2005, esp. Lepofsky et al., 2005; Turner & Peacock, 2005 therein; Gottesfeld, 1994a, b; Kirk, 1986; Turner, 2005). In addition, Kat Anderson's great study of Californian Native plant management practices (Anderson, 2005b) not only covers the extreme southern end of the region, but describes many techniques used throughout.

Increasing numbers of court decisions, local policies, local schools, and other institutions of all kinds have recognized the superiority of much Indigenous management, and have given more and more rights and management charges to Native peoples. A major review book is the edited volume *Plants, People, and Places*, edited by Nancy Turner (2020). Many chapters here recount the legal and policy side (see Lepofsky et al., 2020), but more interesting for our purposes here are the studies of plant use in context. Many of these are by Indigenous authors, including Jeannette Armstrong (2020) on the Syilx (Okanagan), Marianne and Ron Ignace and Ignace (2020) on the Secwepemc, and Leigh Joseph (2020) on Salish management of fritillary bulbs. These, and Deborah Curran and Val Napoleon's chapter in Turner's book (2020), provide important sources on traditional sustainable management. Particularly ironic is John Lutz' story (2020) of the astonishment of early British settlers at the park-like landscape they first saw around what is now Victoria. They compared what they observed to English gardens. They thought it was an Edenic wilderness; of course, its appearance was due to intensive management by local Salish people—it was, quite literally, a garden, though without domestic plants involved.

Jay Miller's statement that "While women did encourage the growth of certain wild plants, they did this unobtrusively" (Miller, 2014: 21) is an odd, possibly naïve, observation. Burning, planting, engaging in massive selective harvesting, and other techniques were far from unobtrusive. These techniques were also very successful, and carefully conducted. In fact, in the same book, Miller more accurately says "this cultivating was intense—much more than a light or selective tending of 'wild' species.... Not quite farming, it was definitely gardening" (Miller, 2014: 131). Miller notes the universality of management by burning, specifically for the Salish groups (Miller, 1988: 88). Burning was general everywhere, except in rainforest areas. Women used digging sticks, and had appropriate prayers and formulae for working with these sticks and with plants in general. These rituals were designed to keep the people constantly aware of reciprocal relationships between them and the growing ones (plants).

Burning was by far the most important method of managing the environment (Agee, 1993, esp.: 56; Anderson, 2005b; Suttles, 1987: 32–35; Stewart et al., 2002). The Northwest forests do not burn easily; some coastal rainforests burned only once every thousand years or so (Lepofsky et al., 2005: 228). However, even they burned occasionally. Clearings, meadows, and brushfields burn much more easily. The old idea that burning had little effect on resources has been disproved by the rapid recent changes observed in environments Native Americans managed by fire. An idea that burning was relatively recent has further been rendered unlikely by pollen studies and Native cultural traditions. Natural burning was common in drier and higher areas and must have played a considerable role in maintaining berry fields. These plants evolved as naturally caused fire-followers over millions of years before people were there to set fires. Nonetheless, pollen records indicate localized burning going back at least several thousand years (Sobel et al., 2013: 37).

Recall that Douglas fir forests, dominant over much of the region, are also fire-following plants (Suttles, 1987: 32). Deliberate burning often kept the succession going. Lightning, rare on the coast, is more common in the mountains, and obviously started fires. Today, climate change is making lightning fires more common in coastal areas. Distinguishing lightning fires from human-set fires is problematic, rendering it more difficult to assess how frequent deliberate burning actually was.

Plateau forests burn much more easily, and were regularly burned to reduce choking undergrowth, increase berry production, eliminate pests, renew the landscape and kill insects and ticks (Ignace & Ignace, 2017: 192; Ross, 2011: 267–272). There was even special spirit power to make one good at burning (Ross, 2011: 271). Groups might burn some of their area (often only a small portion) every three to five years (Lepofsky et al., 2005: 228). Burning for berries was simple and productive (Joyce Lecompte-Mastenbrook, ongoing research; Ross, 2011, esp. 344ff; many other writers).

Root crops also grow largely in burned-over prairie and grassland habitats. Fire suppression has made them rare or absent in many areas, and forests reclaim unburned areas rapidly. The oak forests found from California and Oregon (commonly) to Washington (especially east and south of Tacoma) to British Columbia (only around Victoria) require burning for maintenance to reduce undergrowth, especially poison oak. Those of BC and Washington are now succumbing rapidly to invasion by plants once limited by burning. The oaks once supplied numerous acorns, a staple in California and probably close to one in the Willamette Valley. Acorns had to be leached; in Oregon they were often buried in mud for the winter, thus becoming "Chinook olives" (Gahr, 2013: 74).

However, in at least one area in southwestern Washington, pollen records show *less* burning in the last 2500 years, after Native cultures became complex (Walsh et al., 2008). This seems to accompany a local wet phase, but it is a confusing finding. Small local fires reduce the chance of major fires; perhaps the latter would have left a clearer signature in the record.

Berries "were closely tended and maintained by fire clearing until government prohibited this practice. Today . . .due to lack of burning, they have changed in such a way that inferior species and subspecies of berries have replaced the quality species of the past" (Art Mathews in Daly, 2005: 221; Gitksan data: applicable everywhere in the Northwest). All sources agree that forests have encroached on berry lands, eliminating or diminishing them. The present authors might chalk this up to "good old days" rhetoric if we had not observed it throughout much of western North America over the last 70-odd years. Almost everywhere, the favorite berry patches of our youth have grown up to forests, brush, or grassland. Invasion by inferior species is bad enough; a further problem (mentioned by Mathews) is the fact that shading and crowding make berry plants produce fewer, smaller, and sourer fruit. All home gardeners who have let trees encroach on their berry patches know this, but it is rarely mentioned in the ecological literature.

A single family might eat 100,000 to 200,000 salal berries in a year (Deur, 2005: 14). Salal is the most common berry bush over much of the coastal Northwest, and its bearing capacity, especially when periodically burned over, is incredible. Thousands of families gathering on that scale would not make a dent in its yield. They were sharing these with bears and birds, among other species. Its berries were dried as cake or fruit leather (Hill-Tout, 1978b, d: 104). It responds dramatically to opening up the forest; growing around the roots of evergreens, it is usually too shaded to bear well, and management can have a major effect.

Turner (2014: 2: 167–176) provides an extremely detailed account—unique in the ethnographic literature—of how much yield was obtained from various plant resources by Indigenous harvesting efforts. Few resources provided high yields. Anyone with experience picking berries, which have changed less under domestication than most plants, will have a good idea of effort involved. Domestic berries yield higher than unmanaged wild stands, but no higher than burned and cultivated wild stands (Turner, 2014; personal observation by ENA).

There were countless things to do in a berry patch besides gathering. It was a prime area to tell stories, play games, take care of children, and otherwise enjoy life (Black Elk & Baker, 2020). Largely a women's space, it provided opportunities for women to get away from the house and have adventures. The old folktales stress the darker side: berry patches attracted bears, who did not always relish human competition. The pickers understood how to live with and negotiate with bears, and the folktales often taught this.

Sugar also occurred in sweet sap and conifer cambium, and from Douglas firs, which sometimes exude a trisaccharide sugar from their twigs and branches; which gathers like snow on the twigs. It was so loved in the interior that it was known as "breast milk" in some groups (Turner, 2014: 1: 215, 2020). Pit-cooking of roots such as camas turned the indigestible inulin to fructose, thus making them sweet; European sweets and starches were accepted quickly and seen as parallel (Turner, 2014: 1 and references cited therein). Balsamroot was similarly prepared not only for food but for medicine; it contains strongly antibiotic chemicals (Turner, 2014: 1: 431).

Root crops were harvested in ways that propagated them: large tubers were taken, small ones replanted; roots might be cut and the stem replanted. Harvesting was done such that weeds were eliminated. Particularly favored crops show what appears to be actual selection from thousands of years of this care. Roots have been harvested for over 11,000 years in the Northwest (Ames, 2005: 94).

Camas (*Camassia quamash* and *C. leichtlinii*; the latter may be simply a form of *C. quamash*) was particularly important, and has been for a long time (Lyons, 2015; Lyons & Ritchie, 2017). It seems to have been at least partially domesticated. It shows several traits characteristic of domestic crops, including several morphological changes (Carney et al., 2021), large showy flowers (compared to its close relatives), large edible structures, and extremely good adaptation to gardens and gardening (personal observation, from garden cultivation by ENA and many others). Douglas Deur says it is a "cultivated plant" in the coastal forests, being maintained there only by human activity. Presumably it was introduced from inland (see Deur & Thompson, 2008: 51; Gahr, 2013: 70). A family of Saanich could harvest over 10,000 bulbs a year, according to (thin but suggestive) figures (Deur & Turner, 2005: 14). Turner and Hebda (2012: 120–121) quote a detailed account by Marguerite Babcock of camas management in southeast Vancouver Island; it was clearly cultivated, and if not domesticated then very close to it. A tribe of a thousand people could easily run through a million camas bulbs in a year (Deur, 2005).

The role of cultivation was underemphasized by early anthropologists, perhaps partly because they wanted to emphasize the sophistication of "hunter-gatherer" societies as a way of disproving evolutionary schemes focused on European superiority. It remains true that Northwest Coast peoples were hunting-gathering societies, without significant domestication (besides tamed wolves/dogs) or agriculture. Most hunter-gatherer societies manage their environment, often intensively. For example, Bruce Pascoe (2014) reports from Australia many of the same aboriginal management techniques used in the Northwest Coast and California.

Deur and Turner (2005) emphasize that women were rarely taken seriously by early ethnographers. Indeed, what was said of the Babine could be said of almost all Northwest Coast peoples: "We do not have reliable accounts of women's lives, property, or social obligations for the early contact or precontact eras. All the available fur trader and missionary accounts are full of contradictions and replete with biases" (Fiske & Patrick, 2000: 238).

Roots were managed by harvesting with digging sticks, thinning stands and allowing selection of particular sizes. Balsamroot (*Balsamorrhiza* spp.) produces edible roots that develop over years into huge, woody, inedible taproots; these were left to reproduce, and very small roots were left to grow. Carrot-sized roots were harvested (Ignace & Ignace, 2017: 190).

Wetland resources such as yellow pond lily nuts (the *wokas* of the Klamath; Colville, 1902) and the wapato or "Indian potato" (*Sagittaria;* see Gahr, 2013: 69) were similarly managed. They too may well have been on the road to domestication. This sort of root management gave the Native people the ability to shift rapidly to cultivating potatoes. The Haida were trading potatoes in bulk with English and "Boston" ships well before 1800. In fact, potatoes have been a staple on the coast about as long as they have in Ireland. The potato was introduced by the Spanish, directly or indirectly from their native Andean home. Surviving heirloom potato varieties of the Ozette-Makah and Tlingit types are still genetically identifiable as South American (Ignace & Ignace, 2017: 515; Turner, 2014: 1: 198–200). Presumably, they derive from direct and early introduction by the Spanish.

The root foods, including other managed ones ranging from glacier lilies (*Erythronium*) to clover (*Trifolium*, especially *T. wormskioldii*), were baked for long periods in earth ovens, to break down the long-chain polysaccharides, especially inulin, into digestible short-chain sugars. Leigh Joseph (2020) provides an important study of fritillary *(Fritillaria* spp.), a lily relative with tiny rice-grain-like bulbs—"rice roots." They are baked and eaten widely in the Northern Hemisphere.

Joseph also provides notes on traditional conservation, from his Indigenous family experience. They are also common foods in Siberia.

As experimenters with cooking such roots may be painfully aware, the long-chain molecules, such as inulin, can cause acute digestive distress if ingested without being broken down. Breaking them down requires steaming in a pit (or modern equivalent) for hours or days. The intense heat and steam do the work. Nancy Turner's books give the best, most complete information on these roots and their preparation; she and her students have devoted lifetimes to the research.

Fruit trees like Oregon crabapple, Indian plum (*Oemleria cerasiformis*, a plum relative with tiny but abundant and wonderfully flavorful fruits), hawthorn, serviceberry, and mountain ash were managed. Techniques included pruning around them, opening land for them, and otherwise encouraging them. They were often transplanted into new areas, and local orchards established. Were the seeds planted? Probably, but we do not know. Hazelnuts are better known; carefully managed throughout the Northwest, with trees tended, pruned, and probably transplanted locally (Anderson, 2005b; Armstrong et al., 2018, 2021).

Fiber crops were also managed. The Secwépemc, for instance, "carefully tended" patches of Indian hemp (*Apocynum cannabinum*) "by removing brush and undergrowth near the patch to ensure straight and thick growth of the perennial hemp shoots" (Ignace & Ignace, 2017: 191). Cutting the shoots for fiber made the new ones grow out straight and tall. This is typical of Indian hemp management all over western North America (see e.g., Anderson, 2005b). Indian hemp was sometimes transplanted to new areas in California and presumably in the Northwest Coast region as well.

The only domesticated plant acknowledged in most of the literature was tobacco. Tobacco cultivation was certainly pre-Columbian, but not necessarily in all groups where it is known (Moss, 2005). One species (*Nicotiana attenuata*) was grown by the Haida and Tlingit and probably their neighbors, and grown widely in California (Harrington, 1932) and Oregon (Gray, 1987: 31, on southwest Oregon). It seems likely that other groups grew it, but stopped doing so before anthropologists reached them. It had been grown by the Tlingit long enough for them to embed it in their origin myth of Raven and his many travels and creations (Swanton, 1909: 89). *N. quadrivalvis,* a plant from the central part of the continent, was occasionally farmed in its area of origin, and may have spread into our region.

Bow staves were taken off living yew trees in the coastal forests, junipers and other trees in the Basin and probably the Plateau. This indicates major forbearance and the working of conscience. Yews and junipers produce few straight branches. Staves are best taken off the lower part of a branch, to get the compression wood. The branch very slowly regrows. With junipers, only one stave in about twenty years can be taken (Wilke, 1988); for yews, the figure is probably about the same. This means that people must be restrained by their conscience. No one can patrol the few trees with straight branches. All over the Northwest, one sees yew trees with the characteristic scars, indicating that people saved their bow trees. For the same reason, yew trees were planted thickly around castles in the British Isles in the Medieval times. Many of them are still growing, now centuries old, providing a thick shade. In that situation, however, the lord of the manor guarded the trees. No one could guard bow trees in the old west.

With regard to managing trees for basket making, Pierotti reports the following story involving the Pomo people of northern California:

> In the 1940s, a prominent plant evolutionary biologist from the University of California and his students were very excited. Working in Lake County, California, they found a growth form of willow they had never seen before, with long, straight shoots rather than the usual branching pattern (Fig. 9.1). As good evolutionary biologists, they tried to figure out what environmental factors might be responsible for this unique form, so they assessed the soil characteristics, looked at moisture patterns, and the physiognomy of the local landscape. None of these tests revealed anything unusual, and they were becoming frustrated when one day they encountered an elderly Pomo woman who was carefully examining the willows. Not wanting to leave any stone unturned, the biologist asked the woman if she knew anything about these particular plants and their unusual form. The woman replied, 'Of course, we have been shaping them this way for centuries. They have to be straight if you are going to make the best baskets' (G. L. Stebbins, personal communication to RP, 1985, as quoted in Benedict et al., 2014).

Kat Anderson (2005a) describes the process in detail. Basketry plants were widely managed in similar ways in the Northwest Coast (Deur & Turner, 2005).

Similar "culturally modified trees" ("CMT's") include cedars used for bark or planks. Cuts were made near the foot of the tree and then about 20 or 30 feet up, the cuts being shallow for bark and deep for planks. The bark was pulled off in long strips. The planks were sometimes split off, but it was easier to let the tree grow until the stresses of growth split the plank off without human effort. Such scars reveal, once again, that people took only enough for their needs, without killing or greatly harming the trees. The scars healed, and bark could be taken again (Earnshaw, 2019; Schlick, 1998: 59). Studies on Vancouver Island show that this sustainable one-strip-at-a-time harvesting has been going on for at least 1100 years (Earnshaw, 2019).

Boas records a no-nonsense conservation belief among the Kwakwaka'wakw: "Even when the young cedar-tree is quite smooth, they do not take all of the cedar-bark, for the people of the olden times said that if they should peel off all the cedar-bark...the young cedar would die, and then another cedar-tree nearby would curse the bark-peeler so that he would also die" (George Hunt, in Boas, 1921: 616–617). Several other conservation-related beliefs and prayers are given in this section of Boas' ethnography. The same reasonable observation is recorded from modern Nuxalk (Turner & Peacock, 2005: 123). Peeling bark was a major and important undertaking, with bark serving every purpose from food to medicine to textiles. On the other hand, overharvesting cedars was known to occur, though in a different group (the Haida) with less cedar to manage (Lacourse et al., 2007).

An obvious question is how many of the cultivation techniques of the Northwest were learned from early European settlers. Wayne Suttles (2005), James MacDonald (2005), and Madonna Moss (2005) have considered the matter. By far the most important management technique, burning, was certainly aboriginal; every Native American group in a burnable environment used this method, as do hunters, foragers, and horticulturalists around the world. Unlike the equivocal situation in California and drier parts of the interior, where natural fire is so common that it makes aboriginal burning difficult to assess, historic and archeologically evidenced burning on the Northwest Coast can usually be safely ascribed to Native activity. Natural fire is exceedingly rare west of the Cascades in Oregon and Washington and the Coast Range crest in Canada (see Lepofsky et al., 2005).

> It is obvious to contemporary ecologists that many habitats in North America are dependent upon fire (sometimes called "fire climax" communities). The use of fire by Indigenous peoples is one clear and unequivocal example of an Indigenous impact that shaped ecological communities in the Americas.... Krech (a critic of Indian knowledge) discusses this phenomenon in a way that confuses the issue. He starts out reasonably by pointing out the illogic of the historical Euro-American "Smoky the Bear" perspective... that all fire is harmful and destructive. Krech then discusses the many ways in which Indians employed fire, such as to clear underbrush, create better grazing conditions for ungulates or better habitats for food plants. Krech then segues without pause into a discussion of Indian-set fires that may have gotten out of control and caused damage, or even had an outcome contrary to that intended. Missing from Krech's chapter on fire is any discussion of modern fire ecology, where it is generally appreciated that fire is a positive influence in maintaining habitats suitable for hunting, plant harvesting, and so forth quoted from (Pierotti, 2011: 173).

Indigenous people were usually careful to burn when weather kept fire from spreading, but this could be difficult. Cultures elsewhere differ considerably in this regard. In Baja California, burning goes to the edge of a recent burn and stops there, so burning is managed to reach such an edge soon (Minnich & Franco-Vizcaino, 2002; ENA, personal observation). In Quintana Roo, the Maya are careful to cut firebreaks, and letting a fire escape is a major sin, roundly condemned by the community (Anderson, 2005a). Conversely, in Madagascar, the Tanala take no pains with fire, allowing it to burn into adjacent forests without check (ENA, personal observation). All these patterns might have existed locally in the Northwest. Fire scars and old burns make it appear that picking specific weather and times of year was done, but we will never know the full story.

Intensive root-digging was certainly pre-contact, and indeed goes back far in the archeological record, and the diggers must have known the effects of their cultivation on the crops. Intensive pruning and transplantation and at least some fertilizing are recorded (Turner & Peacock, 2005: 118). Clearing, weeding, fencing, sowing, transplanting, mulching, and fertilizing are all documented (Turner, 2014: 2: 193), and seem likely to be aboriginal in origin. Archeology is rapidly revealing plant and shellfish management in pre-settler times.

In short, resources were abundant before white settlement, and had not become notably less so in the thousands of years of Native occupation. This speaks well for Indigenous management. Fish were maintained in the smallest streams. Some plant resources were increased and improved by cultivation.

This is not to say that management was perfect. Accounts of overhunting, and of unpredictable fluctuations in various resources, leading to famine were abundant, as will appear below. Good management was based on hard-won experience. Also, human population densities remained low, limited by food shortage and other factors. The environment did not have to feed millions. Country care was excellent, pervasive, and fine-tuned, with many implications for better management in the future, but it was not some dream-like ideal, but a real and challenging environment in which detailed knowledge, like that described above from Moose Jackson, was crucial for survival.

References

Agee, J. K. (1993). *Fire ecology of pacific northwest forests*. Island Press.
Alberta Society of Professional Biologists. (1986). *Native people and renewable resource management*. Alberta Society of Professional Biologists.
Alvard, M. (1995). Interspecific prey choice by Amazonian hunters. *Current Anthropology, 36*, 789–818.
Ames, K. (2005). Intensification of food production on the northwest coast and elsewhere. In D. Deur & N. Turner (Eds.), *Keeping it living: Traditions of plant use and cultivation on the Northwest Coast of North America* (pp. 67–100). University of British Columbia Press.
Amoss, P. (1984). A little more than kin, and less than kind: The ambiguous northwest coast dog. In J. Miller & C. M. Eastman (Eds.), *The Tsimshian and their neighbors* (pp. 292–305). University of Washington Press.
Anderson, E. N. (2005a). *Political ecology of a Yucatec Maya community*. University of Arizona Press.
Anderson, M. K. (2005b). *Tending the wild: Native American knowledge and the management of California's natural resources*. University of California Press.
Anderson, E. N. (2014). *Caring for place*. Left Coast Press.
Anderson, E. N., & Medina Tzuc, F. (2005). *Animals and the Maya in Southeast Mexico*. University of Arizona Press.
Anza-Burgess, K., Lepofsky, D., & Yang, D. (2020). 'A part of the people': Human-dog relationships among the northern coast Salish of British Columbia. *Journal of Ethnobiology, 40*, 434–450.
Armstrong, J. (2020). Living from the land: Food security and food sovereignty today and into the future. In N. J. Turner (Ed.), *Plants, people and places: The roles of ethnobotany and ethnoecology in indigenous peoples' land rights in Canada and Beyond* (pp. 36–50). McGill-Queen's University Press.
Armstrong, C. G., Dixon, W. M., & Turner, N. J. (2018). Management and traditional production of beaked hazelnut (k'áp'xw-az', *Corylus cornuta*, Betulaceae) in British Columbia. *Human Ecology, 46*, 547–559.
Armstrong, C. G., Miller, J. E. D., McAlvay, A. C., Ritchie, P. M., & Lepofsky, D. (2021). Historical indigenous land-use explains plant functional trait diversity. *Ecology and Society, 26*, 6.
Arnold, D. F. (2008). *The Fisherman's frontier: People and Salmon in Southeast Alaska*. University of Washington Press.
Atleo, E. R. (2011). *Principles of Tsawalk: An indigenous approach to global crisis*. University of British Columbia Press.
Augustine, S., Lepofsky, D., Smith, N., & Cardinal, N. (2016). The clam garden network: Linking ancient mariculture to modern shellfish management and cultural reconnections. In *Paper, Society for Applied Anthropology, annual conference*.
Barbeau, M. (1929). *Totem Poles of the Gitksan, Upper Skeena River, British Columbia* (National Museum of Canada, Bulletin) (Vol. 61). National Museums of Canada.
Barbeau, M. (1950). *Totem poles*. National Museum of Canada.
Barbeau, M., & Beynon, W. (1987). *Tsimshian narratives*. Canadian Museum of Civilization, Mercury Series, Paper 3.
Beavert, V. R. (2017). *The gift of knowledge: Ttnúwit Átawish Nch'inchi'imamí*. University of Washington Press.

References

Beckerman, S., Valentine, P., & Eller, E. (2002). Conservation and native Amazonians: Why some do and some don't. *Antropologica, 96*, 31–51.

Benedict, M., Kindscher, K., & Pierotti, R. (2014). Learning from the land: Incorporating indigenous perspectives into the plant sciences. In C. Quave (Ed.), *Strategies for teaching in the plant sciences* (pp. 135–154). Springer Publications.

Bernick, K. (2003). A stitch in time: Recovering the antiquity of a coast Salish basket type. In R. G. Matson, G. Coupland, & Q. Mackie (Eds.), *Emerging from the mist: Studies in northwest coast culture history* (pp. 230–243). University of British Columbia Press.

Black Elk, L., & Baker, J. M. (2020). From traplines to pipelines: Oil sands and the pollution of berries and sacred lands from Northern Alberta to North Dakota. In N. J. Turner (Ed.), *Plants, people and places: The roles of ethnobotany and ethnoecology in indigenous peoples' land rights in Canada and Beyond* (pp. 173–187). McGill-Queen's University Press.

Boas, F. (1901). *Kathlamet texts*. Smithsonian Institution, Bureau of American Ethnology, Bulletin 26.

Boas, F. (1910). *Kwakiutl tales. Columbia University contributions to anthropology* (Vol. 2). Columbia University Press.

Boas, F. (1917). *Folk-Tales of Salish and Sahaptin tribes*. American Folk-Lore Society.

Boas, F. (1921). *Ethnology of the Kwakiutl* (Vol. 2). United States Government, Bureau of American Ethnology, annual report for 1913-1914.

Boas, F. (2006). *Indian myths and legends from the North Pacific Coast of America*. [German original 1895.] Trans. D. Bertz. Ed. R. Bouchard & D. Kennedy. Talon.

Boas, F., & Hunt, G. (1905). *Kwakiutl texts, part 2*. American Museum of Natural History, Memoir V, part 2.

Boelscher, M. (1988). *The curtain within: Haida social and mythical discourse*. Unversity of British Columbia Press.

Blukis Onat, A. R. (2002). Resource cultivation on the Northwest Coast of North America. *Journal of Northwest Anthropology, 36*, 125–144.

Busch, B. C. (1987). *The war against the seals: A history of the north American seal fishery*. McGill/Queens' University Press.

Butler, V. L. (2005). *Sustainable use of animal resources on the northwest coast?* Presentation, American Anthropological Association, annual conference.

Butler, V. L., & Campbell, S. K. (2004). Resource intensification and resource depression in the Pacific northwest of North America: A zooarcheological review. *Journal of World Prehistory, 18*(4), 327–405.

Carney, M., Tushingham, S., McLaughlin, T., & d'Alpoim Guedes, J. (2021). Harvesting strategies as evidence for 4000 years of camas (*Camassia quamash*) Management in the North American Columbia Plateau. *Royal Society Open Science, 8*. https://doi.org/10.1098/rsos.202213

Carothers, C. (2012). Enduring ties: Salmon and the Alutiiq/Sugpiaq peoples of the Kodiak archipelago, Alaska. In B. J. Colombi & J. F. Brooks (Eds.), *Keystone nations: Indigenous peoples and Salmon across the North Pacific* (pp. 133–160). School of American Research Press.

Carothers, C., et al. (2021). Indigenous peoples and Salmon stewardship: A critical relationship. *Ecology and Society, 26*(1), 16.

Colson, E. (1953). *The Makah Indians: A study of an Indian tribe in Modern American Society*. University of Minnesota Press.

Colville, F. V. (1902). Wokas, a primitive food of the Klamath Indians. *United States National Museum, Report, 1902*, 725–739.

Coté, C. (2010). *Spirits of our whaling ancestors: Revitalizing Makah and Nuu-chah-nulth traditions*. University of Washington Press.

Coupland, G., Clark, T., & Palmer, A. (2009). Hierarchy, communalism, and the spatial order of northwest coast plank houses: A comparative study. *American Antiquity, 74*, 77–106.

Crockford, S. J., & Frederick, G. (2011). Neoglacial sea ice and life history flexibility in ringed and fur seals. In T. Braje & T. Rick (Eds.), *Human impact on seals, sea lions and sea otters:*

Integrating archaeology and ecology in the Northeast Pacific. University of California Press. https://doi.org/10.1525/california/9780520267268.003.0004

Croes, D. (2003). Northwest coast wet-site artifacts: A key to understanding resource procurement, storage, management, and exchange. In R. G. Matson, G. Coupland, & Q. Mackie (Eds.), *Emerging from the mist: Studies in northwest coast culture history* (pp. 51–75). University of British Columbia Press.

Cruikshank, J. (1998). *The social life of stories: Narrative and knowledge in the Yukon territory*. University of Nebraska Press.

Cummings, B. J. (2020). *The river that made Seattle: A human and natural history of the Duwamish*. University of Washington Press.

Curran, D., & Napoleon, V. (2020). Ethnoecology and indigenous legal traditions in environmental governance. In N. J. Turner (Ed.), *Plants, people and places: The roles of ethnobotany and ethnoecology in indigenous peoples' land rights in Canada and beyond* (pp. 269–281). McGill-Queen's University Press.

Daly, R. (2005). *Our box was full: An ethnography for the Delgamuukw plaintiffs*. University of British Columbia Press.

Deur, D., & Turner, N. (Eds.). (2005). *Keeping it living: Traditions of plant use and cultivation on the northwest coast of North America*. University of British Columbia Press.

Deur, D. (2005). Tending the garden, making the soil: Northwest coast estuarine gardens as engineered environments. In D. Deur & N. J. Turner (Eds.), *Keeping it living: Traditions of plant use and cultivation on the northwest coast of North America* (pp. 296–327). University of Washington Press.

Deur, D., & Thompson, M. T. (2008). South wind's journeys: A Tillamook epic reconstructed from several sources. In M. T. Thompson & S. Egesdal (Eds.), *Salish myths and legends: One people's stories* (pp. 2–59). University of Nebraska Press.

Drucker, P. (1951). *The northern and central Nootkan tribes*. Smithsonian Institution, Bureau of American Ethnology, Bulletin 144.

Earnshaw, J. K. (2019). Cultural forests in cross section: Clear-cuts reveal 1,100 years of bark harvesting on Vancouver Island, British Columbia. *American Antiquity, 84*, 516–530.

Eels, M. (1985). *The Indians of Puget Sound: The notebooks of Myron Eels*. University of Washington Press.

Festa-Bianchet, M., & Côté, S. (2008). *Mountain goats: Ecology, behavior, and conservation of an alpine ungulate*. Island Press.

Fiske, J., & Patrick, B. (2000). *Cis Dideen Kat: When the plumes rise*. University of British Columbia Press.

Frey, R. (2001). *Landscape traveled by coyote and crane: The World of the Schitsu'umsh (Coeur d'Alene) Indians*. University of Washington Press.

Frison, G. (2004). *Survival by hunting: Prehistoric human hunters and animal prey*. University of California Press.

Gahr, D. A. T. (2013). Ethnobiology: Nonfishing subsistence and production. In R. T. Boyd, K. M. Ames, & T. A. Johnson (Eds.), *Chinookan peoples of the lower Columbia* (pp. 63–79). University of Washington Press.

George, E. (2003). *Living on the edge: Nuu-Chah-Nulth history from an Ahousaht Chief's perspective*. Sono Nis Press.

Glavin, T. (1998). *A death feast in Dimlahamid*. New Star Books.

Good, T. P., Ellis, J., Annett, C., & Pierotti, R. (2000). Bounded hybrid superiority: Effects of mate choice, habitat selection, and diet in an avian hybrid zone. *Evolution, 54*, 1774–1783.

Gottesfeld, L. M. J. (1994a). Conservation, territory, and traditional beliefs: An analysis of Gitksan and Wet'suwet'en subsistence, Northwest British Columbia, Canada. *Human Ecology, 22*, 443–465.

Gottesfeld, L. M. J. (1994b). Wet'suwet'en ethnobotany: Traditional plant uses. *Journal of Ethnobiology, 14*, 185–210.

References

Graeber, D., & Wengrow, D. (2021). *The dawn of everything: A new history of humanity.* Farrar, Straus & Giroux.

Gray, D. J. (1987). *The Takelma and their Athapascan neighbors.* University of Oregon Anthropological Papers, No. 37.

Guédon, M.-F. (1974). *People of Tetlin, why are you singing?* National Museum of Man, Mercury Series, Ethnology Division, #9.

Gunther, E. (1926). An analysis of the first Salmon ceremony. *American Anthropologist, 28,* 605–617.

Gustafson, P. (1980). *Salish weaving.* University of Washington Press.

Harrington, J. P. (1932). *Tobacco among the Karuk Indians of California.* United States Government, Bureau of American Ethnology, Bulletin 94.

Harkin, M. E. (2007). Swallowing wealth: Northwest coast beliefs and ecological practices. In M. E. Harkin & D. R. Lewis (Eds.), *Native Americans and the environment: Perspectives on the ecological Indian* (pp. 211–232). University of Nebraska Press.

Helm, J. (2000). *The people of Denendeh: Ethnohistory of the Indians of Canada's northwest territories.* University of Iowa Press.

Henson, L. H., Balkenhol, N., Gustas, R., Adams, M., Walkus, J., Housty, W. G., Stronen, A. V., Moody, J., Service, C., Reece, D., VonHoldt, B. M., McKechnie, I., Koop, B. F., & Durimont, C. (2021). "Convergent geographic patterns between grizzly bear population genetic structure and indigenous language groups in coastal British Columbia, Canada." Ecology and Society, 26 (7), https://doi.org/10.5751/ES-12443-260307.

Hill-Tout, C. (1978a). The Salish people. In R. Maud (Ed.), *The Thompson and the Okanagan* (Vol. 1, p. 14). Talonbooks.

Hill-Tout, C. (1978b). The Salish people. In R. Maud (Ed.), *The Squamish and the Lillooet* (Vol. 2, p. 14). Talonbooks.

Hill-Tout, C. (1978c). *The Salish people. Vol 3: The Mainland Halkomelem* (p. 14). Talonbooks.

Hill-Tout, C. (1978d). *The Salish people. Vol 4: The Sechelt and the south-eastern tribes of Vancouver Island.* Talonbooks.

Hughes, J. D. (1983). *American Indian ecology.* Texas Western Press.

Hunn, E., & Selam, J. (1990). *Nch'i-Wana, the big river.* University of Washington Press.

Hunn, E., Johnson, D. R., Russell, P. N., & Thornton, T. F. (2003). Huna Tlingit traditional environmental knowledge, conservation, and the management of a 'wilderness' park. *Current Anthropology, 44*(Supplement), S79–S104.

Ignace, M. B. (1998). Shuswap. In D. E. Walker (Ed.), *Handbook of North American Indians: Plateau* (Vol. 12, pp. 203–219). Smithsonian Institution.

Ignace, M., & Ignace, R. (2017). *Secwépemc people, land and Laws.* McGill-Queen's University Press.

Ignace, M., & Ignace, R. (2020). A place called Pípsell: An indigenous cultural keystone place, mining, and Secwépemc law. In N. J. Turner (Ed.), *Plants, people and places: The roles of ethnobotany and ethnoecology in indigenous peoples' land rights in Canada and beyond* (pp. 131–150). McGill-Queen's University Press.

Jackley, J., Gardner, L., Djunaedi, A. F., & Salomon, A. K. (2016). Ancient clam gardens, traditional management portfolios, and the resilience of coupled Human-Ocean systems. *Ecology and Society, 21,* 20.

Jacobs, M. (1945). *Kalapuya texts* (p. 11). University of Washington, Publications in Anthropology.

Jochelson, W. (1926). *The Yukaghir and the Yukaghirized Tungus.* Memoir of the American Museum of Natural History, XIII, Reports of the Jesup North Pacific Expedition, IX.

Johnsen, D. B. (2009). Salmon, science, and reciprocity on the northwest coast. *Ecology and Society, 14*(2), 43.

Jolles, C. Z. (2002). *Faith, food and family in a Yup'ik whaling community.* University of Washington Press.

Jonaitis, A. (1999). *The Yuquot whalers' shrine.* University of Washington Press.

Joseph, L. (2020). 'Passing it on': Renewal of indigenous plant knowledge systems and indigenous approaches to education. In N. J. Turner (Ed.), *Plants, people and places: The roles of ethnobotany and ethnoecology in indigenous peoples' land rights in Canada and beyond* (pp. 386–401). McGill-Queen's University Press.

Kan, S. (1989). *Symbolic immortality: The Tlingit potlatch of the nineteenth century*. Smithsonian Institution Press.

Kasten, E. (2012). Koryak Salmon fishery: Remembrances of the past, Perspetives for the future. In B. J. Colombi & J. F. Brooks (Eds.), *Keystone nations: Indigenous peoples and Salmon across the North Pacific* (pp. 65–88). School of American Research Press.

Kay, C. E., & Simmons, R. T. (Eds.). (2002). *Wilderness and political ecology: Aboriginal influences and the original state of nature*. University of Utah Press.

Kennedy, D., & Bouchard, R. (1998a). Lillooet. In D. E. Walker (Ed.), *Handbook of North American Indians: Plateau* (Vol. 12, pp. 174–190). Smithsonian Institution.

Kennedy, D., & Bouchard, R. (1998b). Northern Okanagan, Lakes, and Colville. In D. E. Walker (Ed.), *Handbook of North American Indians: Plateau* (Vol. 12, pp. 238–252). Smithsonian Institution.

Kirk, R. (1986). *Wisdom of the elders*. Douglas & McIntyre.

Koester, D. (2012). Shades of deep Salmon: Fish, fishing, and Itelmen cultural history. In B. J. Colombi & J. F. Brooks (Eds.), *Keystone nations: Indigenous peoples and Salmon across the North Pacific* (pp. 47–64). School of American Research Press.

Krech, S. (1999). *The ecological Indian: Myth and reality*. W. W. Norton.

Lacourse, T., Mathewes, R. W., & Hebda, R. J. (2007). Paleoecological analyses of Lake sediments reveal prehistoric human impact on forests at Anthony Island UNESCO world heritage site, Queen Charlotte Islands (Haida Gwaii), Canada. *Quarternary Research, 68*, 177–183.

Langdon, S. (2016). Tlingit relations with Salmon in Southeast Alaska: Concepts, innovations and interventions. In *Paper, Society for Applied Anthropology, Annual Conference*.

Lepofsky, D., Hallett, D., Lertzman, D., Mathewes, R., McHalsie, A., & Washbrook, K. (2005). Documenting precontact plant management on the Northwest Coast: An example of prescribed burning in the Central and Upper Fraser Valley, British Columbia. In D. Deur & N. Turner (Eds.), *Keeping it living: Traditions of plant use and cultivation on the Northwest Coast of North America* (pp. 218–239). University of British Columbia Press.

Lepofsky, D., Smith, N. F., Cardinal, N., Harper, J., Morris, M., Gitla, E., Bouchard, R., Kennedy, D. I. D., Salomon, A. K., Puckett, M., & Rowell, K. (2015). Ancient shellfish mariculture on the northwest coast of North America. *American Antiquity, 80*, 236–259.

Lepofsky, D., Armstrong, C. G., Mathews, D., & Greening, S. (2020). Understanding the past for the future: Archaeology, plants, and first nations' land use and rights. In N. J. Turner (Ed.), *Plants, people and places: The roles of ethnobotany and ethnoecology in indigenous peoples' land rights in Canada and beyond* (pp. 86–106). McGill-Queen's University Press.

Lewis, M., & Clark, W. (1990). In G. Moulton (Ed.), *The journals of the Lewis and Clark expedition* (Vol. 6). University of Nebraska Press.

Lutz, J. S. (2020). Preparing Eden: Indigenous land use and European settlement on southern Vancouver Island. In N. J. Turner (Ed.), *Plants, people and places: The roles of ethnobotany and ethnoecology in indigenous peoples' land rights in Canada and beyond* (pp. 107–130). McGill-Queen's University Press.

Lyons, K. (2015). Recognizing the archaeological signatures of resident fisheries. In P.-L. Yu (Ed.), *Rivers, fish, and the people: Tradition, science, and historical ecology of fisheries in the American west* (pp. 96–126). University of Utah Press.

Lyons, N., & Ritchie, M. (2017). The archaeology of camas production and exchange on the northwest coast: Ith evidence from a Sts'ailes (Chehalis) village on the Harrison River, British Columbia. *Journal of Ethnobiology, 37*, 346–367.

MacDonald, J. (2005). Cultivating in the Northwest: Early accounts of Tsimshian Horticulture. In D. Deur & N. Turner (Eds.), *Keeping it living: Traditions of plant use and cultivation on the Northwest Coast of North America* (pp. 240–273). University of British Columbia Press.

References

MacMillan, A. D. (1999). *Since the time of the transformers: The ancient heritage of the Nuu-chah-nulth, Ditidaht, and Makah*. University of British Columbia Press.

Martin, C. (1978). *Keepers of the game*. University of California Press.

Martin, P. S., & Szuter, C. R. (1999). War zones and game sinks in Lewis and Clark's west. *Conservation Biology, 13*, 36–45.

Martin, P. S., & Szuter, C. R. (2002). Game parks before and after Lewis and Clark: Reply to Lyman and Wolverton. *Conservation Biology, 16*, 244–247.

Matson, R. G., Coupland, G., & Mackie, Q. (Eds.). (2003). *Emerging from the mist: Studies in northwest coast culture history*. University of British Columbia Press.

McClellan, C. (1975). *My old people say*. National Museum of Man, Publications in Ethnology 6.

McIlwraith, T. F. (1948). *The Bella Coola Indians* (Vol. 2). University of Toronto Press.

McKechnie, I., Moss, M. L., & Crockford, S. J. (2020). Domestic dogs and wild canids on the northwest coast of North America: Animal husbandry in a region without agriculture? *Journal of Anthropological Archaeology, 60*, 101209.

Menzies, C. M. (Ed.). (2006). *Traditional ecological knowledge and natural resource management*. University of Nebraska Press.

Menzies, C. R. (2010). Dm sibilaa'nm da laxyuubm Gitxaala: Picking Abalone in Gitxaala Territory. *Human Organization, 69*, 213–220.

Menzies, C. R. (2012). The disturbed environment: The indigenous cultivation of Salmon. In B. J. Colombi & J. F. Brooks (Eds.), *Keystone nations: Indigenous peoples and Salmon across the North Pacific* (pp. 161–182). School of American Research Press.

Menzies, C. R. (2016). *People of the saltwater: An ethnography of Git lax m'oon*. University of Nebraska Press.

Miller, J. (1988). *Shamanic odyssey: The Lushootseed Salish journey to the land of the dead in terms of death, potency, and cooperating shamans in North America*. Ballena Press.

Miller, J. (1998). Middle Columbia River Salishans. In D. Walker (Ed.), *Handbook of North American Indians: Plateau* (Vol. 12, pp. 253–270). Smithsonian Institution.

Miller, J. (1999). *Lushootseed culture and the shamanic odyssey: An anchored radiance*. University of Nebraska Press.

Miller, J. (2012). Lamprey 'Eels' in the greater northwest: A survey of tribal sources, experiences, and sciences. *Journal of Northwest Anthropology, 46*, 65–84.

Miller, J. (2014). *Rescues, rants, and researches: A review of Jay Miller's writings on northwest Indian cultures*. Northwest Anthropology, Memoir 9.

Minnich, R. A., & Franco-Vizcaino, E. (2002). Divergence in Californian vegetation and fire regimes induced by differences in fire management across the U.S. Mexico boundary. In L. Fernandez & R. T. Carson (Eds.), *Both sides of the border: Transboundary environmental management issues facing Mexico and the United States* (pp. 385–402). Kluwer.

Mishler, C., & Simeone, W. E. (2004). *Han: People of the river*. University of Alaska Press.

Morin, J. (2015). *Tsleil-Waututh nation's history, culture and aboriginal interests in Eastern Burrard Inlet*. Prepared for Gowling Lafleur Henderson LLP 1 First Canadian Place 100 King Street, Suite 1600 Toronto, Ontario M5X 1G5. Retrieved from https://twnsacredtrust.ca/wp-content/uploads/2015/05/Morin-Expert-Report-PUBLIC-VERSION-sm.pdf

Morrell, M. (1985). *The Gitksan and Wet'suwet'en Fishery in the Skeena River System*. Gitksan-Wet'suwet'en Tribal Council.

Moss, M. L. (2005). Tlingit horticulture: An indigenous or introduced development? In D. Deur & N. Turner (Eds.), *Keeping it living: Traditions of plant use and cultivation on the northwest coast of North America* (pp. 274–295). University of Washington Press.

Moss, M. L. (2007). Haida and Tlingit use of seabirds from the Forrester Islands, Southeast Alaska. *Journal of Ethnobiology, 27*, 28–45.

Moss, M. L. (2011). *Northwest coast: Archaeology as deep history*. Society for American Archaeology.

Myers, R. A., & Worm, B. (2004). Rapid worldwide depletion of predatory fish communities. *Nature, 423*, 283–290.

Nadasdy, P. (2003). *Hunters and bureaucrats*. University of British Columbia Press.
Nadasdy, P. (2007). The gift of the animals: The ontology of hunting and human-animal sociality. *American Ethnologist, 34*, 25–47.
Nelson, R. K. (1973). *Hunters of the northern Forest*. University of Chicago Press.
Pascoe, B. (2014). *Black emu, dark seeds: Agriculture or accident?* Magabala Books Aboriginal Corporation.
People of 'Ksan. (1980). *Gathering what the great nature provided*. University of Washington Press.
Phinney, A. (1934). *Nez Percé Texts* (pp. 1–497). Columbia University Contributions to Anthropology.
Pierotti, R. (2010). Sustainability of natural populations: Lessons from indigenous knowledge. *Human Dimensions of Wildlife, 15*, 274–287.
Pierotti, R. (2011). *Indigenous knowledge, ecology, and evolutionary biology*. Routledge.
Pierotti, R., & Fogg, B. R. (2017). *The first domestication: How wolves and humans coevolved*. Yale University Press.
Preston, W. (1996). Serpent in Eden: Dispersal of foreign diseases into pre-mission California. *Journal of California and Great Basin anthropology, 18*, 2–37.
Pryce, P. (1999). *Keeping the lakes' way*. University of Toronto Press.
Ramsey, J. (1977). *Coyote was going there: Indian literature of the Oregon country*. University of Washington Press.
Ranco, D. (2007). The ecological Indian and the politics of representation: Critiquing the ecological Indian in the age of ecocide. In M. E. Harkin & D. R. Lewis (Eds.), *Native Americans and the environment: Perspectives on the ecological Indian* (pp. 32–51). University of Nebraska Press.
Reichel-Dolmatoff, G. (1996). *The forest within: The world-view of the Tukano Amazonian Indians*. Themis, Imprint of Green Books, Foxhole.
Reid, J. L. (2015). *The sea is my country: The maritime world of the Makah*. Yale University Press.
Renker, A. M., & Gunther, E. (1990). Makah. In W. Suttles (Ed.), *Handbook of North American Indians: Northwest coast* (Vol. 7, pp. 422–430). Smithsonian Institution Press.
Reyes, L. L. (2002). *White Grizzly Bear's legacy: Learning to be Indian*. University of Washington Press.
Reynolds, N. D., & Romano, M. D. (2013). Traditional ecological knowledge: Reconstructing historical run timing and spawning distribution of eulachon through tribal Oral history. *Journal of Northwest Anthropology, 47*, 47–70.
Rick, T. C., & Erlandson, J. M. (Eds.). (2008). *Human impacts on ancient marine ecosystems: A global perspective*. University of California Press.
Rick, T. C., & Erlandson, J. M. (2009). Coastal exploitation. *Science, 325*, 952–953.
Ross, J. A. (1998). Spokane. In D. Walker (Ed.), *Handbook of North American Indians: Plateau* (Vol. 12, pp. 271–282). Smithsonian Institution.
Ross, J. A. (2011). *The Spokan Indians*. Michael J. Ross.
Royle, T. C. A. (2021). The use and cultural importance of suckers (Catostomidae Cope, 1871) among the indigenous peoples of northwestern North America: An ethnographic overview. *Journal of Northwest Anthropology, 55*, 299–326.
Sapir, E. (1990). *The collected works of Edward Sapir. VII. Wishram texts and ethnography*. Mouton de Gruyter.
Schlick, M. D. (1998). Handsome things: Basketry arts of the plateau. In S. E. Harless (Ed.), *Native arts of the Columbia plateau: The Doris Swayze bounds collection* (pp. 57–70). High Desert Museum and University of Washington Press.
Schreiber, D. (2008). 'A Liberal and paternal Spirit': Indian agents and native fisheries in Canada. *Ethnohistory, 55*, 87–118.
Simeon, A. (1977). *The she-wolf of Tsla-a-wat: Indian stories for the young*. J. J. Douglas.
Smith, M. W. (1940). *The Puyallup-Nisqually*. Columbia University Press.

References

Sobel, E. A., Ames, K. A., & Losey, R. (2013). Environment and archaeology of the lower Columbia. In R. T. Boyd, K. M. Ames, & T. A. Johnson (Eds.), *Chinookan peoples of the lower Columbia* (pp. 23–41). University of Washington Press.

Sproat, G. M. (1987). *The Nootka: Scenes and studies of savage life*. (Orig. edn. 1868.). Sono Nis Press.

Stewart, O. C., Lewis, H., & Anderson, M. K. (2002). *Forgotten fires: Native Americans and the transient wilderness*. University of Oklahoma Press.

Sullivan, R. (2000). *A whale Hunt: Two years on the Olympic peninsula with the Makah and their canoe*. Scribners.

Suttles, W. (1987). *Coast Salish essays*. University of Washington Press.

Suttles, W. (2005). Coast Salish resource managmgement: Incipient agriculture? In D. Deur & N. Turner (Eds.), *Keeping it living: Traditions of plant use and cultivation on the Northwest Coast of North America* (pp. 181–193). University of Washington Press/University of British Columbia Press.

Swanton, J. R. (1909). *Tlingit myths and texts*. Bureau of American Ethnology, Bulletin 39.

Swezey, S. L., & Heizer, R. F. (1977). Ritual management of Salmonid fish resources in California. *Journal of California Anthropology, 4*, 6–29.

Tanner, A. (1979). *Bringing home animals*. St. Martin's Press.

Taylor, H., & Grabert, G. (Eds.). (1984). *Western Washington Indian socio-economics: Papers in honor of Angelo Anastasio*. Western Washington University.

Teit, J. (1898). *Traditions of the Thompson River Indians of British Columbia*. American Folklore Society by Houghton Mifflin.

Teit, J. (1909). *The Shuswap* (Vol. II). Memoir of the American Museum of Natural History, The Jesup North Pacific Expedition, part VII.

Teit, J. (1912). *Mythology of the Thompson Indians* (Vol. 12). American Museum of Natural History, Memoir; Reports of the Jesup North Pacific Expedition, VIII.

Thompson, M. T., & Egesdal, S. M. (2008). *Salish myths and legends: One people's stories*. University of Nebraska Press.

Thornton, T. F. (2008). *Being and place among the Tlingit*. University of Washington Press.

Toniello, G., Lepofsky, D., Lertzman-Lepofsky, G., Salomon, A. K., & Rowell, K. (2019). 11,500 y of human-clam relationships provide long-term context for intertidal management in the Salish Sea, British Columbia. *Proceedings of the National Academy of Sciences, 116*, 22106–22114.

Trosper, R. L. (2009). *Resilience, reciprocity, and ecological economics: Northwest coast sustainability*. Routledge.

Turner, N. J. (2005). *The Earth's blanket*. University of Washington Press.

Turner, N. J. (2014). *Ancient pathways, ancestral knowledge: Ethnobotany and ecological wisdom of indigenous peoples of northwestern North America* (Vol. 2). McGill-Queen's University Press.

Turner, N. J. (2020). 'That was our candy!' Sweet foods in indigenous peoples' traditional diets in northwestern North America. *Journal of Ethnobiology, 40*, 305–328.

Turner, N. J., & Hebda, R. J. (2012). *Saanich ethnobotany: Culturally important plants of the WSÁNEĆ people*. Royal BC Museum Publishing.

Turner, N., & Peacock, S. (2005). Solving the perennial paradox: Ethnobotanical evidence for plant resource management on the Northwest Coast. In D. Deur & N. Turner (Eds.), *Keeping it living: Traditions of plant use and cultivation on the Northwest Coast of North America* (pp. 101–150). University of Washington Press.

Viveiros de Castro, E. (2015). *The relative native: Essays on indigenous conceptual worlds*. HAU Books.

Walker, D. E., Jr., & Matthews, D. N. (1994). *Blood of the monster: The Nez Perce Coyote Cycle*. N.p.: High Plains Publishing Co.

Walsh, M. K., Whitlock, C., & Bartlein, P. J. (2008). A 14,300-year-long record of fire-vegetation-climate linkage at Battle Ground Lake, Southwestern Washington. *Quaternary Research, 70*, 251–264.

Wilke, P. (1988). Bow staves harvested from Juniper trees by Indians of Nevada. *Journal of California and Great Basin anthropology, 10*, 3–31.

Willerslev, R. (2007). *Soul hunters: Hunting, animism, and personhood among the Siberian Yukaghirs*. University of California Press.

Williams, J. (2006). *Clam gardens: Aboriginal mariculture on Canada's west coast*. New Star Books, Transmontanus 15 [series].

Wilson, E. (2012). The oil company, the fish, and the Nivkhi: The cultural value of Sakhalin Salmon. In B. J. Colombi & J. F. Brooks (Eds.), *Keystone nations: Indigenous peoples and Salmon across the North Pacific* (pp. 25–46). School of American Research Press.

Worster, D. (1994). *Nature's economy: A history of ecological ideas* (2nd ed.). Cambridge University Press.

Chapter 7
White Settler Contact and Its Consequences

Hard Times

This topic is difficult to address, largely because after 500 years, the problem still exists, even though recent court decisions concerning treaty and land rights in both the USA and Canada seem to be mitigating some of the most serious impacts. What is important to keep in mind is that the Nations are, as they often say, "still here." The Nations have learned to master *Survivance*, a combination of survival combined with resistance to assimilation and conquest, as defined by Anishinaabe writer Gerald Vizenor (1994). Survivance is meant to represent the struggle to preserve Indigenous concepts, cultures and ways of thinking and understanding in the face of colonial oppression.

The existence of local dissention and rivalries among Indigenous groups provides some understanding of the ease of European conquest, as well as the influences of the noxious troika, "guns, germs and steel" (Diamond, 1997). From Cortes and the Pizarros forward, Europeans carefully and deliberately took advantage of Native rivalries to set Indigenous peoples against each other. In cases when Indigenous people could unite and stay united, they generally held their own. The successes of the Tlingit and Tsimshian in early fighting stand as examples in the Northwest, but solidarity never reached beyond the "tribal" level, and the Native people never forged a united front. In contrast, the alliance of nations put together by Tecumseh around the turn of the nineteenth century, actually led to serious resistance and garnered support from Canada, which turned out to be a major casual factor in the War of 1812 (Gilbert, 1989). The unity showed by the Nez Perce in the Nez Perce War did not save them in the short run; they were harassed by other tribes and eventually betrayed by Blackfoot warriors. Today the Nez Perce are still a strong political force in their ancient homelands and have turned their Idaho reservation into the major refuge for wolves within that state.

© The Author(s), under exclusive license to Springer Nature Switzerland AG 2022
E. N. Anderson, R. Pierotti, *Respect and Responsibility in Pacific Coast Indigenous Nations*, Studies in Human Ecology and Adaptation 13,
https://doi.org/10.1007/978-3-031-15586-4_7

The total population of the Northwest Coast in precontact times was at least 200,000–250,000, very likely more. This was quite high, given the resource base. The lavish salmon runs occasionally failed, leading to periodic famines, which the peoples worked to prevent through rituals and their own conservation practices. No other resource was abundant along the entire coast in the way salmon were, except—locally—herring and ooligans, and harbor seals, and whales during migrations. Those moderated the impacts of famine, along with other foods gathered from land and sea.

Disease, White settlers, local wars, and massacres reduced the population in many areas by 1900, as established by Robert Boyd (Boyd, 1999; Boyd & Gregory, 2007; Trafzer, 1997). In one particularly horrendous case, the peoples of the lower Columbia River area, as Boyd relates, "plummeted from in excess of 15,000 to just over 500 survivors" between about 1770 and 1855 (Boyd, 2013a: 247, b), a decline of 97%. Among harmful events, from California north into Washington, was the malaria epidemic of 1830–1833 (Boyd, 1999; Boyd et al., 2013). Conversely, some of the tribes of the interior, where malaria was rare or absent, may have lost "only" 75% or so between 1770 and 1900. They continued to lose population in the early twentieth century, when the 1918–1919 and 1928 influenza epidemics wiped out whole communities. Fortunately, these last epidemics reached them along with modern medicine (Helm, 2000: 120–123, 192–219). Boyd notes depression and despair after epidemics; the impacts of these conditions have been underestimated in the past. Thus passed the first, or microbial, phase of the European conquest of the Americas (Crosby, 1972; Martin, 1978; Mann, 2005). One rule seems to hold for the Pacific NW. The further from settler cities and farmlands a people were, the greater their chances of surviving as an intact society. Despite introduced disease, the lies and broken promises, Indigenous nations in both the USA and Canada along the Pacific coast seem to have won major legal victories, from Boldt to Delgamukw legal decisions, which established fishing rights guaranteed through treaty and the validity of Indigenous Knowledge and traditions, which have led to the current Musqueam claim on much of the city of Vancouver, Pacific Northwest First Indigenous nations are establishing themselves as a political force going into the twenty-first century..

Thanks to Covid-19, among other tragedies, even Europeans now know how utterly devastating the loss of even one or two loved ones can be to a community, particularly to children. The loss of 95% of one's society is, to modern Americans and Europeans, unimaginably horrible, except to those who have gone through genocides. Europeans can no longer remember the fourteenth-century plague, which had a profound impact on European philosophy and science (Pierotti, 2004, 2006). The prevailing worldview in Europe prior to the mid-fourteenth century was mythic and symbolic, rooted in an idea of cyclical time; placing more emphasis on links between human and nonhuman aspects of the world than did worldviews arising after the Black Death (Pierotti, 2006). The lack of ability to deal with the resulting death and devastation created both widespread panic and subsequent culture-wide depression (Gottfried, 1983). The impact of massive, inexplicable loss of life on a society cannot be overestimated (Pierotti, 2011: 45–46).

Such experiences clearly colored the often-gloomy stories of western North American Native people. One recourse, but surprisingly rare, was to blame the whites, who had inadvertently or deliberately introduced the disease, and who sometimes threatened openly to unleash it on the Native population (Boyd, 2013a). This blame was a contributing factor to some killings of settlers. Nonetheless, unexplained deaths were assigned to witchcraft, and as in white societies, witches were sought out and killed. Often these victims were medicine persons, since it was assumed that those who had curing power might well have killing power too, and because the failure of these individuals to cope with these diseases created great resentment and fear (Martin, 1978). The loss of these healers made the illness situation worse.

Boyd and others rightly stress disease as the major killer, but the roles of war, massacre, hard usage, denial of hunting and gathering options, the consequences of poor quality "rations" distributed to conquered peoples, and other directly brutal behavior deserve more emphasis than they usually receive. Disease may have been less important, especially early in contact, than has been alleged in most of the literature, and declines were often due to direct or indirect violence (Cameron et al., 2015). Boyd certainly establishes the importance of disease in the Northwest, but it is certainly true that other causes of mortality resulting from European invasions were also important.

Conflict was no stranger to the Northwest coast, even before Europeans arrived, however their arrival seemed to catalyze even more violence. "Chiefs used violence strategically, attacking rivals to seize slaves, resources and land and sea spaces" (Reid, 2015: 57–87). The Makah were one of the most aggressive nations and their control of the northwest corner of the Olympic Peninsula at Cape Flattery gave them strategic advantage in control, especially over the marine areas extending north and west from Cape Flattery and they were close to Vancouver Island than to most of Washington State. The pugnacious nature of the northwest coastal peoples, along with their remote and hard to reach locations meant that they were not massacred the way many inland peoples were (see below). The Makah were important trading partners along the Strait of Juan de Fuca, and were the primary Indigenous nation trading with Victoria, the eventual capital of British Columbia. If they had been in Canada, they would be spoken of today in the same terms of the Sioux and Comanche, however eventually the USA broke them down as it established control over Washington state, passing laws that reduced or even eliminated their abilities to whale, seal, and fish for halibut, which had given them economic dominance during the nineteenth century until after the US Civil War. They were probably the greatest marine power, both economic and military, that Indigenous America ever produced (Reid, 2015). Today, however, few Americans are aware of this history, and treat the Makah as another group of forgotten Indigenous people, barely aware of how powerful they were from 1700 to 1870.

Small wars around the edges of the region—the Rogue River War in southwest Oregon, the Modoc War in northeast California, the Nez Perce war of 1877 fought from Idaho to Montana, the Chilcotin War in central British Columbia, and others—led to additional casualties (Cozzens, 2002). These were wholly one-sided; and usually started by openly genocidal settlers. Governments collaborated to varying degrees in massacres and certainly allowed leaders of these massacres to masquerade as military men, as, for example, "Colonel" John Chivington at the Sand Creek massacre in Colorado in 1864. The Bannock War of 1878 in Idaho was notably one-sided as to fatalities, and followed the outright massacre of Shoshone at Bear River on Jan. 29, 1863, which led to about 400 deaths (Wikipedia, "Bear River Massacre"). These conflicts resemble genocide more than actual war. There was sometimes serious fighting, but often victory by troops was followed by indiscriminate killing. Genocidal killing in California extended into the northwest sector of that state, as told in *An American Genocide* (Madley, 2016). The Yurok and Karuk, in a rare example of successful resistance, took to the hills, and killed numerous genocide-planning gold-miners in the 1850s, though they were inevitably defeated in the end (Madley, 2016: 182, 235). In spite of defeat, they fought hard enough to keep a large chunk of their territory, which, even today, remain less damaged than most of California.

Among Northwest Coast groups, the Rogue River War of the 1850s was particularly harsh (Beckham, 1996; Madley, 2016; Youst & Seaburg, 2002). The Athapaskan and Takelman peoples of the Rogue River drainage initially resisted invasion of their lands, and when miners and settlers flooded in, violence quickly spiraled out of control. There was real combat—it was not pure massacre—but the hopelessly outnumbered Rogue people soon lost, and genocidal European invaders moved in. Local "exterminators"—the word used at the time—with the often-grudging aid of the US Army killed many Rogue River Indigenous people, except for those that retreated into the deep, almost impenetrable, forests of the Klamath Knot or were sheltered by local sympathetic families such as the Applegates. A few survivors were moved to remote reservations, often forced to move on foot. An area that had supported at least 10,000 people was left with virtually no Native American inhabitants.

The results of genocide up and down the coast were not only reduced populations, but vastly reduced landholdings, especially in British Columbia and California, where some well-distributed nations were forced into "reserves" or "rancherias" or became landless. Various Native nations have reservations in the USA, and usually reserves are much smaller in Canada, however land claims are being taken more seriously and First Nations are gaining more influence over Crown (federal) lands. Broken treaties were the norm, where they existed. Enforcement of treaty rights lagged until the 1970s.

British Columbia thought it was clever in the nineteenth century by refusing to negotiate or sign treaties, however this has backfired on them because BC First Nations never ceded land to Canada, which means that land claims are more likely to have validity in much of BC, where few settlers actually live. The uneasiness this generates can easily be seen in the book Death Feast in Dimlahamid (Glavin, 1998), where settlers fear that they will be the ones forced to relocate. This is extremely unlikely and the First Nations have tried to reassure settlers that they will not be removed. The question remains, however, what will happen in future generations after the current settler generations have passed.

Some recent writers, Native and white, have criticized the "myth of the vanishing Indian" for obvious reasons, because the Indians did not, in fact, vanish (see e.g., Harmon, 1998; Pierotti, 2010, 2011; Pierotti & Wildcat, 1999). As of the 1890s, Europeans generally assumed that Indigenous nations as separate peoples might cease to exist in a very few years, especially after Wounded Knee and the collapse of the *Ghost Dance* movement. By 1900, massacres had largely stopped, but disease was still rampant. Also, Settler societies were doing everything they could to relocate and terminate Indigenous culture and assimilate the survivors to Anglo-American norms, especially in Canada. This included the notorious kidnapping of children for boarding schools, followed by Termination policies in the USA from 1948 to 1963 and forced assimilation in Canada. Many groups did vanish as linguistic and cultural entities, including the Tsetsaut, the Chemakum (Chimakum), and several Chinookan groups. Despite this effort, many individuals resisted termination and this seemed to inflame peoples who feared the loss of their land. Many groups elsewhere on the continent were completely gone before 1800, such as the Beothuk of Labrador and several groups in Florida and Texas. Whites who in the late nineteenth century forecast the final extinction of the "race" believed they had every reason to do so.

In contrast to these settler assumptions, many Indigenous people viewed the arrival of settlers as a temporary phenomenon and that eventually they would disappear because they would destroy themselves because of their lack of honor and connection to the real world. The goal of these people was to survive, which is the true meaning of "we are still here," as well of the basis of the idea of Survivance (survival combined with resistance; Vizenor, 1994). The goal was to survive, while bringing as many of our people into the future, along with as many species of our *more than human* relatives as we could.

The survival of the last 5% of the Native Americans of the Northwest resulted from determination and perseverance on the part of the survivors, and their few White allies. The fact that the Indians did not vanish is primarily due to the incredible toughness of the survivors and the power they recognized was retained in their culture ("survivance"). Even though the potlatch ceremony was driven underground, it persisted, as the Sun Dance and the Sweatlodge did with Plains nations. The Canadian government was still pushing effective termination as recently as 1980. The Berger Report in the 1970s revealed the determination of Athapascan and Inuit peoples, who they told the government they valued their cultures more than they did economic growth and being part of the twentieth-century economy. Anthropologists and historians have collected or reconstructed several stories of Native American individuals who not only survived but led their people through such experiences. Notable examples for the Northwest include the story of Jimmy Sewid (Sewid, 1969; cf. Ford, 1941); Daisy Sewid (1969), Agnes Alfred (2004), and Annie Miner Peterson (Youst & Seaburg, 2002). The first three are recorded autobiographies, the last an amazing job of reconstructing a long and eventful life from scattered and fragmentary records. (See also Atleo, 2004, 2011; George, 2003; Reyes, 2002, 2006, 2016, for later accounts written by Native scholars themselves.)

Changing white settler attitudes were also involved. "Indian lovers" were widely hated and despised in the 1870s and 1880s, but remained numerous. The missionaries were purblind in their intolerance of Native ways, but at times sheltered their charges from the worst effects of colonization, while also removing Indigenous children to boarding schools or adoption, in an attempt to destroy Indigenous cultures through assimilation. The process is far from over; however, the trend continues to be reasonably good. Indigenous people still die in disproportionate numbers, and are subject to disproportionate amounts of violence, but today there is no chance of them dying out, and pervasive racism is no longer part of Canadian government policy as it was until repeal of specific measures in 1982.

Some anthropologists were leaders in this change, despite errors and some insensitivity. Unfortunately, in the late nineteenth and early twentieth centuries, many pro-Indian individuals, including some anthropologists, backed breaking up US reservations into individual allotments through the Dawes Act, in a misguided attempt to make the Indians into Anglo-style small farmers on family-owned holdings.

> Federal Indian policy during the period from 1870 to 1900 marked a departure from earlier policies that were dominated by removal, treaties, reservations, and war. The new policy focused specifically on breaking up reservations and tribal lands by granting land allotments to individual Native Americans and encouraging them to take up agriculture. It was reasoned that if a person adopted "White" clothing and ways, and was responsible for their own farm, they would gradually drop their "Indian-ness" and be assimilated into White American culture (https://www.archives.gov/milestone-documents/dawes-act).

Alice Fletcher, a late nineteenth-century anthropologist, was particularly active in promoting allotment among the Nez Perce (Tonkovich, 2012, 2016). The rationale for allotment was given by Commissioner of Indian Affairs John D. C. Atkins in 1886: Native Americans "must be imbued with the *exalting egotism* of American civilization so that he will say 'I' instead of "We,' and "this is mine' instead of 'This is ours'" (Fisher, 2010: 90; emphasis in original). A textbook used in Oregon schools in the first half of the twentieth century said, "the Indian vanished because he could not learn the ways of the white man. He could not survive in competition with the dominant race" (Fisher, 2010: 144). This line reads like an exaggeration from a Hollywood movie, but the man was perfectly serious. More to the point, it is completely self-serving and self-justifying. A whole litany of such quotes can be found in *Shadow Tribe* (Fisher, 2010).

Allotment proved disastrous. Native peoples lost large proportions of their land (see e.g., Fisher, 2010 for Columbia River tribes; Stern, 1966 for the Klamath). Fletcher lived to see and regret the catastrophe that her work produced, as the Nez Perce quickly lost their best land to aggressive, dishonest whites. The million and a half acres of the Siletz Reservation in Oregon dwindled to a few scattered plots, and the Grande Ronde suffered similarly (Youst & Seaburg, 2002). Ironically, today Indian gaming is saving tribal lands: casino earnings are now being used to buy back formerly Indian land.

Conflicts between governments and First Nations declined in frequency after the cultural damage caused by the Dawes Act and the Massacre at Wounded Knee, but starvation, exposure, and other forms of mistreatment continued to kill many Native people on and off reservations. Residential schools for children proved death-traps, because despite their supposed charity, White teachers and administrators cared little about the students except in an abstract sense, ignoring both physical and psychological problems faced by students, which led to terrible mortality rates, some of which are just coming to light as of 2021 (Niezen, 2017 https://www.scientificamerican.com/article/canadas-residential-schools-were-a-horror/).

Alcoholism, which often leads to suicide in these communities, has taken a grim toll. Suicide became common, especially among young people who were being told that their ways of life and cultural traditions were evil, because they did not conform to the strictures of missionary Christianity. This led them to feel they did not belong anywhere. Sexual abuse, especially in schools run by the Roman Catholic Church, which has been revealed as hotbeds of child molesting (still being exposed in the twenty-first century), reached epidemic levels in several cases. These impacts of conquest and racism deserve to be placed with disease as major causes of fatalities. The Anishinaabe and Cree peoples in Canada suffered high rates of sexual and physical abuse, child stealing, and even murder, as described in the writing of Richard Wagamese, Edward Metatawabin, Tomson Highway, and Robert Arthur Alexie, summarized well in Ronald Niezen's *Truth and Indignation* (2017).

These and other residential schools routinely underfed the children or fed them on non-nutritious food, predisposing them to diabetes; they even experimented with how little they could feed the children while keeping them alive (Mosby & Galloway, 2017a, 2017b). They often involved situations where children were violently abused by beatings and rampant sexual molestation (see testimonies in Niezen, 2017). Teaching focused on the English language and remedial working skills. The Catholic church, both in Canada and in Rome, knew about sexual molestation in Catholic schools throughout North America and Europe for decades. To be fair, in places like Newfoundland, having exterminated the Beothuk two centuries earlier, Catholic brothers and priests also sexually and physically abused Irish boys until the 1970s (https://www.heritage.nf.ca/articles/politics/wells-government-mt-cashel.php). The HRC and the Anglican churches, along with the Canadian government have now made apologies, but few amends (Niezen, 2017). Native languages were banned in these schools. A whole "stolen generation" (to borrow an Australian phrase) has resulted. Brutalization led very often to alcoholism among survivors of the residential schools (see writings by Wagamese and others listed above). Raphael Lemkin's classic definition of genocide includes raising children away from and ignorant of their culture, which he knew to be a common feature of settler societies (see e.g., Madley, 2016; Tatz & Higgins, 2016).

In British Columbia, churches and missionaries got the potlatch outlawed from 1884 to 1953 as a heathen institution incompatible with the ethics of the modern state; apparently the missionary mind regards mass organized sexual abuse of minors as more ethical than the potlatch. Ironically, the first legal potlatch in 1953 was the one organized by (white) anthropologist Wilson Duff and Kwakiutl artists to inaugurate the totem pole park at the British Columbia Provincial Museum (now Royal British Columbia Museum; see Hawker, 2003: 138).

Native populations began recovering their numbers, but not necessarily their cultures, around 1900; even now we are unsure how close these are to pre-invasion levels. One thing to keep in mind is how many tribal nations lost 90% or more of their populations to disease. Given that estimates for North America north of Mexico range from 2 to 3 million to more than 20 million, this runs the gamut, since 2–3 million is still only about 10% of at least 20 million (Mann, 2005). Marriage into settler societies has contributed to this. To this day, sizable parts of the Northwest (and other parts of North America) have fewer people than in 1700.

Native peoples were rapidly dispossessed of their fisheries and other resources, despite having these guaranteed through treaties that promised them rights to hunt, fish, and gather as long as grass grows green and rivers run. Canadian Cherokee writer Thomas King parodied this in his comic tour de force, "Green Grass, Running Water" (1994). Individual Indians were urged (or in the case of Canada under Pierre Trudeau), ordered to "follow the white man" and take to farming or menial blue-collar labor during Termination in the USA. When they succeeded as farmers their lands were routinely seized by greedy whites; one well-studied case study is the life of Arthur Wellington Clah (Brock, 2011), who initially succeeded very well in the settler world, but was chased off parcel after parcel until he died in poverty. (He figures in this book in a less direct but very important way: his son Henry Tate, and his grandson [via a daughter] William Beynon, became the great chroniclers of Tsimshian society.)

In the 1960s through 1980s the previously very successful Native fishery in British Columbia was systematically severely compromised by right-wing governments, and damaged by shady practices by white fishermen (Lutz, 2008; ENA, personal observation). Some of the erosion was even supported by the Liberals, usually pro-environment, but more pro-economic growth. Native fishers lost more and more rights, while fishery policies favored larger, more modern boats, which were mostly in the hands of White fishers; loan policies made this steadily worse. On the American side, similar competition occurred, with perhaps less open government sympathy. The Columbia River, once the richest salmon stream, has suffered from both dispossession of Native rights and destruction of the vast majority of its fish resources by dams (Hunn & Selam, 1990), and may lack anadromous fish of any kind by 2100. (Excellent studies include Dupris et al., 2006; Ulrich, 2001; see also Grijalva, 2008 for the general problem of environmental justice and Native Americans. For an account of the salmon wars in Puget Sound see Heffernan, 2012). Canadian First Nations still struggle for similar recognition (see Coté, 2010).

Forced acculturation and active repression of Native traditions further devastated the cultures. Fortunately, both the people and their cultures began to get some measure of attention around 1900, allowing a reversal of population decline, and, to some extent, of cultural decline in spite of continued hostility in many Settler quarters. A standard history on the US side of the border is Alexandra Harmon's *Indians in the Making* (1998), which discusses the creation of the category "Indians" and of particular tribes. A British Columbia equivalent is John Lutz' *Makúk: A New History of Indian-White Relations* (2008). For notably temperate, dispassionate, thoughtful Indigenous views, see Richard Atleo's *Principles of Tsawalk* (2011), or Joseph Marshall, *On Behalf of the Wolf and the First Peoples* (1995). An excellent early account of culture change, adaptation, and native vs. settler education is

George Pettitt's study of the Pettitt (1950). Clear writing and sober, understated tone make all these books noteworthy. Harmon is better for exploring political machinations that created "tribes" and blood quanta; Lutz' book is outstanding for its superb history of Indian labor, with discussion of the ways the Indians were forced out of logging, fishing, trapping, and other activities that formerly gave them a good living. (That process was unfolding when ENA did research in British Columbia in 1984–1987; it was not pleasant to watch.)

The modern "tribes" and "nations" are to an extent creations of settler government policy. From the start, the settler groups needed clear polities and clear leaders to negotiate with. Finding fluid polities and leadership systems, the settlers simply created clear-cut "tribes and chiefs." Sometimes "tried to create" is more accurate; groups remained cantankerously independent and fluid. In much of the USA, there are no tribes identified by the US government by the names they use to refer to themselves. Many Northwest "tribes" do not even have official Indigenous names at all; they are "Flathead," "Nez Perce," and the like. Incorrect names were applied by mistake, or even on occasion by enemy groups. (There is some attempt to reverse the process; the Nez Perce, for instance, are increasingly using the Indigenous name Ni-Mi-Puu.) The process has been described many times (Fisher, 2010; Harmon, 1998; Lutz, 2008). It has left us with most unclear understandings of earlier arrangements, at least on the US side of the border, where change and population decline came earlier, and many tribes have been merged on shared reservations. The high population density and extensive trade reveal that there was organization of life above the village level. Modern "tribes" are usually defined by language, but sometimes simply by where they eventually were allowed to settle.

Native Americans were not even citizens until 1924 in the USA, and until 1961 (and not fully until 1985) in Canada. Fearing American invasion, the British Parliament passed the British North America Act to unite the five eastern colonies (Nova Scotia, New Brunswick, Prince Edward Island, Quebec, and Ontario) to form the new nation Canada. British Columbia was separate, not incorporated into Canada until 1871. No Indian Nation was invited for discussion, or to obtain consent for the formation of a country upon its land. Nor was any Indian Nation even informed of such events happening. In order to expand from sea to sea, the new Canada sent out Indian Agents to make treaties with the various Indian Nations in other provinces, claiming they did so in the Queen's name. As a result, the Nations assumed their treaties were with the British Crown and many continue to believe so today. Ironically as mentioned above, British Columbia has no treaties because they felt they could simply take land, which may cost them greatly as land claims progress. Canada committed a massive fraud but has consistently used post-confederation Treaties Number 1–11 as evidence of having Title to the Indian lands (for this and what follows, see Lutz, 2008).

A major issue was the Canadian government's attempt to define "Status" vs. "Non-Status" Indians. By definition, those who did not sign treaties were not "Indians." This issue continued with the nonstatus criterion remaining intact until Canada got its own Constitution in 1982. In fact, in 1969 a White Paper presented by Pierre Trudeau's "liberal' government, proposed elimination of Indian status altogether (https://indigenousfoundations.arts.ubc.ca/the_white_paper_1969/). This was a very troubling political effort to assimilate Indigenous peoples and remove any special status; starting with an effort to remove Aboriginal status from Indian women who married non-Indian men or even non-status Indian men. Their children were also no longer considered Indians. This was not changed until the 1980s. In BC it was even more noxious, with any Indian who went to university or lived off reservation losing their status.

Fortunately, the Supreme Court of Canada overturned a lower court's decision in Calder v Attorney General of BC, recognizing land rights based on aboriginal title. In 1982, the 1867 BNA was repatriated to Canada from London, and the totality of treaties, proclamations, and other British laws governing the British relationship with the many sovereign aboriginal nations in British North America were incorporated into the Canadian Constitution. This incorporation was vital, because failure to achieve this would have left Britain exposed to disputes with native nations with whom it had signed treaties. The Royal Proclamation, together with the many treaties, now is now part of Canada's Constitution. Under increased pressure from the First Nations, Section 35 (1) was added to the Constitution to recognize and affirm the "existing" aboriginal and treaty rights of the aboriginal people in Canada. In 1985, The Indian Act was amended to remove discriminatory sections because of intense pressure regarding human rights violations pointed out by United Nations to Canada in 1982. Ironically, though, British Columbia, the last Province to join Canada, defied the national government by giving Indians provincial voting rights in 1949. By this time, the land was almost completely alienated and Indigenous cultures damaged.

In the USA, at least some large reservations and solid treaty rights obtained. Canada has been less accepting of Native title and rights. The Indians in British Columbia have only tiny reserves and few subsistence rights. On the American side, until the 1970s having treaties and reservations has not made the Native people much better off than their Canadian neighbors. A classic study concerns the Okanagan, whose territory was split in half by the border. Canadians thought the Canadian Okanagan were better off, US writers thought those on the US side were, but in fact the Okanagan were in the same situation on both sides (Carstens, 1991).

Quite different findings occur for tribes split by the Alaska-Canada border, indicating changes for the better in Canada and for the worse in at least one US state. "In recent years...the Canadian Han have enjoyed far better subsistence hunting and fishing rights than their Alaskan counterparts" (Mishler & Simeone, 2004: pp. xxii–xxiii). Conditions for Indigenous hunters, and for the game they hunt, have rapidly deteriorated in Alaska under a series of Republican governors, their campaigns heavily funded by the oil industry. They represent affluent White constituencies, including rich sport hunters.

Conversely, the Canadian situation improved dramatically when courts began recognizing Native rights in the 1990s. In 1997, in the case of *Delgamuukw* vs. *the Queen*, a British Columbia court handed down an infamous decision, denying the existence of Native land title, basically because the Natives had no written records. Their obvious occupation of the land for over 16,000 years did not count. This ruling was overturned in a historic ruling by the Canadian Supreme Court in 1997, which gave legal status to oral accounts from Indigenous people provided they could establish proof of their existence prior to the existence of Canada itself (Glavin, 1998; Persky, 1998).

Native title was at last recognized, at least under some circumstances. This led to the Nisga'a signing a treaty in 2000 that ceded official title of their lands in exchange for money and use rights. Other tribes still contest their cases. It should be noted that Great Britain had recognized at least some native claims from the beginning of British rule in 1763. Much of the nation was covered under signed treaties with England, which is why even today there is discussion of Crown Lands throughout Canada. However, British Columbia had been a separate political entity (not officially part of Canada) for many years, and had not done so, except for a few in the far northeast and southwest, involving very small areas of land.

On the US side, reservations were bones of contention, because treaties were often unfair, and the US Congress rarely ratified treaties made with "Indians." This led to feelings of betrayal, and consequently to some violent outbreaks and escapes, by people who had been promised larger and better lands. The reservations were treated largely as concentration camps until the 1930s, when John Collier was appointed head of the Bureau of Indian Affairs by Franklin D. Roosevelt. Collier encouraged the tribes to form their own governments. After FDR's death, however, in the late 1940s, the US Congress tried terminating reservations. One example was the "termination" of the Klamath tribe in the 1950s (Stern, 1966). The reservation was privatized; almost all the land was taken even before the Indians got their shares, who were cheated out of most of what was left. Local swindlers even reportedly used the ancient trick of giving a man a bottle of whiskey to get him to sign his name on a blank sheet of paper, and then filling in a deed of sale over his signature. Devastating poverty and social breakdown ensued. The Klamath have been working ever since to get a bit of their land and water back.

Repression of culture and language were comparable in both countries. Canadian Native peoples can expect (though they do not always get) much better health care and other services than most US residents can expect because health care is national, whereas in the USA, recognized Indians have the Indian Health Service (HIS) which is underfunded and highly variable in the treatment people receive. Racism remains widespread and extremely virulent in both countries. Readers should not be fooled by the genteel parlor-liberalism of Vancouver and Seattle; many Whites in the Northwest, especially in rural areas, are unreconstructed. Typical is the answer ENA got when he asked one rural British Columbian why he thought the Indians created so much fine art. His reply was, "They are too lazy to work." This is because whites think art, hunting, and fishing are recreational activities rather than actual labor. An illustration can be observed in the virulent hostility directed at the Makah by "conservationists" and "whale-lovers" when the Makah tried to establish their treaty-guaranteed whaling rights in the 1990s (Coté, 2010; Reid, 2015; Sullivan, 2000). Treaty rights to fishing and whaling were guaranteed in the USA and eventually enforced (though not until the 1970s), although the Makah are still fighting various US "Conservation" groups such as the Sea Shepherd Society to have their full whaling rights restored (Sullivan, 2000).

Land claims, treaty rights, and aboriginal title have been the subject of many lawsuits, some of which have been quite successful (e.g., the Boldt Decision, which restored treaty-based salmon fishing rights to tribes in Washington state; https://www.historylink.org/file/21084). Thus, in the last 30 years, the situation has improved somewhat for Indigenous communities. Treaty issues are so fiendishly complex that discussion must remain outside the scope of this book, even though this book owes its existence to the issue of Indigenous rights, land claims, and the ability of First Nations to control the lands they originally occupied. A very complex example in Canada involves the Musqueam First Nation, who have overlapping and/or shared territory with its First Nation neighbors: Kwikwetlem, Squamish, Tsawwassen, and Tsleil-Waututh (https://www.bctreaty.ca/musqueam-nation, accessed 22 May 2022) was recently given control over federally held lands in the city of Vancouver. What this means in practice is being worked out, with essentially all federally held lands currently not in use within the city reverting to the First Nations, e.g., the old Post Office was repatriated to the First Nations, who are developing it as a shopping area under their ownership. From farther north, Ian Gill provides a particularly good story of the movement that inspired the present book, in his book *All That We Say Is Ours* (2009), a biography and history of the Haida struggle to maintain land claims and fisheries.

A great deal of traditional culture survives, but the languages are almost gone everywhere. Very few tribes have any language speakers under 50 years of age, and those few have, at best, only a few remaining fluent speakers. Only among the most remote groups of the far north are there viable linguistic communities of young people. Some 40% of the Indigenous languages of North America are gone, with another 40% spoken only by older people. Stories, knowledge, art, and experience may persist after the language is gone, but slowly become impoverished. Experience from more heavily impacted parts of the continent shows that more languages and stories will die out, unless current attempts at reclaiming traditions and educating the young are much better funded. In the USA, tribal casino money is funding many such efforts. Language and cultural losses represent environmental disasters as well as humanistic ones. Knowledge of the environment and of managing it is encoded in the language used to talk about it. When the language goes, the knowledge diminishes (Hunn & Selam, 1990; Glavin, 1998).

Language revitalization movements rarely work, partly because they tend to be school-based, often using the familiar rote-drill method that convinced my generation that we could not learn languages. This is not always the case, however as reported by Stiles (1997). Comanche writer Barbra Meek in *We Are Our Language* (2010) provides an account of one of the more hopeful ones, and a full review of the relevant literature concerning the Kaska Dena First Nations from northern BC. Another success is the lifelong work of Nora and Richard Dauenhauer (e.g., 1987, 1990) among the Tlingit, which is exemplary. Similarly, the white linguist Thom Hess inspired the brilliant and indomitable Vi Hilbert, a Lushootseed woman who became not only the most active but also the most inspiring leader of an expanding community of Lushootseed learners, both Indigenous and settler. The whole story of Lushootseed revival, which goes well beyond that team, is told by Laurel Sercombe (2021).

The rural environments in which many Native people live give the lie to the blatantly false claim that people can learn only one language well. (There are Asianist scholars who know over 50 languages.) In fact, the more languages one knows, the better one is at learning more languages, and also at thinking and learning in general. Native Americans who have become fluent in both Native and settler tongues have tended to flourish (an observation confirmed by our research and experience).

It also remains to be seen whether heritage languages learned as second languages in school will preserve the (formerly) accompanying environmental knowledge. Teachers are aware of the concern, but traditional ecological knowledge is generally passed on in the bush, or similar settings, and in some cases can only be effectively passed on in such contexts (e.g., Ridington, 1988; Nadasdy, 2003). The knowledge may be more easily passed on in English in the bush than in a heritage language learned in a settler-style classroom. Pierotti has been involved with a number of his First Nations students recording and developing language retention with their own grandparents and great-grandparents (Dewey, 2007; Switzler, 2012). Now that language programs are under the control of Tribal nations and elders are often involved in teaching the languages and their nuances (e.g., White Hat, 2012) things are looking much better for the future.

Racism continued through the twentieth century, and even the successful Sin-Aikst (Lakes) scholar and artist Lawney Reyes had bitter memories of insults, persecution, and ill treatment in youth (Reyes, 2002). Observation showed that in the 1980s that British Columbia newspapers, when highlighting "social problems" such as crime, substance abuse, and dependence on welfare, usually picked a Native family to foreground in any story. The corresponding Washington state media almost always picked a Black one. The proportions of Black and Native people in these polities were similar in both and quite small. White Anglos had by far the most "social problems," in actual numbers, in both communities.

Conditions on the reservations and reserves are slowly improving, especially now that Laguna Pueblo member Deb Haaland has been named the first Indian Secretary of the Interior by President Joe Biden in 2021. Bureaucracy enters, insidiously, in even the best-intentioned situations. Displacement of Native peoples was followed by waste and misuse of resources by settlers of all sorts. The Canadian government has intervened to save many Northwest Coast groups from starvation and disease, as well as from the alcoholism and violence that followed destruction of livelihood and culture (Anderson, 1992). Even the resulting welfare system has its negative effects. In their detailed major study of bureaucracy and law in a colonial Canadian context, Jo-Anne Fiske and Betty Patrick state: "It [i.e., the welfare system] sustained the hierarchy of state/nation relations that circulates limited resources within the nation while disempowering the extended family...and...has aggravated relations of dependency" (Fiske & Patrick, 2000: 119). In the US court decisions have established that First Nations function as the legal equivalent of state governments, a ruling that came about because of clashes over casinos. This considerably reduced the virulence of debate. Even the most benign extension of alcoholism treatment, nursing services, and the like, however, has the effect of rubbing in the "problems" and "difficult situation" of Indigenous peoples, which is not appreciated (Fiske & Patrick, 2000: 182).

As a result of action taken by anthropologists and Indigenous peoples, some languages and cultures have been recorded and available for study. Some Native people have denounced such recording, because of concerns about "cultural appropriation." One strong form of refutation can be found in the marvelous image of Nuu-chah-nulth elder Hugh Watts reading to his grandchildren from the texts that Edward Sapir collected from Hugh's great-grandfather, Sayach'apis, in the early twentieth century. This photo is presented by Charlotte Coté, also a descendent of Sayach'apis (Coté, 2010: 83; Sapir & Swadesh, 1955). Sayach'apis' texts represent one of the greatest literary documents in any culture worldwide, especially considering that his language was reduced almost to the vanishing point in his time. We are more than lucky to have them, and Sapir's warm and respectful biography of Sayach'apis (Sapir, 1922)—a tribute and acknowledgment far ahead of its time, and even of much of ours.

Modern histories tell of a slow but steady revival of traditions, reclaiming of lands, and winning civil rights legislation. Many groups are writing their own histories (e.g., Karson, 2006; White Hat, 2012). First Nations are slowly winning more control of their resources and rights to manage their own lands (e.g., Louise Takeda's sensitive account of Haida Gwaii efforts, 2015, and Charles Wilkinson's of the Siletz, 2010). A notable example of survival through sheer toughness is that of the Jamestown S'Klallam (Peck, 2021), of the central northern Olympic Peninsula, who were forced into reservations in the 1850s. They walked out, went home, built the community of Jamestown on unclaimed land, and slowly worked their way into local and ultimately national recognition by sheer indomitable effort.

There and elsewhere on the Northwest Coast, cultures are reviving. The brilliant art in all media is the most visible marker of a major reclamation of community, lifeways, and pride (for s examples of First Nations animation look at https://www.youtube.com/watch?v=74Y38Oy4AM4 and https://www.youtube.com/watch?v=l2IWV24CWHk&list=PL6E0864154436817B. Northwest Coast art was one thing that was appreciated right from the beginning of contact. Markets for it developed in the early nineteenth century, and have grown steadily and rapidly since. Totem pole making has not only revived, but spread to some Salishan and other communities that did not traditionally carve them.

Health care is far better than it once was. Canada has a model comprehensive medical care system that reaches even remote communities. It may not be perfect, but may be the best, except for New Zealand, that a large nation with dispersed minority communities has developed. The USA lags behind, but has improved steadily, and has a separate federally-funded Indian Health Service (see https://www.ihs.gov/). Good nutritious food is available widely, though still sparse and expensive in isolated communities. Education, and educational opportunities, are very different from boarding school times. Overall, life is better for many Native people than it was when ENA carried out field research, but many people are still left behind.

Resource Mismanagement Since 1800

The Northwest has been as devastated as the rest of the world by exploitation over the last 150 years. Overlogging has decimated many forests. Yet another stream or lake going out of production of fish almost every year. Loss of berries, roots, and game continues.

Modern agriculture, especially the diverse small-farm agriculture of the Willamette and Lower Fraser Valleys and several other areas, is far more productive in terms of biomass, but in terms of community function, than anything Indigenous, but only through establishment of monocultures and the use of pesticides and herbicides. Industrial farming pays in a few areas, notably the Columbia River valley. However, these successful enterprises cover only a small fraction of the Northwest. Elsewhere, often the best that settler societies can do is log the forests or run cattle on range with low carrying capacity.

Moreover, rapid urbanization is rapidly eliminating farmland, though some First Nations are resisting. The only large tracts of good soil with a long growing season in far western Canada are along the lower Fraser River, and in tracts on the east-facing coast of Vancouver Island. These have been the scene of explosive urban growth, eliminating almost a third of the Lower Fraser farmland. Much of the rest is casually maintained, the feeling being that urbanization is inevitable in spite of conservation attempts and local actions.

Counter-trends—restored fish, successful farming, sustainable forestry—have often come from counter-traditions. Either the Native Americans have regained some control of their resources (as in Haida Gwaii) or "counter-cultural" settler groups have had their way. Of these latter, small-scale and mixed farming remains fairly widespread, but organic farming, community forestry, fish restoration by dam removal, and other successful interventions are happening with increased frequency, e.g., the taking down of the Elwha Dam (https://www.nps.gov/olym/learn/nature/dam-removal.htm).

The "settler" mentality has been unkindly but accurately summarized as "rape, ruin and run." This mentality is not confined to Anglo-Americans (Anderson, 2014). It is typical of new settlers of pioneer fringes throughout the world. It has been detected archeologically in the settlement of Polynesia and in the Bronze Age and Iron Age Mediterranean. It characterized the Japanese spread into Hokkaido. The Russians and Chinese both acted it out in Siberia. But the Anglo-Americans had an extreme form of the ideology, one that typically considered all nature to be an enemy, to be utterly destroyed as soon as possible and replaced with a Europe-derived artificial landscape (Anderson, 2014; Pierotti, 2011). They had the tools: guns, steam engines, fish wheels, and much more.

Another and more insidious and deadly difference from the Native Americans was and is that the settlers were interested in only a very few resources which they exploited without thought to the future. They depleted the game, mined gold, coal and other minerals, and logged the forests. They concentrated on salmon and later on other fish, but could not understand the appeal of eulachon. Gone was the sensitive Indigenous total-landscape management that concerned itself with berries, roots, bark, grasses, small animals, and other resources. The white settlers not only drew down the immediately saleable resources with little thought of sustainability; they destroyed the other resources without even considering them.

Indigenous management of fish was displaced over time and replaced by Anglo-American management (excellent histories exist: Arnold, 2008; Harris, 2008; Reyes, 2016; Schreiber, 2008). Salmon are at least a concern even to Whites, but lampreys, sturgeon, suckers, and other species have almost disappeared without much notice. Sturgeon survive only in a few of the biggest rivers. Dams, overfishing, pollution, and the other usual factors have wiped out the salmon from much of their original range and reduced them to low abundance everywhere. The "unkindest cut" has been farming Atlantic salmon, *Salmo salar* (Schreiber, 2004). Diseases and parasites—notably "fish lice," parasitic copepods—from these fish escape and decimate the local salmon, which have little natural resistance. A small salmon-farming industry in the waters east of Vancouver Island has led to extermination of pink salmon throughout that area—millions of wild self-reproducing salmon sacrificed for high-cost alien ones. Native people have objected to fish farming (Schreiber, 2002; Schreiber & Newell, 2006—the latter contrasts correct but "spiritual" Native knowledge with ignorance and mismanagement by White managers).

The fate of the Northwest Coast fisheries shows something about ownership regimes. The White settlers prefer private property, and, failing that, national government ownership and control: National Parks and Forests in the USA, Crown lands or provincial and national parks in Canada. The national governments lease out a good deal of non-private land outside parks for forestry or mining by giant corporations. The Native people, by contrast, universally vested property rights in localized kinship groups—communities, in a word.

For farming, individual ownership has its points. Fisheries do not work that way. Runs cannot be owned by individual humans. Native control allowed exquisitely careful management of stocks—using the most efficient methods to get the exact number of fish that could be safely taken, and then stopping the fishery when enough were taken. Western control has led to individual or national development that was incompatible with fish—dams, for instance—and, until recently, to virtually uncontrolled overfishing at all levels. The combination has been destructive. As an example, one of the major problems damaging salmon runs in many areas is uncontrolled clear-cutting which destroys stream structure and ecology. Ironically, fishermen blame Indians and loggers blame environmentalists, when in fact they are each other's worst enemies, but no industry representatives will acknowledge this fact. A brilliant but unfortunately unpublished thesis by Sara Breslow (2011) describes in detail the conflicts in the Skagit Delta between farmers, loggers, fishers, environmentalists, and government agencies—each blaming the others for all the problems, with the result that nobody does anything adequate to stop irreversible decline. State and national governments do not have the necessary focus, attention, or priority structure to resolve such conflicts, which are at heart, ecological, and in these conflicts it is often only the Native Americans that understand ecology at a gut level. Settler societies have other things to do, like using the rivers to generate hydroelectric power or to disperse and dispose of contaminants. They are also subject to voter pressure by fishermen desperate to take just a few more and let the future take care of itself. There is simply no way to prioritize the fishery while simultaneously regulating it tightly enough to preserve it. To conserve a fishery, people have to care deeply about the species involved (Anderson, 1996).

Andrew Fisher's excellent study of the devastation of the Columbia River people and their salmon is a stunning history (Fisher, 2010). Riverine fish were basic to Indigenous cultures, but were depleted first by overfishing, and later by the huge dams built along the river and its tributaries. Some of these had "fish ladders" to allow the fish to swim on upstream, but many of these ladders were designed badly, so that very few fish could successfully manage the trip.

Recently, the tendency is to take down dams. As noted, the old dams on the Elwha River, on the Olympic Peninsula, were taken out in the early twenty-first century, leading to rapid recovery of resources, including some fish. Providing future hope is that, unknown to many people, dams have relatively short lifespans, and must be decommissioned, usually within 50 years of their construction. They are the ultimate settler project: they last only as long as the people who build them, leaving future generations to deal with the mess they have created. Dam removal is slated for the Klamath River in California, and even proposed for the Columbia, but all these projects are held up by local opposition from the few beneficiaries.

In contrast, kinship and community dominate traditional Indigenous management practices. Extended kinship groups (widely called "clans," though not always in the technical anthropological usage) are religiously and ritually constituted. Succession to leadership titles in them is by means of potlatch involving wealth donations and feasting, and leaders can be removed if they fail to manage properly (Trosper, 2009). Clans in turn own and control resources. As was observed as long ago as 1826, an individual of authority in a clan, when hunting, "is particularly careful neither killing too many himself nor allowing any to do so" (William Brown, quoted in Fiske & Patrick, 2000: 128). In this group—the Babine—as in others, clans maintained the fishing weirs and owned fishing stations. By introducing individual rather than communally owned nets, Canadian settlers ruined this arrangement, which of course had unfortunate knock-on effects on the entire kinship system as well as on resource management. Enforcing conservation is almost automatic in a clan-owned weir but almost impossible in a world of individually-owned nets.

Alaska still has enough Native and local control to keep some fisheries healthy, but mining and oil interests are ever present. Oregon has, ironically, saved some of its fish thanks to powerful sport-fishing lobbies. The sport fishers are not dependent on fishing for survival, as are the commercial fishermen, and are more willing to let some fish escape today to make sure there will be fish for next year. Some practice "catch and release."

California's bit of Northwest Coast scenery, the Klamath drainage, provides a particularly revealing case. The Klamath Basin at the headwaters was settled early on by farmers and stockmen, displacing the Indians in the genocidal Modoc War (*Hell with the Fire Out,* Quinn, 1997). More recently, drought has reduced the Klamath flows, while water use continues to expand in the farming areas because of climate change. There is no longer enough water for both farms and fish. This situation came to a head in 2003 and again in 2019 and 2021. The US Bureau of Reclamation had disposal rights over most of the upstream water, and was subjected to intense lobbying by the upriver farmers and downriver fishers. The farmers are largely well-to-do Republicans. The fishermen downriver are less affluent and very often vote Democrat, and include large Native American groups. 2003 and 2019 being years of Republican national government, the water went to the farmers, and the fish died (see Doremus & Tarlock, 2008; Petrie, 2020; Service, 2003 for balanced accounts; Williams, 2003 for an unabashedly pro-fish position; Carolan, 2004 for a broader overview, but he incredibly misses the political dimension, thus his account is of rather limited worth). The historic drought of 2021 came after Donald Trump's racially and environmentally polarizing presidency, and conflicts between farmers and Native Americans became dangerously serious (Chabria, 2021). Finally, state government and the power companies entered the picture, resulting in a resolution to take out the hydropower dams that are destroying the Klamath's fish runs. This will substantially solve the downriver problem, but global climate change, which seems particularly serious in the area, may make it a hollow victory.

Until quite recent times, and locally even since, decisions always went to Whites over "Indians"—racism being severe in the Northwest, particularly in the nineteenth and early twentieth centuries—but, above that, to giant corporate interests over everyone and everything else. Despite Woody Guthrie's song *Roll on Columbia,* the Columbia was dammed, not so much for irrigation as for hydroelectric power. Much of this, in turn, went to ALCOA to produce aluminum for the burgeoning aircraft industry in western Washington and the manufacture of canned soft drinks. Dramatic changes began, starting with Judge George Hugo Boldt's decision in Washington state in 1974 to enforce the treaty rights, and giving the Native people an equal share of the fish resource. Since then, the USA and Canada have had to recognize at least some Native title to, or at least administrative control over their former lands. Other decisions have restored community management regimes and ownership systems in both nations.

Significantly, the classic claim that humans go for the main chance—material or financial maximization—is massively disproved by the history of the Pacific Northwest. Throughout the last 200 years, White authorities have decided for regimes and stakeholders that damaged the resource base with extremely low returns to themselves or society. The potatoes and barley raised by the White farmers up the Klamath were worth less than the fish would have been if the runs had been maintained. The hydropower gained from damming the Klamath and many other smaller rivers is, again, less valuable than the fish would have been. Destructive logging of forests that could have been—and sometimes actually once were—sustainably managed, has destroyed whole towns. In addition, forests that regulated runoff are gone, allowing flooding of many coastal communities. Political power—from votes, campaign donations, skin color, and political position—trumps financial benefit except to the giant corporations, and even then, they benefit often only in the short run.

An interesting case is the evolution of the MacMillan Bloedel timber corporation. H. R. MacMillan and Charles Bloedel brought scientific sustainable forestry practices to British Columbia and Washington, respectively. They appear to have been quite idealistic and to have run their companies and forests responsibly. Eventually they joined forces. At first all went well, but even before MacMillan died in 1976, the logging had become destructive, often called "forest mining" because trees were treated as a resource to draw down rather than to harvest sustainably. By the 1980s, after decades of corporate mergers and political shenanigans, "Mac and Blo" became a worldwide byword for irresponsible logging and merciless treatment of local communities. Things have improved since, especially since Mac and Blo fell on hard times and was taken over by Weyerhauser in 1999. However, the sustainable forestry of the original founders remains a dream. In contrast, the typical result of "Settler" management is ravaged timberlands that will not support logging or anything else for a very long time. Local communities have been ruined by this "cut-and-run" approach. (The literature on this is huge; see Vaillant, 2005 for a dramatic but accurate popular account).

Signing much of southern Alaska's timberlands over to Native Corporations has not been a full cure-all, and success was hard to find (Dombrowski, 2002). Native people were cajoled or cheated into signing away their rights for very slight compensation, as in the bad old days (cf. Stern, 1966). Outside the region, Thomas Davis chronicles a far more successful Indigenization of forest management among the Menominee in Wisconsin. The forest managed by the Menominee Nation is indistinguishable from natural forests, while allowing the harvest of more than two billion board feet of lumber over the last century (Davis, 2000; Trosper, 2007) showing success through selective management is possible.

Today, outside of national parks and reserves, tree farming and selective harvesting continue, but the old-growth forests remain under attack by timber executives employing a discount-based economic philosophy, tree farms are not always managed sustainably, and forest resources are still under pressure. Legislation and better management slow the decline, but have not stopped it. Contemplating satellite photographs of the region (see Google Maps) is a sobering experience.

Efforts to slow and reverse this pioneer-settlement process began early. Overfishing was targeted well before the end of the nineteenth century. Fish wheels, which could exterminate a huge run in a year, were banned very rapidly. After this early victory, conservation of the fish resource was a slow, difficult fight, with new threats (from dams to farmed salmon) constantly wiping out gains from years of dedicated work. As noted above, salmon are in serious trouble, but at least everyone cares about them and wants them; other resources that once were enormously valued, are neglected, and allowed to decline.

Game animals have done much better, because they are less valuable to settlers from a monetary perspective, and sport hunters provide a lobby for sustainable management. Parks and reserves protect the animals. There has even been some recent reintroduction of extirpated wildlife, such as the reintroduction of fishers (*Pekania pennanti*) to the North Cascades. Conservation has brought back the sea otter, almost exterminated from the coast by the fur trade, but unfortunately it has not thrived in the upper Northwest, for several reasons, even though the species thrives in California and the Aleutian Islands.

Environmentalism, conservation, and even basic sustainable management are major values in the region, but are heavily concentrated among the urban middle and upper classes of the settler community. Rural inhabitants of the Northwest remain widely opposed to such ideas, which they perceive—without evidence—as limiting their freedom and can be devastating to their incomes. Class conflict and rural distrust of cities, based on fear, ignorance, and racism, the unholy troika of the contemporary worldwide right, add bitter personal resentment to what might otherwise be a reasonable dialogue.

An interesting comparison comes from the Russian coast closest to the USA. Anna Kerttula (2000) describes Russian mismanagement and waste of resources in easternmost Siberia. The Yup'ik villagers there had been sustainably and efficiently hunting sea mammals, birds, and fish while the Chukchi herded reindeer on the tundra. This is a new economy for the Maritime group of Chukchi, who began as sea mammal hunters, and only switched to reindeer through conquest and colonization (Rytkheu, 2019). Russians brought wasteful hunting and herding practices (throwing away much of the meat). They also damaged the tundra seriously by construction and driving heavy vehicles, and caused rampant pollution. This was totally uneconomic; it was subsidized by the state as a way of "civilizing" the "natives," whom the Russians treated as inferiors. When the USSR collapsed, this artificial and destructive economy collapsed with it, and the local people were thrown back on their traditional subsistence economy—much damaged by Russian environmental abuse. It is interesting to see how clearly foolish and culture-bound the Russians' allegedly superior practices appear to Americans, who are so used to Anglo settler mismanagement that they tend to overlook its problems, and how much it is driven by culture rather than economics. This comparison illustrates why Pierotti and collaborator Pierotti and Wildcat (2000) titled their oft-cited paper "The Third Alternative," because both Socialist and Capitalist Regimes show similar disregard for the environment and sustainability. The viable strategy is to employ an Indigenous view as The Third Alternative.

Co-management of resources has fortunately entered the picture. Local Native communities cooperate with government agencies (Natcher et al., 2005; see above). One notable area is the west coast of Vancouver Island (Goetze, 2006; Pinkerton, 1989; Pinkerton & Weinstein, 1995). This does not always work well, especially when the government ignores Native input—as is often the case (e.g., Nadasdy, 2003). Obviously, co-management works only if it is, in fact, co-management and not simply consultation, which means First Nations are allowed to speak, but their opinions are simply ignored.

Perhaps it is best to leave the last word to Snuqualmi Charlie, talking to Arthur Ballard in the 1920s. He concluded a long story of Moon's creation and transformation of the world:

> Moon said, 'Fish shall run up these rivers; they shall belong to each people on its own river. You shall make your own living from the fish, deer and other wild game.'
>
> I am an Indian today. Moon has given us fish and game. The white people have come and overwhelmed us. We may not kill a deer nor catch a fish forbidden by white men to be taken. I should like any of these lawmakers to tell me if Moon or Sun has set him here to forbid our people to kill game given to us by Moon and Sun (Ballard, 1929: 80).

References

Alfred, A. (2004). *Paddling to where I stand: Agnes Alfred, Qwiqwasutinuxw noblewoman*. Trans. D. Sewid-smith, ed. Martine Reid. Seattle: University of Washington Press.

Anderson, E. N. (1992). A healing place: Ethnographic notes on a treatment center. *Alcoholism Treatment Quarterly, 9*(3/4), 1–21.

Anderson, E. N. (1996). *Ecologies of the heart*. Oxford University Press.

Anderson, E. N. (2014). *Caring for place*. Left Coast Press.

Arnold, D. F. (2008). *The Fisherman's frontier: People and Salmon in Southeast Alaska*. University of Washington Press.

Atleo, E. R. (2004). *Tsawalk: A Nuu-chah-nulth worldview*. University of British Columbia Press.

Atleo, E. R. (2011). *Principles of Tsawalk: An indigenous approach to global crisis*. University of British Columbia Press.

Ballard, A. C. (1929). Mythology of southern Puget Sound. *University of Washington Publications in Anthropology, 3*, 31–250.

Beckham, S. D. (1996). *Requiem for a people*. (Orig. publ. 1971). Oregon State University Press.

Boyd, R. T. (1999). *The coming of the spirit of pestilence: Introduced infectious diseases and population decline among northwest coast Indians, 1774-1874*. University of Washington Press.

Boyd, R. T. (2013a). Lower Chinookan disease and demography. In R. T. Boyd, K. M. Ames, & T. A. Johnson (Eds.), *Chinookan peoples of the lower Columbia* (pp. 229–249). University of Washington Press.

Boyd, R. T. (2013b). Lower Columbia Chinookan ceremonialism. In R. T. Boyd, K. M. Ames, & T. A. Johnson (Eds.), *Chinookan peoples of the lower Columbia* (pp. 181–198). University of Washington Press.

Boyd, R., & Gregory, C. D. (2007). Disease and demography in the plateau. *Journal of Northwest Anthropology, 41*, 37–70.

Boyd, R. T., Ames, K. M., & Johnson, T. A. (Eds.). (2013). *Chinookan peoples of the lower Columbia*. University of Washington Press.

Breslow, S. J. (2011). *Salmon habitat restoration, farmland preservation, and environmental drama in the Skagit River valley*. Ph. D. dissertation, Department of Anthropology, University of Washington.

Brock, P. (2011). *The many voyages of Arthur Wellington Clah*. University of British Columbia Press.

Cameron, C. M., Kelton, P., & Swedlund, A. C. (Eds.). (2015). *Beyond germs: Native depopulation in North America*. University of Arizona Press.

Carolan, M. (2004). Ontological politics: Mapping a complex environmental problem. *Environmental Values, 13*, 497–522.

Carstens, P. (1991). *The Queen's people: A study of hegemony, coercion, and accommodation among the Okanagan of Canada*. University of Toronto Press.

Chabria, A. (2021, June 25). *Racism surfaces in water war*. Los Angeles Times, A1, A12.
Coté, C. (2010). *Spirits of our whaling ancestors: Revitalizing Makah and Nuu-chah-nulth traditions*. University of Washington Press.
Cozzens, P. (2002). *Eyewitnesses to the Indian wars, 1865-1890. Vol. 2: The wars for the Pacific Northwest*. Stackpole Books.
Crosby, A. W. (1972). *The Columbian exchange: Biological and cultural consequences of 1492*. Greenwood Press.
Dauenhauer, N. M., & Dauenhauer, R. (1987). *Haa Shuká, our ancestors: Tlingit Oral narratives*. University of Washington.
Dauenhauer, N. M., & Dauenhauer, R. (1990). *Haa Tuwunáagu Yís, for healing our Spirit: Tlingit oratory*. University of Washington.
Davis, T. (2000). *Sustaining the forest, the people, and the spirit*. SUNY Press.
Dewey, M. (2007). *Language retention practices and modern technology*. Indigenous Nations Studies Program, University of Kansas.
Diamond, J. (1997). *Guns, germs and steel*. W. W. Norton.
Dombrowski, K. (2002). The praxis of indigenism and Alaska native timber politics. *American Anthropologist, 104*, 1062–1073.
Doremus, H., & Tarlock, A. D. (2008). *Water war in the Klamath Basin: Macho law, combat biology, and dirty politics*. Island Press.
Dupris, J., Hill, K., & Rodgers, W. (2006). *The Si'lailo way: Indians, Salmon and Dams on the Columbia River*. Carolina Academic Press.
Fisher, A. H. (2010). *Shadow tribe: The making of Columbia River Indian identity*. University of Washington Press.
Fiske, J., & Patrick, B. (2000). *Cis Dideen Kat: When the plumes rise*. University of British Columbia Press.
Ford, C. S. (1941). *Smoke from their fires: The life of a Kwakiutl chief*. Yale University Press.
George, E. (2003). *Living on the edge: Nuu-Chah-Nulth history from an Ahousaht Chief's perspective*. Sono Nis Press.
Gilbert, B. (1989). *God gave us this country: Tekamthi and the first American civil war*. Atheneum Press.
Gill, I. (2009). *All that we say is ours: Guujaaw and the reawakening of the Haida nation*. Douglas & McIntyre.
Glavin, T. (1998). *A death feast in Dimlahamid*. New Star Books.
Goetze, T. (2006). Empowered co-management: Towards power-sharing and indigenous rights in Clayoquot sound, BC. *Anthropologica, 47*(2), 247–266.
Gottfried, R. (1983). *The black death: Natural and human disaster in medieval Europe*. Robert Hale.
Grijalva, J. M. (2008). *Closing the circle: Environmental justice in Indian country*. Carolina Academic Press.
Harmon, A. (1998). *Indians in the making: Ethnic relations and Indian identities around Puget Sound*. University of California Press.
Harris, D. (2008). *Landing native fisheries: Indian reserves and fishing rights in British Columbia, 1849-1925*. University of British Columbia Press.
Hawker, R. W. (2003). *Tales of ghosts: First nations art in British Columbia, 1922-1961*. University of British Columbia Press.
Hefferman, T. (2012). *Where the Salmon run: Life and legacy of Billy Frank Jr*. University of Washington Press.
Helm, J. (2000). *The people of Denendeh: ethnohistory of the Indians of Canada's northwest territories*. University of Iowa Press.
Hunn, E., & Selam, J. (1990). *Nch'i-Wana, the big river*. University of Washington Press.
Karson, J. (Ed.). (2006). *Wiyáxyxt/as days go by/wiyáakaaʔawn: Our history, our land, and our people: The Cayuse, Umatilla, and Walla Walla*. Tamástslikt Culturtal Institute and Oregon Historical Society Press.

References

Kerttula, A. M. (2000). *Antler on the sea: Yup'ik and Chukchi of the Russian Far East*. Cornell University Press.

King, T. (1994). *Green grass, running water*. Bantam.

Lutz, J. S. (2008). *Makúk: A new history of Aboriginal-White relations*. University of British Columbia Press.

Madley, B. (2016). *An American genocide: The United States and the California Indian catastrophe, 1846–1873*. Yale University Press.

Mann, C. C. (2005). *1491: New revelations of America before Columbus*. Knopf.

Marshall, J. (1995). *On behalf of the wolf and the first peoples*. University of New Mexico Press.

Martin, C. (1978). *Keepers of the game*. University of California Press.

Meek, B. (2010). *We are our language: An ethnography of language revitalization in a northern Athabaskan community*. University of Arizona Press.

Mishler, C., & Simeone, W. E. (2004). *Han: People of the river*. University of Alaska Press.

Mosby, I., & Galloway, T. (2017a). 'The abiding condition was hunger:' Assessing the long-term biological and health effects of malnutrition and hunger in Canada's residential schools. *British Journal of Canadian Studies, Special Issue on Health and Residential Schools, 30*(2), 147–162.

Mosby, I., & Galloway, T. (2017b). 'Hunger was never absent': How residential school diets shaped current patterns of diabetes among indigenous peoples in Canada. *Canadian Medical Association Journal, 189*(32), E1042–E1045.

Nadasdy, P. (2003). *Hunters and bureaucrats*. University of British Columbia Press.

Natcher, D. C., Davis, S., & Hickey, C. G. (2005). Co-management: Managing relationships, not resources. *Human Organization, 64*, 240–250.

Niezen, R. (2017). *Truth and indignation: Canada's truth and reconciliation on Indian residential schools*. University of Toronto Press.

Peck, A. M. (2021). 'We Didn't go anywhere': Restoring Jamestown S'Klallam presence, combating settler colonial amnesia, and engaging with non-native in Western Washington. *Journal of Northwest Anthropology, 55*, 105–134.

Persky, S. (1998). *Delgamuukw: The supreme court of Canada decision on aboriginal title*. David Suzuki Foundation, Greystone Books, Douglas and McIntyre.

Petrie, M. (2020, November–December). *Riches to rags* (pp. 44–46). Ducks Unlimited.

Pettitt, G. (1950). *The Quileute of La Push, 1775-1945* (Vol. 14, p. 1). University of California, Anthropological Papers.

Pierotti, R. (2004). Animal disease as an environmental factor. In S. Krech & C. Merchant (Eds.), *Encyclopedia of world environmental history*. Berkshire Publishing.

Pierotti, R. (2006). The role of animal disease in history. In *Encyclopedia of world history*. Berkshire.

Pierotti, R. (2010). Sustainability of natural populations: Lessons from indigenous knowledge. *Human Dimensions of Wildlife, 15*, 274–287.

Pierotti, R. (2011). *Indigenous knowledge, ecology, and evolutionary biology*. Routledge.

Pierotti, R., & Wildcat, D. R. (1999). Traditional knowledge, culturally-based world-views and Western science. In D. Posey (Ed.), *Cultural and spiritual values of biodiversity* (pp. 192–199). United Nations Environment Programme.

Pierotti, R., & Wildcat, D. (2000). Traditional ecological knowledge: The third alternative. *Ecological Applications, 10*, 1333–1340.

Pinkerton, E. (Ed.). (1989). *cooperative management of local fisheries: New directions for improved management and community development*. University of British Columbia.

Pinkerton, E., & Weinstein, M. (1995). *Fisheries that work: Sustainability through community-based management*. David Suzuki Foundation.

Quinn, A. (1997). *Hell with the fire out: A history of the Modoc war*. Faber & Faber.

Reid, J. L. (2015). *The sea is my country: The maritime world of the Makah*. Yale University Press.

Reyes, L. L. (2002). *White grizzly Bear's legacy: Learning to be Indian*. University of Washington Press.

Reyes, L. L. (2006). *Bernie Whitebear: An urban Indian's quest for justice*. University of Arizona Press.
Reyes, L. L. (2016). *The last fish war: Survival on the rivers*. Chin Music Press.
Ridington, R. (1988). *Trail to heaven: Knowledge and narrative in a northern native community*. University of Iowa Press.
Rytkheu, Y. (2019). *When the whales leave*. Trans. by I. Y. Chavasse. Milkweed Editions.
Sapir, E. (1922). Sayach'apis, a Nootka Trader. In E. C. Parsons (Ed.), *American Indian life* (pp. 297–323). University of Nebraska Press.
Sapir, E., & Swadesh, M. (1955). *Native accounts of Nootka ethnography* (p. 1). Publications of the Indiana University Center for Anthropology.
Schreiber, D. (2002). 'Our food sits on the table': Wealth, resistance and salmon farming in two first nations communities. *American Indian Quarterly, 26*, 360–377.
Schreiber, D. (2004). *The social construction of Salmon farming in British Columbia*. PhD Dissertation, University of British Columbia.
Schreiber, D. (2008). 'A Liberal and paternal Spirit': Indian agents and native fisheries in Canada. *Ethnohistory, 55*, 87–118.
Schreiber, D., & Newell, D. (2006). Negotiating TEK in BC Salmon farming: Learning from each other or managing tradition and eliminating contention? *BC Studies, 150*, 79–102.
Sercombe, L. (2021). History of Lushootseed language instruction. *Journal of Northwest Anthropology, 55*, 23–41.
Service. (2003). 'Combat biology' on the Klamath. *Science, 300*, 36–39.
Sewid, J. (1969). *Guests never leave hungry: The autobiography of James Sewid, a Kwakiutl Indian*. Yale University Press.
Stern, T. (1966). *The Klamath tribe: A people and their reservation*. University of Washington Press.
Stiles, D. B. (1997). Four successful indigenous language programs (p. 16). In *Teaching indigenous languages*. Retrieved from https://files.eric.ed.gov/fulltext/ED415079.pdf
Sullivan, R. (2000). *A whale hunt: Two years on the Olympic peninsula with the Makah and their canoe*. Scribners.
Switzler, V. (2012). *That is all I have to say: An argument for immersion as a language revitalization method in the warm springs community*. University of Kansas.
Takeda, L. (2015). *Islands' spirit rising: Reclaiming the forests of Haida Gwaii*. University of British Columbia Press.
Tatz, C., & Higgins, W. (2016). *The magnitude of genocide*. Praeger, an imprint of ABC-CLIO.
Tonkovich, N. (2012). *The allotment plot*. University of Nebraska Press.
Tonkovich, N. (2016). *Dividing the reservation: Alice C. Fletcher's Nez Perce allotment diaries and letters* (pp. 1889–1892). Washington State University Press.
Trafzer, C. E. (1997). *Death stalks the Yakama*. Michigan State University Press.
Trosper, R. L. (2007). Indigenous influence on Forest management on the Menominee Indian reservation. *Forest Ecology and Management, 249*, 134–139.
Trosper, R. L. (2009). *Resilience, reciprocity, and ecological economics: Northwest coast sustainability*. Routledge.
Vaillant, J. (2005). *Golden spruce: A true story of myth, madness, and greed*. W. W. Norton.
Vizenor, G. (1994). *Manifest manners: Postindian masters of survivance*. Wesleyan University Press.
White Hat, A. (2012). *Zuya (Life's journey): Oral teachings from rosebud*. University of Utah Press.
Wilkinson, C. (2010). *The people are dancing again: The history of the Siletz tribe of Western Oregon*. University of Washington Press.
Williams, T. (2003, March). Salmon stakes. Audubon, pp. 42–52.
Youst, L., & Seaburg, W. (2002). *Coquelle Thompson, Athabaskan witness*. University of Oklahoma Press.

Chapter 8
The Ideology Behind It All

General Principles

Northwest Coast resource management depends on a specific ontology, a worldview of how the universe is constructed with a distinctive metaphysics and epistemology that generate an equally distinctive set of ethical principles. These concepts deal with how animals (including humans), plants, mountains, glaciers, rivers, and other beings fit into the world, usually considered to be sentient creatures with immortal and unendingly reincarnated souls. These can be summarized as:

> A common general philosophy and concept of community appears to be shared by all of the Indigenous peoples of North America, which includes: 1) respect for nonhuman entities as individuals, 2) the existence of bonds between humans and nonhumans, including incorporation of nonhumans into ethical codes of behavior, and 3) the recognition of humans as part of the ecological system" (Pierotti, 2011: 198–199).

These principles fully apply on the Northwest Coast, where they are considerably elaborated, producing a philosophical tradition grounded in spirituality.

These views and others, some from everyday experience and some from shamanistic illuminations, comprise a whole cosmovision, or ontology—far more than a monotheistic "religion" in the ordinary sense, far more than a mere worldview or creation story. It is different from the Western scientific view although it shares some epistemological principles, such as "connectedness" (cf. ecology), relatedness (evolution), and the overall importance of reciprocity (Anderson, 1996; Kawagley, 1995; Trosper, 2009; Pierotti, 2011). This perspective differs from Western religious stories, which are constructed by priests and theologians. A cosmovision based entirely on logical developments from direct, embodied, experiential knowledge is an amazing and wondrous creation. (See Turner's superb overview, 2014: 224–350, which defies summary but emphasizes the personhood of plants and landscapes including mountains and rocks.)

Conservation was based on the simple principle of "leave some for others" and "make sure that the resources will persist for seven generations (or more) after you pass." Ronald Trosper (2009: 18) has emphasized the importance of reincarnation, including reincarnation of humans as nonhumans and vice versa; this makes sustainability for an indefinite period more economically rational. More to the point, core Northwest Coast societies hold that animals will not reincarnate for those who do not treat them with respect (Pierotti, 2010). They go elsewhere. Overhunting and waste are the key marks of disrespect (see the following chapter), so reincarnation does not mean that game is automatically there. The Nuu-chah-nulth phrase was *7uḥ-mowa-shitl*, "keep some and not take all" (George, 2003: 74; the 7 represents an initial catch in the voice). This is John Locke's point: that individual users, even if they are owners of his beloved "private property," have an obligation to leave "enough and as good" (Locke, 1924 [1698]: 132) for others when drawing on a resource base. There are, however, several crucial beliefs that underlie this, which are the keys to conservation in Northwest Coast ideology, and all the resources of the collective representation of community are deployed to make people accept them.

A good place to begin is with seven "fundamental truths" of Northwest Coast ecological thinking, as abstracted for the Heiltsuk Nation:

> 1. Creation: We the coastal first peoples have been in our respective territories (homelands) since the beginning of time. (This might be interpreted as "since the beginning of our peoples as identifiable cultural and linguistic entities," since in fact much migration has occurred, and many of the origin myths actually recount it.)2. Connection to nature: We are all one and our lives are interconnected.
>
> 3. Respect: All life has equal value. We acknowledge and respect that all plants and animals have a life force.
>
> 4. Knowledge: Our traditional knowledge of sustainable resource use and management is reflected in our intimate relationship with nature and its predictable seasonal cycles and indicators of renewal of life and subsistence.
>
> 5. Stewardship: We are stewards of the land and sea from which we live, knowing that our health as a people and our society is intricately tied to the heath of the land and waters.
>
> 6. Sharing: We have a responsibility to share and support to provide strength and make others stronger in order for our world to survive.
>
> 7. Adapting to change: Environmental, demographic, socio-political and cultural changes have occurred since the creator placed us in our homelands and we have continuously adapted to and survived these changes (Brown & Brown, 2009: folder 2)

More specific rules for food gathering have also been presented by Inez Bill of the Tulalip of northwest Washington state:

> 1) Taking and gathering only what you need so Mother Nature can regenerate her gifts to us; 2) Remembering to not waste any of our traditional food; 3) Sharing what you gather with family, friends and elders that are not able to go out and gather whenever possible; 4) Including prayer and giving thanks when gathering; 5) Preparing local native foods at gatherings; 6) Preparing food with a good heart and mind so when you serve your meal, people will enjoy their meal, and 7) Providing nourishment for our people and their spirits, but also the spirit of our ancestors. We will strive to continue this way of life (quoted from Krohn & Segrest, 2010: 42).

There is some obvious modernization and settler society influence here, i.e., Mother Nature, that may divert attention from two critically important traditional ideas: prayer and thanks while gathering, and having a "good heart and mind" when preparing and distributing. Throughout the Northwest Coast and onward to a great deal of North America and Siberia, these are vital points for maintaining good relations with the spirits of plants and animals, and indeed with the spirit powers in general. One may contrast the Euro-American food rules for the Muckleshoot school, found a few pages later in the same folder: "No trans-fat and hydrogenated oil. No high fructose corn syrup," etc. (Krohn & Segrest, 2010: 49). The contrast of spiritual and chemical is significant on many levels. For one thing, they clearly imply natural compared with processed foods. The same points about traditional foodways are made at more length by many First Nations authors in *Wisdom Engaged* (edited by L. M. Johnson, 2019).

Rodney Frey, writing on the Coeur d'Alene of northern Idaho (with some modesty and tentativeness) also identifies five basic principles:

>...the understanding that the landscape is spiritually created and endowed...that the landscape is inhabited by a multitude of 'Peoples,' all of whom share in a common kinship.... That... [humans have an] ethic of sharing [which includes the animal people too]... that...the gifts [of nature] are also to be respected and not abused...and...one is to show thanks for what is received... The fifth and final teaching encompasses...the ethic of competition" (Frey, 2001: 9–12).

The last is particularly interesting. Humans do not simply wait around for the gifts to come to them. They must work, compete with each other and with other beings, and compete with the powers of weather and geography. Life is not easy, and nothing comes without major effort.

Similar points about food—contrasting the modern international diet with traditional fish and berries—are made by Charlotte Coté, Nuu-chah-nulth scholar, in her book *A Drum in One Hand, a Sockeye in the Other* (2022). She stresses the dependence of Nuu-chah-nulth philosophy and cosmology on the idea of unity through interconnection with the world (*ḥačatakma ćawaak* or *ḥišukʔiš čawaak*, depending on dialect), which in turn leads to *ʔiisaak*, "respect," with implications wider and deeper than the English word (Coté, 2022: 8, 146). The word *čawaak*, as *Tsawalk*, is eponymous of Richard Atleo's book *Principles of Tsawalk* (2004), on the Nuu-chah-nulth cosmology and philosophy surrounding the term. Also stemming from unity is *ʔuʔaaɫuk*, "to take care of," and *tiičʔaqƛ*, "holistic health," both with wider connotations than usual English usage, but similar to extensions of those phrases in the literature on healing and natural lifeways. Respect receives more attention below (Chap. 10).

Nuu-chah-nulth anthropologist Richard Atleo, whose title is Umeek (2011: 143–144), sees unity as basic, reconciling complementary polarities. He sees individual identity as "an insignificant leaf" (144) in face of basic unity. He elaborates a philosophy of *haḥuuɫi*, "land" in the sense of communally owned and managed place, deriving from the idea of unity. He has also pointed out that our common origin with other life forms in the creation and transformation of the world makes mutual respect obligatory: "If the salmon are not properly respected and recognized they cannot properly respect and recognize their human counterparts of creation" (Atleo, 2005: viii). He contrasts the usual empiricism of the post-Locke west with "[t]he indigenous worldview [which] holds that real truth can only be known in and from the spiritual realm, which is the reality upon which empiricism is based, founded upon, and sourced" (Atleo, 2005: x). This appears to be widely the case, remarked on by shamanistic peoples from South America to Siberia (see, e.g., Kohn, 2013; Laugrand & Oosten, 2015; Viveiros de Castro, 2015). The Dene, for instance, recognize that whites believe that "knowledge can be removed from its context of production," but the Dene do not believe this (Legat, 2012: 30). Another way of thinking about this is that Native people regard specific events as part of an ongoing pattern which emerges from ecological processes (Pierotti, 2011). When an individual animal is encountered it is recognized that the humans who are present experience the event as being in a state which is unfolding, and the humans must understand it as part of an emerging pattern (Lopez, 1992: 168). This can be thought of as requiring a story to examine the phenomenon, rather than a snapshot or a data set (Pierotti, 2020: 6). However, many people do not realize that science is also a process, even though it may not always recognize the importance of phenomena that are not represented as data in tables and figures. This context includes the usual moral rules. One individual wanted so much to know about the caribou that he lived with them and eventually turned into a caribou, unable to regain humanity (Legat, 2012: 89).

Andy Schooner, Nuxalk elder, narrated the story of the world. He began with the mountains, which traveled around and talked, looking for their permanent home. One that was far up the coast near Port Simpson knew he had to go to Bella Coola, the Nuxalk center. That mountain was the last to leave that part of the coast and settle in Bella Coola. Andy Schooner continued with tales of greedy eagles, a flood myth (variant of a story with worldwide versions, possibly recounting—among other things—the massive melting that accompanied the last glacial maximum), and then closing with the local brass band—telling all with the same clear, direct realism. The wandering mountains are as real and immediate as the local band (Storie, 1973: 61, 86). Such an ontology calls for respect for all beings. The mountains are as sentient, below the surface, as Andy Schooner's musician friends, and as deserving of our care.

One major component was attachment to place, as was argued by Vine Deloria (1995) and Martin (2001), and interpreted to explain why almost all Indigenous knowledge is designed to be applied locally, rather than globally as Western science attempts to do (Pierotti, 2011). Groups had been where they were from time immemorial. They were related to the landscape by deep and complex bonds through careful multigenerational observation carried out for centuries, if not millennia (Pierotti, 2011). On the Skagit River, the local version of a universal Northwest Coast story of a boy left alone who must re-create his slain tribe includes a passage that says it all: the people he revives from bones "had no sense…so the boy made brains for them from the very soil of that place" (Miller, 2014: 70–71). The full implications of that brief sentence includes everything in this book.

Recent collaborations of settler-background scientists with Native American authorities have been notably important in describing Indigenous worlds. These include Nancy Turner's work with Native coauthors, and the collaboration of

Eugene Hunn and James Selam (1990). Just out of our region, but very close and sharing the same ideology, are the Yup'ik of Alaska, whose ethical relations with the animal persons have been described in exquisite detail by Ann Fienup-Riordan (1994, 2005) and Yup'ik scholar and educator Angayuqaq Oscar Kawagley (1995; collected works available at https://www.uaf.edu/ankn/publications/collective-works-of-angay/). Many Northwest Coast groups remain sadly little known, but at least we have excellent work by many ethnographers for representative groups from every part of the region. No other part of the world except the American Southwest and the central desert of Australia has been so well documented in terms of ecological and environmental beliefs. Few if any accounts of local knowledge and conservation in a modern American town are as good as the work of Turner, Hunn, or Fienup-Riordan.

According to Theresa O'Neill:

> Western philosophy marginalizes morality and spirituality as optional while 'perceiving practical principles of processes of production, reproduction and power relations as basic to human life everywhere....' Western notions of the self, based on economics and the individual as opposed to notions of the self, based on spirituality and the group underlie conflicts that arise in efforts to create autonomous and/or parallel legal orders (Fiske & Patrick, 2000: 236–237).

This is overdrawn—Western thought does not consider morality optional (except perhaps in economics). However, there is a point here. Both the fusion of spiritual, ethical, and practical in Native Northwest Coast Indigenous thought and the application of all three to both human and nonhuman persons are different from anything in the Western tradition (see also Anderson, 1996; Pierotti, 2011).

The account of the *balhats*—the Babine potlatch—includes the note that it was said to exist to allow negotiation over land instead of fighting, a far better way of dealing with conflict than Hobbes' concept of the elevation of a king (see also Trosper, 2009). They quote the early anthropologist Pliny Earle Goddard on potlatching in general: "There are perhaps two main principles involved; first that all events of social or political interest must be publicly witnessed; and second that those who perform personal or social service must be publicly recompensed. There is a third more general social law that all guests on all occasions must be fed" (Fiske & Patrick, 2000: 218, quoting Goddard, 1934: 1311.)

Hunting and Conserving

Progressing from north to south, conservation ideology can be documented throughout the Northwest. (Once again, we emphasize that we are seeking broad generalizations across a wide range of cultures, *not* implying that these cultures are "the same" in any way.) For example, statements about the Han Athapaskans of Alaska can also stand as general throughout the region:

> Traditional Han religious beliefs were based on the conviction that the universe is full of power. This power is immense and lodged in all things from a mountain to an animal, an arrow, a gun, a stone, or menstrual blood. Since all things possess power, people believed the world was alive; within this animate world they warily negotiated their way by enlisting animal allies and observing a multitude of taboos or rules...at one time animals and humans were the same category of being, able to talk to one another and exchange shape or form....
> Hunters must not only have respectful intentions toward their prey but also show proper respect toward its remains. If a hunter does not properly dispose of the animal's carcass, the animal's spirit will take offense and make itself unavailable... (Mishler & Simeone, 2004: 123, 126; see also Pierotti, 2010).

The Han seem to have had a more intense belief that the world was alive, and consequently more wariness, than many groups, but otherwise this description could go for any northwest North American or northeast Siberian people.

The related Koyukon, living somewhat northwest of the Han, have a similar knowledge system (Nelson, 1983). Like other groups described in this book, Koyukon see the world as having originally been inhabited by animals ancestral to modern ones but to some degree human or (like the sasquatch) humanoid. As they prepared the world for the future, including the coming of humans, they shaped themselves or were shaped by transformers into their modern forms. Animals possess spirits that respond to human good or bad treatment.

The stories about this have implicit conservation messages: these animals must be respected as ancestral people, not just walking meat. Animals regularly communicate with people, or show signs that carry messages (Nelson, 1983: 27–29). "Not only the animals, but also the plants, the earth and landforms, the air, weather, and sky are spiritually invested. For each, the hunter knows and array of respectful gestures and deferential taboos.... Violations against them will offend and alienate their spirits, bring in bad luck or illness...." (Nelson, 1983: 31). Animals are not killed out, nor is waste permitted (Nelson, 1983: 220–224). The "principles of the Koyukon worldview," as he lists them, are: "The watchful world"—spirits continually watch, and so do animals; "natural and supernatural worlds are inseparable"; natural beings have spirits that require consideration and respect, and that can punish, foretell the future, and otherwise interact with humans (Nelson, 1983: 225, 227, 228–237). Relations with the natural world vary in relation to the age and gender of the human(s) involved.

Nelson (1983: 201) warns against generalizing Koyukon attitudes without evidence, but there is plenty of evidence for the generality of the patterns we have picked out. Specifics, however, set the Koyukon apart in many ways, including their conservation of game in contrast to many eastern Canadian groups influenced by intensive fur trade (Nelson, 1983: 201). These latter groups often over trap, have trapping territories originally established by the fur trade, and are less concerned with subsistence quests, than with having more cash in hand.

Also from the Yukon realm are statements quoted by Catherine McClellan in her comprehensive work *Part of the Land, Part of the Water* (1987):

> "Oldtime Yukon hunters point out that, just as one should not say anything bad about animals, so one should never say what animal one is planning to hunt" (p. 260). Her consultant Tim Smith told her that his father warned him about this; punishment, at least

missing your shots, occurred if you talked. Also, "oldtimers...killed game only when people needed it. They also tried never to kill animals that were with young...'Their belief is never to waste. So they never kill just for the pleasure of killing. And when they killed, they made sure they used all of it.'" (McClellan, 1987: 261, quoting Virginia Smarch of Teslin).

Children were supposed to stop after one trout caught. Tom Peters told her: "'They used to look after the game and the fish. We used to look after them because this is what we live on'" (McClellan, 1987: 261). She continues with a long version of the Bear Mother story (see below), used to teach respect for animals.

The Dene of north-central Canada say: "The Tłı̨chǫ elders attribute the disappearance of wildlife to industrial development and to a lack of respect, which inevitably causes destruction" (Legat, 2012: 30). The usual respect rules hold, though today this Dene group does not call animals by the same term as human persons.

Conservation was done everywhere by a combination of forbearance—never take all, never take too many—and always act in a respectful manner.

> The animals sharing the land with the Upper Tanana Indians were considered almost like partners of Indian daily life. As the source of food and as living beings they had to be treated with respect. Children were taught early not to disturb nests or eggs, not to bother helpless babies of any kind." Taking young animals was avoided, and "one was not allowed to kill anything without use. To kill something one would not eat was a waste and a 'sin' sanctioned by bad weather, bad luck or worse.... The same closeness characterises the relationship between the hunter and the animal who let himself be killed. For instance, if someone kills a wolf, he is not allowed to touch the skin or the body without first apologizing to the wolf. It is always better to explain to your victim that you had to kill it in order to feed your family. Men and animals also share the same basic feelings and sometimes belong to the same kind of society (Guédon, 1974: 29).

This complex of ideas is essentially universal in the region.

The pattern is worldwide, extending to the Yuit of St. Lawrence Island. Though slightly outside our region, they display the basic principles, and R. Apatki provided a classic statement:

> Whatever they first caught, they show them to...God....
> They return the bones, whatever they are to the fields.
> Those [bones] that belong to the sea, they went to throw away down to sea....
> They light small lamps. There they sacrifice. The small ceremonial fires are some sort of altars....
> They [the ancestors] had taken very good care of it [St. Lawrence Island], just like they honor it, those people.
> They take good care of the hunting grounds, these places—all hunting ground. Respectfully.
> They don't go to places until when it is time to go there.
> They don't put some things [trash and the like] on hunting grounds. They respectfully use them.... (Jolles, 2002: 296–297; the whole text is well worth looking up).

Jolles provides further accounts of respectful treatment of kills. Estelle Oozevaseuk told her in 1989 of giving water to seal kills, a universal custom among Inupiat and Inuit peoples and their neighbors (Pelly, 2001), and giving bits of food to fox kills and others (Jolles, 2002: 298–299). A whole complex of respectful behaviors is described: special games, special ways of dividing kills, special stories. Reverential ceremonies for major game animals, notably whales, go on all year. As an example, the Sun Dance, perhaps the major ceremony among Plains' peoples, began essentially as a buffalo-hunting ceremony, and "religious dimensions of these cultures...condensed in ritual symbols, constituted the way the people related to animals that were hunted and killed for food (Harrod, 2000: 79). Jolles effectively shows how closely the respect for animals tracks the very respectful, structured behavior traditional within the human community that recognizes their dependence on their nonhuman relatives.

Similar beliefs are shared widely in the Subarctic. Adrian Tanner (1979) and Robert Brightman (1993) for the Cree, Frédéric Laugrand and Jarich Oosten (2015) for the Inuit, and Georg Steller for the Itelmen of Kamchatka (Steller 2003:211, 249) all provide similar but extremely detailed accounts of treatment of animals in these hunting societies. Hunters address slain animals respectfully, sometimes apologizing, offering them drink or special foods, and always offering thanks for the gift of their lives.

> All these societies share ideas of reciprocity: the game animals and fish give themselves to hunters, the hunters provide respect and offerings. They carry out often inconspicuous but deadly serious small rituals for slain animals. A highly traditional Cree elder said: "There was a story of a young hunter killing more than what he needed.... And it finally came where the moose was scarce.... So one of the Elders was a Shaman. So the Shaman went into a ceremony and they contacted the moose spirit. Said, 'moose come you're not coming around our area anymore? We're going hungry here, we can't find you.' The moose spirit told him, 'well my bones are scattered all over, you're disrespecting me. You're not taking care of me'" (Danto et al., 2022: 69).

That short statement perfectly summarizes this chapter.

These peoples are not above using trickery—snares and traps are universal—and they are not necessarily conservationist, but the ideology of offering and gratitude remains. They are not Shepard Krech's sardonically titled "ecologically noble savage" (Krech, 1999; the phrase comes from an inaccurate and biased article by Kent Redford, 1990, an opponent of involving Indigenous peoples in conservation efforts), but they keep the game and fish abundant by careful management (see detailed refutations of Krech in Trosper, 2009 and Pierotti, 2011, Chap. 8). The confusion in Krech is because he does not know the difference between "environmental" and "ecological"; Indians are very good at ecology, but did not originally have "environmentalist" attitudes, lacking any sense of separation from "nature" in the first place, and thus do not always relate to such attitudes now (Pierotti, 2011; see also Chap. 16).

Such a difference can be more explicitly observed in Robert Sullivan's (2000) account of the controversy over the attempt by the Makah Nation to re-establish their whaling tradition, where local conservationists acted in a profoundly racist fashion, including death threats. Respect includes ensuring that all of the prey animal is consumed, making it suffer as little as possible, treating the slain body with courtesy, and offering water or other gifts to its spirit. The soul or spirit is immortal and will be reincarnated as another animal. Laugrand and Oosten (2015) provide many stories of respect, and of disrespect followed by condign punishment by the spirits, which would cause sickness to disrespectful individuals.

As noted by George Emmons (1991: 102, from his perceptive observations in the 1880s) for the Tlingit:

> "The Tlingit never took life unnecessarily, having a positive belief in the existence of a spirit...in all nature [i.e., in every natural thing]. Even inanimate objects possessed for him something more than the mere material form. The shadow cast by the tree in the sunshine" was a spirit. Winter wind was an ice spirit. Animals had their spirits, and "were always treated with great respect. The hunter or fisherman appealed to them before capture; and after their death, followed certain observances that were supposed to propitiate their spirits. The folktales tell of punishment of those who even spoke disrespectfully or slightingly of animals."

Emmons summarizes numerous tales and observances, including bathing, changing clothes, and fasting before hunting trips; disrespect led to bad luck or worse (Emmons, 1991: 103). Some of these observances are still known.

> [Salmon] "were always cut from the vent, along the belly to the throat. This was believed necessary, lest the fish or their spirits feel offended and desert the stream. There is a tradition among the Stikine that once, in a salmon stream up the river..., a fish was cut along the side by mistake. Whereupon all of the others immediately disappeared." A shaman fixed the situation, after apologies to the fish (Emmons, 1991: 143).

An early settler ethnographer, Charles Hill-Tout, thus describes the ideology of the Lillooet: "When the sockeye salmon (*lauwa*) run commenced, the first salmon caught was brought reverently and ceremoniously upon the arms of the fisherman, who never touches it with his hands, to the *wqatceoqaloc,* or seer (the term meaning 'he went to see'), who always conducts the (First) salmon ceremonies among the Lillooet.":

> Nothing that the Indian of this region eats is regarded by him as mere food and nothing more. Not a single plant, animal or fish, or other object upon which he feeds, is looked upon in this light, or as something he has secured for himself by his own wit and skill. He regards it rather as something which has been voluntarily and compassionately placed in his hands by the goodwill and consent of the spirit of the object itself, or by the intercession and magic of the culture heroes, to be retained and used by him only upon the fulfillment of certain conditions. These conditions include respect and reverent care in the killing or plucking of the animal or plant, and proper treatment of the parts he has no use for.... (Hill-Tout, 1978b: 117, orig. 1905).

He later (p. 129) provides a description of the sturgeon ceremony, intended to pay reverent respect to the sturgeon people. He notes that they "endow every object and agency in their environment with conscious power and being," but that many or most of these spirits are "capricious" or "malevolent," necessitating healing power (Hill-Tout, 1978a, c: 49, d).

Marianne Boelscher Ignace and Ron Ignace explain the Secwépemc (Shuswap) foundational belief thus:

> ...central to the relationship between an animal and the fisher or hunter who 'bags' the animal is the concept that the animal gives itself to the fisher or hunter.... Plants as sentient beings also give themselves.... Therefore, the harvesting of all things in nature presupposes

> prayer that thanks the Tqelt Kúkwpi7 (Creator) for providing the animals or plants...as well as thinking the animals or plants for giving themselves.... All parts of the Secwépemc land and environment are ...thought of as a 'sentient landscape.' (Ignace & Ignace, 2017: 382).

This belief appears to be universal throughout northwestern North America and eastern Siberia. More specific to the Shuswap is a tradition of painting one's face red or black at power places, such as lakes and mythic spots, but other groups have similar ways of paying respect to spiritual sites. This general custom extends through much of east Asia, where such sites may be tied with sacred cloths, circled clockwise on foot, and otherwise revered.

Frederica de Laguna summarizes the Tlingit ideology of hunting in this manner:

> Hunting and fishing were not simply subsistence activities, but moral and religious occupations, for the Tlingit was killing creatures with souls akin to his own. No animal, especially no little bird, should be slain needlessly, nor mocked, nor should the body be wasted. Rather, the hunter would pray to the dead animal and to his own 'spirit above,' explaining his need and asking forgiveness. The dead creature was thanked in song, perhaps honored with eagle down (like a noble guest); certain essential parts...were interred, returned to the water, or cremated, to insure reincarnation of the animal (De Laguna, 1990: 209).

The pattern seems to be consistent and we could provide many similar examples. These peoples showed respect for the animals, plants, and geographical features that they depended upon for their survival in landscapes that were sometimes unpredictable. This was an economic philosophy that did not expect progress or profit, but resilience and survival as their rewards (Pierotti, 2011). The pattern extends far into Asia, where ENA has encountered it in Mongolia.

A long set of Tsimshian rules for hunting is detailed by Franz Boas (1916: 448–451), including the familiar ones of eating or using all parts of the animal and burning the few useless or discarded parts, so that the animals and fish are reincarnated. Among the Yukon Indigenous groups, every animal had a spirit that stayed with it until it was killed, and then "would enter the body of a new animal on earth, just as the soul or spirit of a person who had died would enter the body of a newborn baby" (McClellan, 1987: 260). The Nuxalk also have a number of complex rules to make sure game animals and fish reincarnate and do not die for good (McIlwraith, 1948: 1: 76–77; these can involve sacrifice, abstention from sex, and prayer; 104–116). The widespread rule that salmon bones must be returned to the river is one example. These include conservation rules that neutralize the problem of believing that game animals can be killed at will because they come back. However, their coming back is required for human persistence, and the philosophy of these nations is designed to maximize the chances of their return.

Purity was another part of respect. The Tsimshian are typical: "Men purified themselves before hunting or fishing. They fasted, bathed, drank the juice of the root of the devil's club, and practiced sexual continence. Animals were said to be offended by unclean persons and to refuse to allow themselves to be caught by them" (Halpin & Seguin, 1990: 271). Males were the hunters, and avoided contact with women before major hunts. Similar restrictions were universal, the extreme being found in Nuu-chah-nulth whaling, which might be said to be the most extreme form of hunting (Sapir, 2004).

A direct personal form of this was manners and etiquette. Formal occasions, especially feasts and potlatches, called and still call for formal behavior. Food is graciously accepted and not refused. People are quiet, especially the young, and everyone listens to the formal and often stunningly eloquent oratorical performances of the elders. Proper respect according to social standing is rendered, and superiors are supportive and kind to ordinary people (see, e.g., People of 'Ksan, 1980: 112–113).

Extreme care about life and the world was universal, as expected in a hunting and fishing world where survival depends on skill in an uncertain environment where resources can fail and deplete stored food, and where social life was often poised on the edge of war. "The world is as sharp as a knife" is a proverb known to Haida (Abbott, 1981) and Tlingit (Kan, 1989: 121).

Plants and animals are "other-than-human persons" (or "more-than-human relatives"), in a phrase apparently introduced in English (although its equivalent in local languages is ancient beyond memory; Marshall, 1995), by A. Irving Hallowell from his studies on the Native peoples of central Canada (1955, 1960). Hallowell also introduced the use of the philosophical term "ontology" to refer to traditional ideas of reality (as well as to modern philosophic ones). Like humans, these other-than-human people have spirits that are conscious and sentient and have agency and language. All have a life force. The Nuu-chah-nulth anthropologist and artist Ḳi-ke-in relates that there are many *ch'ihaa*, spirits, in the world, which draw closer in winter, and that humans and other beings have a life force, *thli-makstii,* which goes ultimately into the stars after death, eventually ending in the Taa'winisim (Milky Way; Ḳi-ke-in, 2013: 28). Plants were people, and as such critical in ceremonies and in the search for guardian spirits (Turner, 2014: 2: 324–350).

Other-than-human persons can take human or humanoid form when in their own world; salmon, for instance, are people who live in houses at the bottom of the ocean, and don their salmon skins only to sacrifice themselves for the benefit of their relatives on shore. This represents Philippe Descola's ontology of shamanism (Descola, 2013), being the same belief that he reports from his South American researches. Descola sees this as part of animism. Laugrand and Oosten (2015) have applied this view to Inuit hunting, and it applies equally well on the Northwest coast. In contrast, Ridington (1988: 71) argues that:

> There is no documentary evidence of foxes who live like people (thus)... we [i.e. Settler society scientists] are compelled to class (such stories) as myth. In our thoughtworld, myth and reality are opposites. Unless we can find some way to understand the reality of mythic thinking, we remain prisoners of our own language, our own thoughtworld... The language of Western social science assumes an object world independent of individual experience. The language of Indian stories assumes that objectivity can only be approached through experience. A hunter encounters his game first in a dream, then in physical reality. In the Indian thoughtworld, stories about talking animals and stories about the summer gathering are equally true, because both describe personal experience. Their truths are complementary.

Pierotti knows this experience at some levels; in his case, he has dreamed of animals that have spoken to him, providing both reassurance and remonstrating him about things he should not hope to accomplish. He recognizes that these are not literally true. However, the messages he has been given always turn out to be correct in the real world. Therefore, the truths are indeed complementary.

The Katzie Salish, living where the most sturgeons spawn, in the lower Fraser drainage, believed the ancestral man at Pitt Lake had two children; the boy became "a white owl-like bird" that gives power (Jenness, 1955: 12); the girl became the original sturgeon. Whenever they caught a sturgeon, it was their beloved sister sacrificing herself for them (Jenness, 1955: 12; Suttles, 1955: 21). Mountain goats were similarly related, stemming from a Katzie turned into a goat by the Transformer, and some grizzlies kept peace with the Katzie, especially if called by a special name rather than the usual word for the species (Suttles, 1955: 23). A later ancestor, Swaneset, shaped the country, producing berry and wapato grounds.

Widely believed among Salishan groups is the idea that all living resources are humanlike in their own homes, donning outer coverings for human use, which might be seen as a way of ensuring that all members regard these other species as their equals, with significant lives, full of meaning on their own terms. Berries are like small babies in their real—spiritual—state (Miller, 1999). For the Sahaptin:

> *waqayšwit* 'life' [is] an animating principle or 'soul' possessed by people as well as animals, plants, and forces of nature. The presence of 'life' implies intelligence, will, and consciousness. This is the basis for a pervasive...morality, in which all living beings should respect one another and are involved with one another in relations of generalized reciprocity (Hunn & French, 1998: 388).

These are not new ideas, and certainly not the product of Native Americans catching up with the environmental movement of the 1960s. Manuel Andrade, collecting Quileute language texts in 1928, recorded a "speech...spontaneously offered" by Jack Ward, "when he found out that the texts which he was dictating would be published." The entire short speech is worth looking up. It is devoted to calling on the Whites "to observe conservancy of the products of the land...." Jack Ward speaks in eloquent Quileute. Andrade translates:

> ...you should take good care of the trees.... Be careful with fire. Make sure to extinguish it when leaving a camping ground. Otherwise, all the animals of the woods may disappear, such as elk, deer, and others. This applies also to the fish.... Let no one...destroy too much trout, steal-head [sic], salmon, and all other kinds of fish. Proud and happy I am knowing that my people, the Indians, are moderate in the use of nature's supplies, never killing wantonly the fish in the waters.... But you, White people, are wasteful. You are not mindful of what you do in your camps, and consequently, many trees are often destroyed by fires. It is heartbreaking to us Indians to see how the country around us has changed...all the animals in the land are beginning to disappear. Much of the fish in the...waters is disappearing. Many of the good trees are disappearing.... (Andrade, 1969 [1931]: 12).

A Bridge to Creation Myths

The basic or original being that became an animal, or was transformed into one by the Transformer, was incarnated in all subsequent animals of that kind, or—alternatively—remained as master of them. Thus, mythic Raven is incarnated in all ravens, or else exists in the spirit world as the guardian of all ravens. The former version of the belief seems to prevail in the north. The latter prevails in some parts of the coast (for instance, among the Puget Sound Salish; Miller, 1999: 10–11). It is the standard form in many areas to the south, including Mexico. This belief reaches to the Maya of Quintana Roo. The deer god *Siip*, Zip in old literature, still has some role in overseeing deer today; the "leaf-litter turkey," a giant ocellated turkey, is the eternal master of wild turkeys, punishing hunters who take too many of them (Llanes Pasos, 1993). Versions of the same belief extend into South America [Blaser, 2009; Descola, 2013]).

The leading character in creation myths was a Transformer who changed ambiguously humanlike persons, and more or less formless supernatural beings, into the humans, animals, trees, and geological formations of today. This story is known from Alaska (e.g., the Han people; Mishler & Simeone, 2004) to California and beyond (see comparisons, mostly for the central coast, in Miller, 1999; and many stories in Boas, 1916, 2006; Thompson & Egesdal, 2008). People who are broadly humanlike but have some animal-like characteristics became the animals they resembled. Formless supernaturals became rivers and mountains.

Usually, the Raven is the transformer on the coast; Coyote is often the major one in the interior, except for the Paiutes and Shoshone (Numic peoples) where Wolf is the older creator sibling, and Coyote the mischief making younger sibling (Pierotti & Fogg, 2017, Chap. 8). (The missionaries originally used the Chinook word for "coyote" for God.) A being without specifically known form does the job farther north and east. The Gitksan Transformer, Wegyet, can be either raven or humanoid, apparently at will; similar shape-shifting is found in several other creator figures throughout North America; e.g., the Blackfoot creator Naapi can be coyote or humanoid, but is more specifically symbolic of processes of change.

Most Northwest Coast groups seem to have had no single most powerful being, but rather a wide range of animal powers and other spirit beings and forces. However, this is one area where cosmologies differ strikingly between local regions. The Tsimshian had a high divinity in the sense of powerful remote being known as *gal* (Miller, 2014: 42), and some Haida seem to have regarded the Shining Heaven as a high god. In the Columbia Plateau region, there was a widespread concept of a high god who oversaw the animal spirits. Coyote may have been the immediate creator of much that we have here, but he was under control of a higher power. (The old idea that "primitive" people lack a high god or "big god" and have only animistic spirits was disproved a century ago; some in fact do have a single most powerful being and others do not; Lowie, 1924.). For the Colville of northeast Washington, the creator was originally a sort of power-in-process, a being whose existence was the flow of his creation. Eventually "the 'chief' who made the world and all its creations decided thereafter that there would be no body, legs, or head, and would instead become Sweatlodge for the continued benefit of human beings" (Miller, 2010: 47; cf. Ray, 1933). It lives on, instantiated in every sweatlodge created since.

Also reported from the Plateau and not from other areas is a concept of sacred spaces with sacred deer:

"Near my home is a sacred area—about one acre—where sacred deer used to live. One adult deer would always present itself only to a young man just before he was about to get married. But only to a young man who, before starting his hunt, had for three days sweated while singing his power song, of course fasted," and been virtuous. He would shoot the deer, give the meat to his intended-wife's mother to redistribute, and make the hide into a robe for the new wife or (in due course) their baby (Ross, 2011: 145, quoting an unnamed elder).

Spirits and Places

Not only plants and animals, but also mountains and rivers, were people, and were owed consideration; "we…talked to the river, so we could get some fish from it without hurting the river at all," as a woman Elder told First Nations student Johnnie Manson (2016). Various spirits, probably tied to the land and living beings, were always present. The Tillamook would burn food that was dropped, because a spirit was thought to have pulled it from the eater's hand, and leftover items were burned as well. Food uncooked but eaten while traveling in a strange place had to be shared with local spirits by having a bit scattered around (Jacobs, 2003: 83), which is a Tillamook expression of a well-nigh worldwide human custom of scattering food items for spiritual reasons.

Such kinship with all beings obviously made conservation a great deal more compelling. Even people who care little for unrelated fellow humans will care for their own kin. This implies that animals are not simply killed by humans; the prey offer themselves voluntarily to their human relatives, but only if the humans showed proper respect, being clean of spirit and body, and acting in a proper manner. Their spirits then reincarnate in new members of the species, just as human souls reincarnate within a descent group.

Today as in the past, in many cultures, a new baby will be examined to see which recently deceased elder has reincarnated (we have often encountered this; see also Trosper, 2009: 18). The Siberian Khanty (Wiget & Balalaeva, 2011) and Yukaghir have the same belief, and the ever-perceptive Ronald Trosper (2009) and Rane Willerslev (2007: 50ff.) point out for the latter that, while the essence of the person is reincarnated, the new child is his or her own person—not simply a recycled relative. This is so strikingly like genetic reality that one assumes the Indigenous peoples noticed that descent produces great similarities in appearance and basic personality, and yet considerable differences in the final result. Like Aristotle realizing that heredity must be a "pattern" conveyed from parent to child, rather than a tiny homunculus or divine intervention, the Northwest Coast people recognized the inheritance of traits and reasoned an explanation based on their worldview.

If the souls of previously hunted animals were treated with respect, new (newly reincarnated) individuals of that species will offer themselves. Of this, much more below, but we must quote Richard Atleo's point that Northwest Coast people pray for the animals to make themselves available, but when European "people pray to God for a supply of meat…the meat has no say in the matter" (Atleo, 2004: 85), presumably because the meat has usually been raised in captivity and has no control over its life. Among the Tillamook, strangers were allowed to fish in local waters, but not if the strangers "'treated fish mean and made them leave'" (Jacobs, 2003: 99).

Under special circumstances, humans can visit the salmon people, or the bear or mountain goat people, and see them in human or humanoid form. Countless Northwest Coast stories recount such adventures (see Andrade, 1969 [1931], with tales of visits to shark people, seal people, and others). Unsurprisingly, there is also a Yukaghir version of this tale as well (Willerslev, 2007: 89–90). Humans can marry animals, plants, or even stars, and many Northwest Coast kinship groups are descended from such unions.

Totem poles typically show the nonhuman ancestors of the sponsors of the pole. Hunting involves the hunter leaving the human realm and entering into animal worlds. This often requires a great deal of ritual, in which the hunter's wife is often also an essential participant, required to abstain from certain actions while her husband is hunting, especially when seeking large animals from whales to deer (Reid, 1981, gives a detailed description of Kwakwaka'wakw ideology in this area). This concept seems to be present in all cultures that practice whaling from the Makah to the Chukchi and Inuit.

Many a local crest derives from such visits. Some members of the Tsimshian Blackfish phratry "met Nagunaks" when they "inadvertently anchored over his house" at the bottom of the sea; "he sent blue cod, one of his slaves, to investigate.... The steersman was annoyed by the splashing of the fish, caught it and broke its fins," which caused the sea-guardian Nagunaks to take them to his house, warn them never to harm animals unnecessarily, and then released them with privileges and songs, sending them off in a fast-speeding copper canoe. They thought they had been gone for four days, but they were gone for four years. The privileges became lineage crests and emblems (Garfield et al., 1950: 42). This is a version of a story known not only throughout the Northwest Coast but through much of North America and Eurasia.

Such stories represent a form of the Native American theory, defined by Eduardo Viveiros de Castro as "'perspectivism': the conception according to which the universe is inhabited by different sorts of persons, both human and nonhuman, which apprehend reality from distinct points of view (Laugrand & Oosten, 2015: 13). This conception was shown to be associated to some others, namely:

(1) The original common condition of both humans and animals is not animality, but rather humanity; (which establishes evolutionary history)

(2) Many animals species [sic], as well as other types of 'nonhuman' beings, have a spiritual component which qualifies them as 'people'; furthermore, these beings see themselves as humans in appearance and in culture, while seeing humans as animals or as spirits;

(3) The visible body of animals is an appearance that hides this anthropomorphic invisible 'essence,' and that can be put on and taken off as a dress or garment;

(4) Interspecific metamorphosis is a fact of 'nature.'

(5)...the notion of animality as a unified domain, globally opposed to that of humanity, seems to be absent from Amerindian cosmologies" (Viveiros de Castro, 2015: 229–230).

Northwest Coast people, however, more often hold that animal persons have two forms, humanoid and nonhuman, both equally their own (see also Ridington, 1988).

Often, animals and other beings—even rivers and mountains—had amorphous or vaguely humanoid forms before a mythic Transformer turned them into what they have now. Whole cycles of these tales abound in Northwest story records (e.g., Jacobs & Jacobs, 1959: 121–151). Closer to Viveiros' idea are the observations of a Tillamook hunter: "Sometimes you come upon the elks holding a dance of their own, just like people. A big buck stands in the center and sings and all the does and calves dance around him in a circle." Dogs could charm elk (Jacobs, 2003: 75, quoting Ellen Center's memories of her father).

At this point we should note that Viveiros' "ontological turn" has been accurately criticized by Grant Arndt (2022) for dealing with abstract Indigenous philosophy and rather neglecting the empirical basis for Indigenous ideas, thus allowing it to be seen as highly different from generalized Western or bioscientific ideas. The reality is that people of different cultures interact with the same world, perceive it in somewhat different ways, analyze and interpret what they perceive in increasingly different ways, and eventually develop culturally constructed representations that differ from each other (Arndt, 2022). Thus, attending to the philosophical side of Indigenous thought maximizes the differences, allowing what Larry Nesper in Arndt's article calls "exoticization" and consequent "depoliticalization" (Nesper comment in Arndt, 2022: 24). A good corrective, moving constantly back and forth from empirical observations to Indigenous constructions, is Charlotte Coté's *A Drum in One Hand, a Sockeye in the Other* (2022), which explains Indigenous philosophy, the real-world grounding of it, the resulting differences in worldview, and the consequent thoughtful and often ironic views of "the whites" in Indigenous society.

Animals have powerful spirits, and are in fact spirit beings, able to move back and forth between our everyday world and the spirit realm. For the Spokan, "[a]nimals are in touch with, and move between, the two worlds. That sensitivity allows certain animals to foretell events, such as weather, death, or the arrival of a guest. They can see into the spirit world.... Through animals, man can obtain a vision into that other world; through them, he finds his spirit power...." (Egesdal, 2008: 140.)

Animals have many powers. In the northern Northwest Coast, the land otter (the river otter, *Lutra canadensis*, is called "land otter" in Northwest Coast English) is recast as the *Kustaka*, a dangerous spirit being, which can lure humans to drowning or near-drowning, and then take over body and soul and turn the person into an "otter-man," a zombie-like being. A shaman could meet it and obtain its power, but this was terrifying (Emmons, 1991: 360, 370–375). This is still a matter of concern, and on Haida Gwaii the fear has spread to the Anglo-Canadian settlers, who can be quite nervous about the *gogeet* (Haida *gagitx*, otter-man; Anderson, 1996 and personal research in Haida Gwaii, 1984–1985). Marianne Boelscher, who spells the name *gogiid* (Boelscher, 1988: 186–198), adds that individuals lost on small islands after canoe accidents were tempted by otters who gave them food that turned out to be wood ticks, or appeared as old friends and relatives. Throwing urine on these "ticks" or apparent friends, or otherwise defiling them, made them take their true form. This could be embarrassing when the individuals who appeared were actually rescuers (Boelscher, 1988: 189). Fear of otters and stories of dangerous but successful encounters with them extend to the Han of Alaska (Mishler & Simeone, 2004: 147) and the Tagish and Tutchone of Yukon (Cruikshank, 1983, with many stories).

Otters move easily in both the great realms of life: land and water (De Laguna, 1990, esp.: 258–276, 727–755; Halpin, 1981; Jonaitis, 1981, 1986). They even enter the underworld, since they live in burrows. Moreover, they play a lot, and even incorporate humans into their play. Most daunting of all, they often try to lure people into the water to play with them, as ENA has personally experienced on several occasions. They love to play and mean nothing but friendship, but it can seem uncanny. Significantly, otters abound in shamanic art (charms and the like) but are largely absent from public art—totem poles and crest art (Jonaitis, 1981, 1986; Anderson personal research), though a few appear on Gitksan poles (Barbeau, 1929: 73, 88). Barbeau recorded a shaman's otter song, for curing, among the Tsimshian (Garfield et al., 1950: 122). The *bukwus* or "wild man" took the place of the otter, to some extent, in Kwakwaka'wakw tradition (Halpin, 1981).

Generally, mink and weasel have some similar magical powers (Albright, 1984). Mink in particular was also a hero of a long series of stories, widespread in the region. For the central Salish groups, the fisher, a very otter-like (though land-based) animal, has magic curing powers, and passing a stuffed fisher over a person is part of healing (Lévi-Strauss, 1982). Fisher skins brought back to life by shamanic power consoled people for the death of a child, among the Katzie (Jenness, 1955: 73).

The dog, as the only aboriginal domestic animal, also moved in the human and the nonhuman realms (Amoss, 1984, summarizes widely found Northwest Coast attitudes on this; for the Yukaghir, Willerslev, 2007: 76), as the otter moves in three worlds. It is singularly absent from myths and art on the Northwest Coast.

The Tłı̨chǫ Dene (somewhat outside the Northwest Coast area) treat dogs with respect but think them unable to survive in the wild—unlike wild animals and the Tłı̨chǫ themselves. The Tłı̨chǫ also tend to think of animals as *very* different in powers and natures from humans, though still being other-than-human persons. The Tłı̨chǫ Dene are descended from a dog who took human form and married a human woman (Legat, 2012: 89), as are the Heiltsuk (Olson, 1940: 189–190), according to origin stories. The most widespread of all Northwest Coast stories concerns a woman who married a dog (this story will be fully considered in Chap. 10).

Salmon are people too, and abundantly present in myth. Something of the empathy between human and salmon people is found in a powerful song recorded by Marius Barbeau and William Beynon—the song the salmon sing as they go upriver:

> I will sing the song of the sky.
>> This is the song of the tired—the salmon panting as they swim up the swift current.
>>> I go around where the water runs into whirlpools.
>>> They talk quickly, as if they are in a hurry.
>>> The sky is turning over. They call me."

This was sung by Tralahaet, a Tsimshian chief (translated by Benjamin Munroe and William Beynon, early twentieth century; see Garfield et al., 1950: 132). Recall that the salmon give birth and die; this poem is about sacrificing one's life for the rising generations.

A similar Kwakwala song of the salmon sacrificing themselves to humans seems related:

> Many salmon are coming ashore with me
> They are coming ashore to you, the post of our heaven.
> They are dancing from the salmon's country to the shore.
> I came to dance before you at the right-hand side of the world, overtowering, outshining, surpassing all; I, the salmon (Colombo, 1983: 34, after Boas).

Knowledge: Empirical and Spirit-Driven

This allows us better understanding of the superficially "strange" or "exotic" ethnozoology of the Athapaskan peoples of northwest Canada. Recent descriptions of Athapaskan groups by Julie Cruikshank (1998, 2005), Jean-Guy Goulet (1998), Marie-Françoise Guédon (1974), Leslie Johnson (2010), Thomas F. McIlwraith (2007), Paul Nadasdy (2003, 2007), Robin Ridington (1981a, b, 1988, 1990), Henry Sharp (1987, 2001), David Smith (1999), and others have presented a very consistent picture of conservation ideology and practice. From somewhat outside our area, but involving closely related Athapaskan peoples, come the accounts of Richard Nelson (1973, 1983), which speak to conservation issues.

The Athapaskans see the world through trails and paths (Johnson, 2010; Legat, 2012; Ridington, 1981b, 1988). They have complex terminologies for types of places, mostly habitats for hunting and gathering. They may name features that seem small and insignificant to outsiders, while leaving whole mountains unnamed—as do many peoples worldwide (Johnson, 2010; one may point out that many substantial peaks remain unnamed today). The trailways were taken by ancestral beings long ago, and the small features were spots critical in those mythic adventures. Even groves that might seem quite recent in growth to the outsider may have originated in timeless pasts.

Salishan groups living close to the Athapaskans also have a worldview rich in maps, travels, sites, and stories about all these, as shown in detailed studies of the Secwepemc by Marianne and Ronald Ignace (2017), Andie Palmer (2005), and others. A classic study in this tradition is *Maps and Dreams* (Brody, 1982), a detailed study of life among the Beaver Cree of Northern British Columbia that discusses how these people were able to prevent a pipeline to transport Alaskan oil being carved through their territory, by revealing traditional land use patterns using maps.

Leslie Johnson has written at length about the differences between this moving-traveler view of the world and more static or totalizing views (Johnson, 2010). She contrasts the Athapaskan view, rich in narrative, personal history, and subsistence knowledge, with the abstract, arbitrary, geometric grids laid down on the land by modern states, the model for which was established in North America by US President Thomas Jefferson who was convinced that the entire USA could be gridded into square mile plots, which would be made available for farming. Jefferson was a gifted individual, but his imagination ran to romantic fantasies, based on a total ignorance of ecology, or apparently how landscapes are structured.

However, once settlement is achieved, even within grids, settlers too can come to know and love particular landscapes. Having grown up in rural Nebraska and Texas almost a century ago, ENA can testify that a few generations of farming and childrearing by the least spiritual of settlers transforms a grid into a richly known, well-loved, story-laden landscape. Pierotti has experienced the same on his 13% (of a square mile) section in northeastern Kansas, where it is possible to create refugia for plants and animals endangered by agricultural development. The real question is: has there been any real advantage to having fenced in small areas of a continuous landscape? It is the living, or "dwelling" (Ingold, 2010; Legat, 2012), that counts. When settlers stay and make a living from the total landscape, as the Native peoples did, they come to appreciate it, as Johnson and other long-resident immigrants have in fact done, even if they will never quite understand it in the way the Indigenous residents did.

Athabaskan knowledge is rarely formally discursive. In other words, people rarely talk about it in general principles (Nelson, 1983). However, people are quite conscious of, and vocal about, the need to manage and to harvest selectively (McIlwraith, 2007). Education was thorough, systematic, and extensive, and stressed careful, respectful management practices (Mandeville, 2009: 199–204), and carried out through a combination of field training and apprenticeship, myths and tales, ordinary storying, and plain old-fashioned outright explanation of relationships and the nature of place-based living.

One summary of an Athapaskan worldview sees the past as "continually pulled through to the present" by stories and experiences: humans constantly learning spiritual and practical knowledge; "all beings take different trails, each utilizing different places within the dè [land, country]," (Legat, 2012: 200). All are related and interact. Stories and trails hold the world together.

Robin Ridington studied the *Dunne-za* ("Beaver") people of British Columbia. He reports: Beneath each mountain or valley lay the body of a giant animal that had been sent below the earth's surface when the great transformer *Saya* taught the Dunne-za of old to be hunters rather than game, a point emphasized by Pierotti in his Chap. 3, "Predators not Prey: 'Wolves of Creation' Rather Than 'Lambs of God'" (2011). The melodic lines of their songs were said to be "the turns of a trail to heaven and the steady rhythm of drum, the imprint of tracks passing over this imaginary terrain" (Ridington, 1981b: 240). The giant underground animals are widely known in the Northwest (Sharp, 1987), and provide the explanation for earthquakes; stories about one of them cluster around the faults running through Seattle. A sisyuł (giant double-headed serpent) caused a landslide or earthquake that opened a river for fish in Nuxalk country (Davis & Saunders, 1980: 87–94); this may refer to an actual earthquake.

Ridington describes finding the game by dreaming where they are, as does Hugh Brody (1982). Camps among the Dunne-za were always set up so that there was bush unbroken by human trails in the direction of the rising sun. People slept with their heads in this direction and received images in their dreams. Dreaming the locations of game is also noted for related nearby groups by Henry Sharp (1987, 2001) and Jean-Guy Goulet (1998). The hunters, among these Athapaskan groups as among other North American Indigenous nations, find that:

> ...the animals could not be killed unless they had previously given themselves to the hunter in his dream images. It was said that the real hunt took place in the dream and the physical hunt was only a realization of what had already taken place (Ridington, 1981b: 241).

For Canadian Athabaskan peoples, animals exist as spirits, as well as in fleshly form. One dreams of these animals, and dreams of their trails. The spirit comes from the sky (Ridington, 1981a, 1988) or some other nebulous place, moving about on earth. The hunter dreams the path and goes to intersect the animal. If successful, he kills it, and the animal's soul returns to spirit land, to take fleshly form again later (as we have seen, above, for other groups). Pierotti (2010, 2011) argues that hunters knew so much information concerning the individual makeup of local populations that they were aware of which individual organisms they should focus on. ENA observed this in Yucatan among Maya hunters (Anderson & Medina Tzuc, 2005). This instruction manifested itself through the dreaming process, as described below.

What is astonishing is that the hunters do, in fact, dream where the moose and deer are, go where their dreams lead, and actually find and kill the animals. On numerous occasions, an Anglo-Canadian hunter has no success using hi-tech hunting methods in a habitat, only to have a Native hunter guide him directly to game, on the basis of a dream. Thanks to the work of Goulet (1998), Pierotti (2010), Ridington (1981a, b, 1988, 1990), Sharp (2001), and David Smith (1999), we have some knowledge of how bioscience can explain this. The Indigenous hunter notices countless cues, often at a pre-attentive level, while he (men do most of the hunting) is out in the forest. Every faint track, nibbled twig, bent branch, distant sound, muddied watercourse, old bedding ground, and unusual bird noise is noted. Trying to notice and evaluate all these cues consciously would overwhelm the mind. They are attended preconsciously or at the bare edge of consciousness. At night, they are all integrated in dreams—and the dreams do, indeed, tell the hunter where the game is most likely to be found. Psychologists now hold that the purpose of dreams is to integrate and organize such pre-attentive knowledge, processing it for use.

The First Nations of the Northwest know every tree, every rock, and every fishing spot in a vast wilderness. They also believe that giant fish in the earth produce earthquakes by twitching their tails (Sharp, 2001) and that animals have spirits that communicate with hunters. Such juxtaposition of stunning empirical knowledge with what outsiders see as flagrant error is perfectly reasonable, because such beliefs make perfect sense in the context of animism. Many, perhaps most, cultures worldwide explained earthquakes by assuming a giant sentient being lives underground. Spirits that communicate with humans are not only the basic postulate that defines "animism"; it is also a perfectly logical and reasonable explanation for what is perceived by hunters in traditional situations. It is certainly more reasonable than the idea of animals as mere mindless "machines," as held by Descartes (1999 [1637]:

71–73; see Pierotti, 2011), which set the tone for "objective" Western biologists after him.

A notable belief, subject of a lengthy, brilliant study by Julie Cruikshank (2005), holds that glaciers hate to have meat fried near them. This belief spread around Glacier Bay among the Tlingit and their neighbors, and has apparently even spread to some local Whites. As Cruikshank points out, people avoid throwing meat into the fire, or otherwise creating a smell of burning meat, because it attracts bears (as McIlwraith, 2007: 114 also reports; see also Schreiber, 2008). Since a glacier is a huge, moving, unpredictable, constantly changing, and dangerous thing, it seems like a grizzly bear to people who believe that all animated and growing things have spirits. Nothing could be more natural than assuming that the glacier will wax savage at the smell of scorching meat, as a grizzly would. Glaciers also hate loud noises. Even non-Indigenous geologists and glaciologists agree that loud noises can trigger avalanches. They agree with the Tlingit that glaciers need and deserve respect.

As Cruikshank says,

> ...conventions associated with kinship and personhood extended to relations with non-human entities, including a sentient landscape that listens and responds to human indiscretion. This concept does not translate well into European languages.... Moreover, the language is rich in verbs and emphasizes activity and motion, making no sharp distinction between animate and inanimate. Hence mountains, glaciers, bodies of water, rocks, and manufactured objects all have qualities of sentience (Cruikshank, 2005: 142).

As with a vast number of Indigenous languages, Tlingit focuses on verbs and heavily inflects them, unlike our noun-focused Indo-European tongues. In many Indigenous languages inflected verbs can express ideas that require entire sentences in European languages. Verbs are more important (and abundant) relative to nouns in Indigenous languages. In the Mohawk language, there are approximately five verbs for every noun (Pierotti, 2011: 79–8), which means the concepts that Europeans consider to be nouns and static, like creation, are considered to be nouns and active processes by Indigenous peoples. This is one reason language preservation is so important in these cultural traditions.

Community, Human and Nonhuman

"Commonality between species" means "we are all related" which means all living beings (Pierotti & Wildcat, 2000; Pierotti, 2011, Chap. 5). Connectedness and relatedness are involved in the clan systems of many Indigenous peoples, where nonhuman organisms are recognized as relatives whom the humans are obliged to treat with respect and honor (Pierotti & Wildcat, 2000: 1333). This implies "respect, kindness, generosity, humility, and wisdom" not only toward and for humans, but for other life forms (Atleo, 2004: 88)

Since animals are persons involved in human kin networks, one can be quite literally brother to a wolf or sister to a whale. This extension of kinship does fade over distance, and a far-off "brother" or "sister" can be no closer than a long-lost third cousin in Anglo society. However, it is taken seriously. Correspondingly, Nuu-chah-nulth are struck by the weakness of kin ties among Euro-Canadians (as several have told ENA).

Society—human, but also those of other social mammals—is reinforced by strong social codes and by bonds that form among individuals. This way of thought includes (1) respect for nonhuman entities as individuals and the existence of bonds between humans and nonhumans, including incorporation of nonhumans into ethical codes of behavior, (2) the importance of local places, and (3) the recognition of humans as part of the ecological system, rather than as separate from and defining that system (Pierotti, 2011: 5). These are generated, and applied, at the level of community. Thus, for humans on the Northwest Coast, this means the local realm of a chiefly lineage or lineage group and associated commoners and others. This would involve around 500 people—fewer in the cold interior and many more around the rich Salish Sea. It also involved all the nonhuman lives in the area as argued by Anderson, who pointed out that to Northwest Coast peoples, "fish, bears, wolves, and eagles were part of the kinship system, part of the community, part of the family structure. Modern urbanite ecologists see these as Other, and romanticize them, but for a Northwest Coast Indian, an alien human was more Other than a local octopus or wolf" (Anderson, 1996: 66).

Communities were tightly organized. This too varied, from the extremely close-knit kin and community groups of the Nuu-chah-nulth to the loose and flexible ones of the interior Athapaskans, but the similarities appear to be more important than the differences. However, even the Nuu-chah-nulth recognized individuals and their needs, balancing them with the collectivity in a very self-conscious manner (see Atleo, 2004: 55–56). The strong awareness that a community is made up of interacting individuals led to a philosophy of individuals-in-society being both connected and related. This integrated harmoniously with the philosophy of people-in-nature—more accurately, not "nature," only different kinds of persons (see Trosper, 2009, on "nature" as problematic concept in this context).

The contrast is with the Western world, where the usual alternative to individualism is "communitarianism," which, often, consists of attempts to justify top-down autocracy. Northwest Coast community morality integrates the individual in the community but does not subject them to the absolute authority of a king, pope, or dictator. Chiefs had real authority, and in some parts of the coast could sometimes be tyrannical, but their power was limited—usually very much so. These were face-to-face communities, strongly ranked, but small enough that the chief knew he had to get along; otherwise, as many observers noted, he would be abandoned or killed.

What has been said of some Plateau groups would apply throughout the Northwest: "On winter nights [children] were an apt audience for elders reciting myths; daily they heard the headman's exhortations and on special occasions the war stories.... From their parents and grandparents, as well as those brought in to speak of their lives, they heard praised the virtues of obedience, of honesty, and of charity to the unfortunate" (Stern, 1998: 406). The "unruly" were punished. Teaching stories were filled with lore on managing and the morality behind it (Turner, 2014, vol. 2, esp.: 244–266).

Thus, the Northwest Coast peoples could be paradoxically freer than most Westerners manage to be. Rampant individualism cannot work in practice in a large group for obvious reasons, and the Western remedy has always been top-down autocratic control—by church if not by centralized government. The ancient Greeks already recognized, and criticized, this outcome, in their pessimistic ideas on the sequence from democracy to aristocracy to tyranny. The alternative of free self-organizing and self-governing communities was not unknown to them. It was a reality later in much of the Celtic and Germanic world until medieval times, and locally later, but it never replaced the dismal alternation between libertarian and tyrannical alternatives. Thomas Hobbes (1950 [1657]) is the locus classicus; he saw *no* alternative to "warre of each against all" except a totalitarian king. If he had known enough about the peoples he called "savages," he would have realized that other possibilities had existed for millennia.

On the other hand, recall what has been said concerning warfare above; typical of the Northwest Coast was superb relations with nonhuman persons, good relations with one's own people, and warfare with other people of different language and homeland. The goal, at least for most, was peace with all species within one's own lands, and warfare outside of that. By contrast, the Western world has long had an ideal—often neglected but never quite forgotten—of peace with one's own species everywhere.

Stewardship

The intense Native American relationship of community and land must be continually stressed, which brings us back to stewardship. A group is localized, and their local world is existentially important. Its mountains, trees, major rocks, and rivers are regarded as transformed beings that are part of local society. They may be actual kin to the humans of the immediate area. This concept is alien to the modern Euro-American settlers, who have been in the area for three centuries or less. However, it is similar to the rootedness of many villages in Europe, with their sacred wells and springs, their megalithic monuments, and their churches built on sites of pre-Christian temples.

The Inuit teach that "[t]he earth grew angry if men out hunting worked too much with stones and turf in the building of their meat stores and hunting depots" (Laugrand & Oosten, 2015: 28). In Mongolia, a girl from a herding family told ENA she wanted to collect pretty rocks, but would not do this, because "moving them for no reason would be disrespectful to the rocks" (ENA, June 2013). Large and striking rock formations are still regarded as living entities on the Northwest Coast as well as throughout much of Asia, and North America. Especially in northern California, but also among the Nuxalk (McIlwraith, 1948: 1: 594–608) and others, some places are inherently powerful. (There are two Thomas F. McIlwraiths, one the great ethnographer of the Nuxalk, 1899–1964, and the other his descendant, who received a Ph.D. from the University of New Mexico in 2007. 1948 references are to the elder; the 2007 and 2012 references to the younger.)

Such designated "power places" are used for contacting spirits. Some places were known for major power, e.g., waterfalls or rapids, where the extreme power of water is amply visible. Power places can be mountaintops, dramatic cliffs or protrusions, waterfalls, stream eddies, or other dramatic spots. Mountaintops and unique rock formations are preferred sites for spirit or vision questing, because of their isolation and extreme conditions of cold, violent weather, spectacular scenery, and the ability to inspire fear. Some of them become "cultural keystone places" (Lepofsky et al., 2017), a newly coined term which also includes vitally important berrying grounds, clam beds, and other important managed subsistence sites.

The Lakes (Sinixt, Sin-Aikst) Salish of southern British Columbia had almost all moved to Washington state (to a reservation with related groups), had mostly merged into other groups, and had been declared extinct in Canada as long ago as 1956. However, when a road was punched through their main surviving cemetery they appeared in force and blockaded the road (Pryce, 1999). A threat to sacred ground brought the tribe together again.

The Vision Quest

Individual moral and spiritual experience, especially the vision quest, is a crucial component of community morality across many North American nations. Vision quests are often all-important life-transforming experiences that all Northwest Coast young people used to experience and that a surprisingly large number still undergo. This is described by the Nuu-chah-nulth scholar Richard Atleo, a traditionally raised Nuu-chah-nulth scholar who is highly educated in European science and philosophy: "*Oosumich* is a secret and personal Nuu-chah-nulth spiritual activity that can involve varying degrees of fasting, cleansing, celibacy, prayer, and isolation for varying lengths of time depending upon the purpose" (Atleo, 2004: 17). It also includes beating oneself with medicinal branches, including those of a plant containing a natural soap, which thus has a cleansing effect (Atleo, 2004: 92). For whaling, it could last eight months, for hunts that only lasted three or four days (Atleo, 2004: 17). "Oo" or *uu* literally means "spiritual mysteries" and is used just to mean "be careful" (Atleo, 2004: 74, 83). It is related to management practices, even controlled burning, which similarly manages the world to reconcile conflicting impulses of destruction and preservation (Atleo, 2011: 133).

> In traditional Nuu-chah-nulth culture...the world of good and evil is known and experienced collectively through the practice of *oosumich*. Consequently, in these communities the collective spiritual experiences of people determine human perceptions of the nature of the world. There can be no equivalent to a Plato, or a Socrates, or some such individual who might create a school of thought about reality that is later shared by some loyal followers. Rather, good and evil are determined by consensus through personal spiritual experiences

that are reflected in the physical realm. Individual experiences are judged in the context of broad community experiences (Atleo, 2004: 37; see also Atleo, 2011).

Socialized conservation comes naturally to the highly social Nuu-chah-nulth, who live in a world of kinship and community that is tight and close even by Native American standards, but it applies generally. Social links are maintained by gratitude as well as respect. The first word Anderson learned in Nuu-chah-nulth was ƛ'eekoo (pronounced, more or less, *tɬeko*) which is most simply translated "thank you!" but which has deeper implications of exchange, interdependence, formal relations, and group morality that cause it to be routinely used by Nuu-chah-nulth who are otherwise strictly English speakers. (See, e.g., Coté, 2010: xi–xii. She spells it *kleko*, which is itself interesting; English speakers find "ƛ" difficult as an initial, and substitute "kl," which has crept into ordinary spelling. Traditional Nuu-chah-nulth speakers have the opposite accent, naturally substituting "ƛ" for "kl," thus, for instance, saying "ten o' ƛock." Coté is Nuu-chah-nulth, and often spells it in the Nuu-chah-nulth way, but in this book, she found the English spelling more convenient. In her later book from 2022, she uses the formal linguistic spelling.) In this case as in many others, the Nuu-chah-nulth tend to pack a huge range of associations into one word, instead of multiplying highly specialized words as English does.

Nuu-chah-nulth ceremonies for hunting and gathering include ʔo:simch (or ʔuusimch), Atleo's *oosimich*: purification rituals involving washing in cold water, scrubbing hard with branches, and going without sleep, often in isolation in the hills (George, 2003: 48). Often this was done to acquire power to aid in activities such as hunting, and went with elaborate face painting and similar art (Black, 1999). All sexual activity must be avoided. Rituals involving face painting reinforce this (Thomas, 2000). The purpose is to make oneself pure, partly to minimize human scent (which matters when hunting land mammals), but also to pay respect to the game, fish, shellfish, berries, roots, and trees; a dirty or impure person is offensive.

The Haida seem more self-consciously individualist, and the Athapaskan groups less communal in lifestyle, but they have the same philosophy, the difference being that the resulting consensus is looser. Athapaskan spirit power and healing power are more individual matters, less public (see Helm, 2000: 271–292 for extended accounts). They are not shared and validated by great ceremonies. The contrast of Nuu-chah-nulth and Euro-Canadians is striking in this regard, as anyone who has observed extensive interactions within and between these groups can testify. Discussion, usually moderate and civil, to reach a collective moral conclusion, contrasts vividly with a tendency toward defiant independence and militant liberty of conscience (Atleo, 2011: 55–56). In the final analysis, the Northwestern collective morality is based on individual experiences, from which the shared principles are then teased out. This contrasts with the rigid top-down morality and power/knowledge of many Western institutions, such as churches, schools, and the state.

The Plateau groups, however, display a key difference: Individuals can indeed use their spiritual experiences on vision quests to start social movements that take on a life of their own. The "prophets" famous in nineteenth-century Plateau history almost certainly have a long, complex history reaching back centuries or millennia.

This makes them more like the ancient Greeks, especially when one remembers that the philosophy of Plato—and still more that of sages like Pythagoras—had a strong spiritual component. Lonely quests for power, involving bathing in cold water, beating oneself with medicinal branches, fasting, and other self-denials, occur among all documented Northwest Native groups (Benedict, 1923; Miller, 1988, 1997, 1999). Such quests or spiritual retreats are undergone throughout life by medicine persons, warriors, and others in need of extra spiritual power. The extreme eight-month retreats of Nuu-chah-nulth and Makah whalers (Atleo, 2011: 101–102; Jonaitis, 1999; Sapir, 2004) show the great danger, potential benefits to the whole community, and consequent special needs for spiritual power of that enterprise. "Indian doctors," shamans or spirit possessors, were primarily curers, but they could kill by sorcery, become invisible, foretell the future, protect people, and find lost persons and objects (see particularly complete account in Ignace & Ignace, 2017: 387–393; Jolles, 2002: 172–173).

Secwépemc children as young as five or six might receive spirit powers. Elders might "doctor" them with animal skins or parts, to give them the power of the animal. Ronald Ignace received grizzly bear and woodworm powers this way. Woodworms, the larvae of wood-boring beetles, are known throughout the Northwest for their ability to drill slowly and patiently through the hardest wood, and thus represent persistence and tenacity. In early adolescence, youths went for the more typical Northwestern vision quest:

> ...a young person had to live in solitude for days, potentially weeks and months, often on more than one stay. Through...fasting and prayer in solitude, individuals found their personal *seméc* (spirit guardian power)... Songs would come to them, given by an animal or an element of nature, like fire or water, that thus showed itself to the person questing and transferred its spirit power (Ignace & Ignace, 2017: 383–384; see long list of spirit powers on p. 386).

The initiate thus had to shift for himself or herself for many days, with only the few resources brought along for the trip. One elder lamented the difficulty of finding remote enough places in this modern world. Getting one's power from natural beings created an intense bonding with the wild and with natural realities.

Bathing in cold water was apparently a universal part of the quest in the Northwest. Under conditions of hunger, cold, stress, and loneliness, the young person would sooner or later have dreams and visions, and these would validate the calling. Stress physiology combined with cultural expectations increase the likelihood that even the least sensitive person will have them.

Vision questing is known from the Northwest to the Plateau (Benedict, 1923; Hunn & Selam, 1990: 237–241) and Oregon (Jacobs, 1945: 345–348 gives a fine Native account). It extends from there throughout the continent; a classic account from the Plains was given by Black Elk (Neihardt, 1932), and there are many others. Vision questing extends across North America, but from the southwest and southern USA on southward into Mexico, where climates are less physically trying, it was often supplemented or replaced by group or social training and initiation under Indigenous religious leaders. In the vast majority of cases, it is undergone around the time of puberty. Often there is one long and arduous quest; sometimes a boy or girl will undergo several quests. Boys usually—but not always—undergo longer and harder quests than girls, and further discussion below will focus on the boys, whose quests are much thoroughly described in the literature. Typical adolescent spirit quests usually involve isolation at an area of great known concentration of spirit power. Power places are not just for adolescent quests; they are of wide use. The Gitksan recognized *spahalait*, shaman power places, and *spanaxnox*, supernatural power spots related to the privileges of names and named groups (Cove, 1987: 183), as well as adolescent quest spots.

The quest was highly individual among the Plains and Plateau nations, was more socially constructed and integrated into collective ceremonies on the Coast, and reduced to a quiet, secret, lonely forest experience among the Athapaskan groups. The Salish peoples are especially well documented. Among them:

> In order to exploit his natural environment successfully a person needed to establish lines of communication with the nonhuman realm. He would do so primarily through a vision or encounter with a more or less individualized, personalized manifestation of the power which pervaded the wild realm of nature.... Contact could be achieved only if the human supplicant were purged of the taint of human existence.... Daily bathing in the river was the cornerstone of this regime (Ames, 1978: 12–13, research with the Nooksack).

The young person had to be away from humans, in a power place, subjecting himself or herself to privation and discomfort. Women were restricted in action, less often allowed to go far off, and thus tended to get lesser powers (Ames, 1978: 13–14). They could, however, achieve clairvoyance.

The Katzie had complex and deep beliefs in these individual powers; Old Pierre provided an extremely full and detailed account of guardian spirit powers (Jenness, 1955: 37, 48–64). These powers were shown, usually revealed cryptically, in the winter dances (Jenness, 1955: 41–47). People danced their powers, but usually in such a manner that viewers had to guess, and were often mystified. A youth is not supposed to reveal or talk about the actual spirit that gave him the vision, but receives a song and dance that sometimes make the spirit identifiable. A youth who moves and growls like a bear, or a wolf, for instance, reveals what sort of spirit he has (Hill-Tout, 1978c: 111–112). Others dance their visions, or simply dance along with everyone.

Openly revealing powers risked their loss. The Quinault had an extensive vision quest knowledge, with many spirits accessed (Olson, 1936: 145–158). A particularly detailed and important account was provided by Billy Mason, whose experience with the spirits was long, intricate, and gripping (Olson, 1936: 155–157). Very few Native people have been so open about sharing such accounts. Another account was provided by Ike Willard, a Shuswap (Secwépemc) elder who had trained as a boy. (His narrative was translated by Charley Draney and Aimee August in Bouchard & Kennedy, 1979: 132–134.) The term translates as "training for power...when boys were about eight or nine years old their fathers or grandfathers took them into the forest and showed them how to kill animals and live off the resources that the forest had to offer... a boy had to...prepare a sweathouse near a stream where a grizzly bear swims." He "was given a digging stick" for roots, and left alone, being checked on rarely; human proximity would scare off spirit powers. "Young girls also went into the forest to train. Sometimes, when an Indian doctor wanted his or her own grandchild to become an Indian doctor, he would give the child something with his own power on it, such as a piece of an animal, and tell the child to find the rest of the animal." "Every night the boy sang and danced." He had to rise and bathe in cold water before sunrise, sweat frequently (by a fire or using a small shelter), and use evergreen boughs with sweating. Things seen in the water during the bath might be power objects. "During the training period, the boy thought about his power and about the kind of man that he would be when he grew up." He ate little, to master learning to resist greed. He jumped among trees and otherwise exercised. "After the boy had found his power and his own special song, he returned to the village... Sometimes the boy would paint a picture of his power on a rock cliff."

Another personal account from the Kalapuya notes the virtually universal belief that illness would follow if the youth did not follow the spirit vision and do what the vision directed him to do with his life (John Hudson in Jacobs, 1945: 56–58; 180–184, 345–348; for detailed accounts of spirits and spirit questing among the Kalapuya; see Jilek, 1982 for the general case). The vision typically tells the youth what power he has—or, in Western psychological terms, gives the youth mental freedom to realize what he really wants to do with his life (Jilek, 1982). For the Plains tribes, it normally gave warrior power or knowledge, but the Northwest Coast groups, with their highly varied range of activities, presented a wider range of possibilities. Woodworking power came from Master Carpenter or an equivalent spirit being, thus proving the deeply religious nature of art in traditional society. Seamanship, fishing, housebuilding, and all manner of other powers could be given.

Since this power comes from the all-important spirit world, it cannot be gainsaid. Thus, if parents had been hoping the boy would become a woodworker, but the spirits make him a hunter, a hunter he shall remain. This is not infrequently used by a boy or girl to validate his or her career choice at the expense of parental planning. Many a college student pressured by parents to "go into medicine," and then learning of this aspect of the spirit quest in an anthropology class, has dearly wished for such an institution in modern America.

Obviously, some spirit gifts were more noble than others, and youths were counseled to be happy with whatever they got. Some delightful stories make the point. Among the Upper Skagit, one youth got only the power to win eating contests. He was embarrassed and ashamed. But then a much more powerful enemy tribe challenged the Upper Skagit people to war or equivalent competition. The Upper Skagit had the good sense to demand the right to pick the type of contest...and of course our poor youth was soon the hero of his people (Collins, 1974).

Conformity was less important than having one's own spirit visions. For the Kwakwaka'wakw (and they are not alone):

> Everyone had to be divided from the familiar, the 'natural,' so that he could see freshly and learn to act according to circumstances and not according to routine. The anti-social hero is typically he who 'looks through' society. It was one of the main aims of Kwakiutl society to use his insight...It would be true...to say of Kwakiutl society that it was based on common dissent. Every member was carefully trained to be able to decide for himself (S. Reid, 1981: 250–251; this was and is as true for women as for men).

The ability to heal was considered to be special, in almost every culture. All the way from Siberia to California (thus bridging the Bering Strait), healing or shamanic power had a special name, separate from other spirit power. A healer received power to travel to the spirit world to find out the cause and cure of illness. The whaling power of the Nuu-chah-nulth was also of a special nature. It involved meditating among human skulls and other *mementi mori*, as part of dealing with the danger of the quest (Jonaitis, 1999; Sapir, 2004).

Spirit dancing holds the community together, reveals and reinforces social structure, spreads resources widely, and acts as religious focus; as such, it maintains a vibrant cultural and social world in what is otherwise a situation of limited possibilities (Ames, 1978). Spirit dancing has proved to function as a powerful social therapy, and has thus been rehabilitated lately for treating adolescent alienation and problematic behavior, and community alienation and aimlessness (Ames, 1978; Jilek, 1982; Jilek & Jilek-Aall, 1985). This is especially true among the Salish groups, because among them the youth returns to society and dances out his vision in a public ceremony. It has been instrumental in reducing alcoholism, suicide, depression, and anomie among young people, especially if combined with other forms of help. It has given people a sense of control and self-efficacy. Literally hundreds, perhaps by now thousands, of Salish youths have had their lives turned around by it (Ames, 1978; Collins, 1974; Jilek, 1982). It continues to be important and beneficial, as Suzanne O'Brien describes in her excellent and comprehensive study of healing in Salish communities, *Coming Full Circle* (2013).

Psychologist Wolfgang Jilek (1982: 1; interview with Dr. Jilek by ENA, 1984) memorably described his findings on this and his conclusion that the spirit dancing worked far better than clinical therapy for the Salish youths. In British Columbia in 1985, Anderson interviewed a Catholic priest who had actually taken part in the dancing. Someone asked "Doesn't the church say that is all from the Devil?" To which the priest replied, quietly and seriously, "Anything that has the effects I've seen must be from God."

A point not often enough made is that the major role of curers in traditional times often involved treating mental illnesses and sicknesses. Infectious diseases were rather few, until the Whites brought the full range of Old-World pathogens. Healers had no way of coping with these, and died along with their patients, which greatly altered cultures (Martin, 1978). In pre-contact times, the few infectious diseases were usually minor and easily treated with herbs. Probably much more common were physical traumas—accidents, animal bites, war wounds, falls—which could be treated by first aid and some dramatic placebo-effective ritual. Conversely, mental and social problems were evidently numerous and serious. Winter confined people in their houses with seemingly endless rain, cold, and dark outside, especially at higher latitudes. Summer brought frantic activity to put away food. Dangers of war and interpersonal conflict were always present. Anyone even slightly prone to depression, aggressiveness, schizophrenia, or other problems would be sorely tested. Thus, healers had to specialize in these conditions. Of course, they interpreted them as spiritual, but their ways of dealing with them were pragmatic. Psychologists such as Jilek have pointed out that traditional psychological therapy was stunningly effective, and was based on principles perfectly comprehensible to a modern psychologist (Anderson, 1992; Anderson & Pinkerton, 1987).

Returning to the wider—lifelong—quest for spirit power, as Richard Atleo says, powerful individual spiritual experiences validated individuals' sense of self, which then fed back into the community's collective experience of spirituality and thus of morality. Morals were constantly being tested against individual transformative experience. Since the latter was, inevitably, highly conditioned by cultural expectation and tradition as well as personal experience, it normally reinforced the moral tenets of the people.

On the other hand, transformation obviously allows for change, and validates ideas about how to cope with new circumstances. This is most obvious on the Plateau, with its many prophets announcing visions about the Whites, the horses they brought, and other new matters. These prophets were lifelong vision questers, who had mystical experiences and intense spirit journeys on a regular basis throughout life. The coastal Salish also had prophet cults, and other groups had individuals who taught new adaptations.

The relevance of this to conservation will appear below, in the consideration of traditional stories, many of which depend on some young person having a spirit vision in which the animal persons tell him or her what they expect and need from humans. The vision is described matter-of-factly, and may be regarded as a perfectly factual account. Often, for instance, a youth meets a game animal and learns to treat it properly, and thus to hunt it successfully. James Teit (1912: 261–264) reports a Nlakapamux story in which a young man marries, has children by, and thus becomes a successful hunter of, mountain goats (such stories are common and will be further discussed below). This and similar stories are believed to be true events that happened in the past. Sometimes, however, many touches reveal that such journeys are spirit journeys rather than physical ones.

Spirit journeys by medicine persons were also regular features of life in the Northwest. Healers sent their souls to the spirit worlds for many purposes. They also used their special spirit powers, especially those peculiar to the curing life. Curing was important, often involving sucking out objects or "pains" magically implanted in a patient by a witch, or else going to the land of the dead to retrieve a lost soul (see Boyd, 2013a, b). Among the Haida, they could also diagnose a situation in which people were "moused" (*kuganaa*), subjected to sorcery involving mice or mouse spirits. Mice could be conjured out of the sorcerer's body (Boelscher, 1988: 192–198). As elsewhere in the world, from Salem to Sudan, accusations of witchcraft and sorcery on the Northwest Coast were political, generally involving accusations of rivals, suspected enemies, or unpopular lineage mates. More directly relevant to ecology were shamans' spirit journeys to find out why fish were not running, or why game was scarce. Typically, the finding involved breaking of a taboo—in particular, someone had been disrespectful to the animal persons, who then absented themselves locally. Such revelations powerfully reinforced social rules.

Shamanism

All this ritual, healing, and animal protection is part of a religious system loosely known as "shamanism." True shamanism is a tightly defined complex of religious ideas and behaviors that centers on Siberia and central Asia. The classic account by Mircea Eliade (1964) is dated and unsatisfactory; see the superb work of Caroline Humphrey (1995, 1996) and the beautifully illustrated work of Pentikäinen et al. (1999) for better overviews. The word "shaman" is the word for a soul-sending individual healer, in the Tungus languages of eastern Siberia. Northwest Coast religion is so similar in some rituals that it can be called "shamanism" without stretching definitions, but as one moves farther east and south, the characteristic shamanistic features gradually fade, and we are left with a religion more focused on individual spirit experience, and less on the spectacular and complex rituals of the shaman per se.

A shaman uses a drum, often a shiny object (possibly for self-hypnosis), and other equipment to send himself or herself into trance, thereby visiting the Land of Gods and Dead or equivalent spirit realm usually to cure illness, reclaim lost souls, explain misfortunes, find lost objects, and otherwise solve problems. Shamans are individual practitioners rather than being part of any organized religion. Shamans receive animal powers, with many animals being specially associated with particular ethnic groups and their shamans. Sometimes the tongues of animals were obtained for power. (There is an enormous literature on shamanism, involving numerous controversies; space forbids mention of sources outside the Northwest Coast. See Boelscher, 1988; De Laguna, 1990; see book-length account: 669–725; Emmons, 1991: 368–397).

Shamanism is based on individual soul travel to the land of gods and dead or to the skyworld or underworld, to find lost souls or lost objects, find causes and cures for illness, and otherwise fix social and personal problems. The shaman usually has an animal familiar, or several of them; however, often other kinds of spirit powers exist. Shamans can normally transform themselves into animals or spirit beings. Ordinary people with special spirit powers can transform also, but with much less versatility. Moreover, animals—especially predators—can transform into humans, or into other shapes. In fact, the transformation, interpenetration, and mutual shape- and soul-shifting between human and natural realms is a key basic tenet of shamanic societies. This problematizes, to say the least, Philippe Descola's claim that "animism" postulates spiritual similarity but physical difference between people and other-than-human persons (Descola, 2013.)

This classic definition applies only to the shamanism of Siberia, far north China, and arctic and subarctic America. Shamanism blends into other types of spirit cults as it moves south. Religious behavior in temperate-zone Asia and the Americas is generally quite different, but shaman-like behavior crops up sporadically as far as southern China and South America, showing possible earlier streams of shamanic-type practice.

Some authors carelessly describe *all* Native American religion as shamanic, even the priestly religion of the high cultures of Mesoamerica and the Andes; this does violence to the above definitions (and to common sense). Others do not even allow the Northwest Coast into the "shamanism" camp. Still others draw the line somewhere in between. Some acrimonious exchanges have taken place. Most studies, including almost all the early literature and important recent books like Jay Miller's *Shamanic Odyssey* (1988; see also Miller, 1999), integrate Northwest Coast religion into the shamanism complex, because of its close relationship with the Siberian forms, including the original "shaman" practices of the Tungus.

Some of the views considered above are found in Indigenous societies throughout the northern hemisphere (spirits in all beings; transformation; respect for all beings and their spirits—see Chap. 14); others are widespread in North America (vision quests, respectful treatment of game—the latter extending into Siberia); others are strictly Northwest Coast (potlatching, complex local proprietary rights, and many specific stories and accompanying moral practices). The Northwest groups have brought together widespread patterns of thought and action and develop them into a comprehensive system of living with and living through plant and animal fellow travelers.

References

Abbott, D. (Ed.). (1981). *The world is as sharp as a knife: An anthology in honour of Wilson Duff*. British Columbia Provincial Museum.

Albright, S. L. (1984). *Tahltan ethnoarchaeology* (p. 15). Simon Fraser University.

Ames, P. (1978). *Coast Salish Spirit dancing: The survival of an ancestral religion*. University of Washington Press.

Amoss, P. (1984). A little more than kin, and less than kind: The ambiguous northwest coast dog. In J. Miller & C. M. Eastman (Eds.), *The Tsimshian and their neighbors* (pp. 292–305). University of Washington Press.

Anderson, E. N. (1992). A healing place: Ethnographic notes on a treatment center. *Alcoholism Treatment Quarterly, 9*(3/4), 1–21.

Anderson, E. N. (1996). *Ecologies of the heart*. Oxford University Press.

Anderson, E. N., & Medina Tzuc, F. (2005). *Animals and the Maya in Southeast Mexico*. University of Arizona Press.

Anderson, E. N., & Pinkerton, E. (1987). *The Kakawis experience* (p. 68). Kakawis Family Development Centre.

Andrade, M. (1969). *Quileute texts*. (Orig. Columbia University Press, 1931.) AMS Press.

Arndt, G. (2022). The Indian's white man: Indigenous knowledge, mutual understanding, and the politics of indigenous reason. *Current Anthropology, 63*, 10–30.

Atleo, E. R. (2005). Preface. In D. Deur & N. J. Turner (Eds.), *Keeping it living: Traditions of plant use and cultivation on the northwest coast of North America* (pp. vii–xi). University of Washington Press.

Atleo, E. R. (2004). *Tsawalk: A Nuu-chah-nulth worldview*. University of British Columbia Press.

Atleo, E. R. (2011). *Principles of Tsawalk: An indigenous approach to global crisis*. University of British Columbia Press.

Barbeau, M. (1929). *Totem poles of the Gitksan, Upper Skeena River, British Columbia*. National Museums of Canada.

References

Benedict, R. F. (1923). *The concept of the Guardian Spirit in North America*. American Anthropological Association Memoir 29.

Black, M. (1999). *Out of the mist: Treasures of the Nuu-chah-nulth chiefs*. Royal British Columbia Museum.

Blaser, M. (2009). The threat of the Yrmo: The political ontology of a sustainable hunting program. *American Anthropologist, 111*, 10–20.

Boas, F. (1916). *Tsimshian mythology* (pp. 27–1037). Smithsonian Institution, Bureau of American Ethnology, Annual Report for 1910.

Boas, F. (2006). *Indian myths and legends from the North Pacific coast of America*. [German original 1895.] Trans. D. Bertz. Ed. R. Bouchard & D. Kennedy. Talon.

Boelscher, M. (1988). *The curtain within: Haida social and mythical discourse*. University of British Columbia Press.

Bouchard, R., & Kennedy, D. (Eds.). (1979). *Shuswap Stories*. CommCept Publishing.

Boyd, R. T. (2013a). Lower Chinookan disease and demography. In R. T. Boyd, K. M. Ames, & T. A. Johnson (Eds.), *Chinookan peoples of the lower Columbia* (pp. 229–249). University of Washington Press.

Boyd, R. T. (2013b). Lower Columbia Chinookan ceremonialism. In R. T. Boyd, K. M. Ames, & T. A. Johnson (Eds.), *Chinookan peoples of the lower Columbia* (pp. 181–198). University of Washington Press.

Brightman, R. (1993). *Grateful prey: Rock Cree human-animal relationships*. University of California Press.

Brody, H. (1982). *Maps and dreams*. Pantheon Books.

Brown, F., & Brown, K. Y. (2009). *Staying the course, staying alive (CD and folder)*. Biodiversity BC.

Collins, J. M. (1974). *Valley of the spirits: The Upper Skagit Indians of Western Washington*. University of British Columbia Press.

Colombo, J. R. (1983). *Songs of the Indians*. Oberon.

Coté, C. (2010). *Spirits of our whaling ancestors: Revitalizing Makah and Nuu-chah-nulth traditions*. University of Washington Press.

Coté, C. (2022). *A drum in one hand, a sockeye in the other: Stories of indigenous food sovereignty from the northwest coast*. University of Washington Press.

Cove, J. (1987). *Shattered images: Dialogues and meditations on Tsimshian narratives*. Carleton University Press.

Cruikshank, J. (1983). *The stolen woman: Female journeys in Tagish and Tutchone*. National Museums of Canada, National Museum of Man, Mercury Series, No. 87.

Cruikshank, J. (1998). *The social life of stories: Narrative and knowledge in the Yukon Territory*. University of Nebraska Press.

Cruikshank, J. (2005). *Do glaciers listen? Local knowledge, colonial encounters, and social imagination*. University of British Columbia Press.

Danto, D., Walsh, R., & Sommerfeld, J. (2022). Learning from those who do: Land-based healing in a Mushkegowuk community. In D. Danto & M. Zangeneh (Eds.), *Indigenous knowledge and mental health: A global perspective* (pp. 63–76). Springer.

Davis, P. W., & Saunders, R. (1980). *Bella Coola texts*. British Columbia Provincial Museum. Heritage Record No. 10.

De Laguna, F. (1990). Tlingit. In W. Suttles (Ed.), *Handbook of North American Indians: Northwest Coast* (Vol. 7, pp. 203–228). Smithsonian Institution Press.

Deloria, V. (1995). *Red earth, white lies: Native Americans and the myth of scientific fact*. Harper & Row.

Descartes, R. (1999). *Discourse on method and related writings*. Trans. by Desmond M. Clarke. Penguin.

Descola, P. (2013). *Beyond nature and culture*. Trans. by Janet Lloyd (French original 2005). University of Chicago Press.

Egesdal, S. (2008). Introduction to 'Coyotes and Buffalo: A traditional spoken story. In M. T. Thompson & S. Egesdal (Eds.), *Salish myths and legends: One people's stories* (pp. 139–146). University of Nebraska Press.
Eliade, M. (1964). *Shamanism: Archaic techniques of ecstasy*. Routledge & Kegan Paul.
Emmons, G. T. (1991). *The Tlingit Indians*. Edited with additions by F. de Laguna. American Museum of Natural History.
Fienup-Riordan, A. (1994). *Boundaries and passages: Rule and ritual in Yup'ik Eskimo Oral tradition*. University of Oklahoma Press.
Fienup-Riordan, A. (2005). *Wise words of the Yup'ik people: We talk to you because we love you*. University of Nebraska Press.
Fiske, J., & Patrick, B. (2000). *Cis Dideen Kat: When the plumes rise*. University of British Columbia Press.
Frey, R. (2001). *Landscape traveled by Coyote and Crane: The World of the Schitsu'umsh (Coeur d'Alene) Indians*. University of Washington Press.
Garfield, V. E., Wingert, P. S., & Barbeau, M. (1950). *The Tsimshian: Their arts and music*. J. J. Augustin.
George, E. (2003). *Living on the edge: Nuu-Chah-Nulth history from an Ahousaht Chief's perspective*. Sono Nis Press.
Goddard, P. E. (1934). *Indians of the Northwest Coast*. American Museum of Natural History.
Goulet, J. A. (1998). *Ways of knowing: Experience, knowledge, and power among the Dene Tha*. University of Nebraska Press.
Guédon, M.-F. (1974). *People of Tetlin, why are you singing?* National Museum of Man, Mercury Series, Ethnology Division, #9.
Hallowell, A. I. (1955). *Culture and experience*. University of Pennsylvania Press.
Hallowell, A. I. (1960). Ojibwa ontology, behavior, and world-view. In S. Diamond (Ed.), *Culture in history: Essays in honor of Paul Radin* (pp. 19–52). Columbia University Press.
Halpin, M. (1981). 'Seeing' in stone: Tsimshian masking and the twin stone masks. In D. Abbott (Ed.), *The world is as Sharp as a knife: An anthology in honour of Wilson Duff* (pp. 269–288). British Columbia Provincial Museum.
Halpin, M., & Seguin, M. (1990). Tsimshian peoples: Southern Tsimshian, coast Tsimshian, Nisha, and Gitksan. In W. Suttles (Ed.), *Handbook of north American Indians: Northwest coast* (Vol. 7, pp. 267–284). Smithsonian Institution Press.
Harrod, H. L. (2000). *The animals came dancing: Native American sacred ecology and animal kinship*. University of Arizona Press.
Helm, J. (2000). *The people of Denendeh: Ethnohistory of the Indians of Canada's northwest territories*. University of Iowa Press.
Hill-Tout, C. (1978a). *The Salish people. Vol. 1: The Thompson and the Okanagan*. Talonbooks.
Hill-Tout, C. (1978b). *The Salish people. Vol. 2: The Squamish and the Lillooet*. Talonbooks.
Hill-Tout, C. (1978c). *The Salish people. Vol 3: The Mainland Halkomelem*. Talonbooks.
Hill-Tout, C. (1978d). *The Salish people. Vol 4: The Sechelt and the south-eastern tribes of Vancouver Island*. Talonbooks.
Hobbes, T. (1950 [1657]). *Leviathan*. E. P. Dutton.
Humphrey, C. (1995). Chiefly and shamanist landscapes in Mongolia. In E. Hirsch & M. O'Hanlon (Eds.), *The anthropology of landscape: Perspectives on space and place* (pp. 135–162). Oxford University Press.
Humphrey, C. (1996). *Shamans and elders: Experience, knowledge, and power among the Daur Mongols*. Oxford University Press.
Hunn, E. S., & French, D. (1998). Western Columbia River Sahaptins. In D. E. Walker (Ed.), *Handbook of north American Indians: Plateau* (Vol. 12, pp. 378–394). Smithsonian Institution.
Hunn, E., & Selam, J. (1990). *Nch'i-Wana, the big river*. University of Washington Press.
Ignace, M., & Ignace, R. (2017). *Secwépemc people, land and Laws*. McGill-Queen's University Press.

References

Ingold, T. (2010). *The perception of the environment: Essays in livelihood, dwelling and skill* (2nd ed.). Routledge.
Jacobs, M. (1945). *Kalapuya texts* (p. 11). University of Washington. Publications in Anthropology.
Jacobs, E. D. (2003). *The Nehalem Tillamook*. Oregon State University Press.
Jacobs, E. D., & Jacobs, M. (1959). *Nehalem Tillamook Tales*. University of Oregon Books.
Jenness, D. (1955). *The faith of a coast Salish Indian*. British Columbia Provincial Museum [now Royal British Columbia Museum], Memoir 3.
Jilek, W. (1982). *Indian healing*. Hancock House.
Jilek, W., & Jilek-Aall, L. (1985). The metamorphosis of 'culture-bound' syndromes. *Social Science and Medicine, 21*, 205–210.
Johnson, L. M. (2010). *Trail of story, Traveller's path: Reflections on ethnoecology and landscape*. AU Press (Athabaska University).
Johnson, L. M. (Ed.). (2019). *Wisdom engaged: Traditional knowledge for northern community well-being*. University of Alberta Press.
Jolles, C. Z. (2002). *Faith, food and family in a Yup'ik whaling community*. University of Washington Press.
Jonaitis, A. (1981). *Tlingit halibut hooks: An analysis of the visual symbols of a rite of passage* (Vol. 57). American Museum of Natural History.
Jonaitis, A. (1986). *Art of the northern Tlingit*. University of Washington Press.
Jonaitis, A. (1999). *The Yuquot whalers' shrine*. University of Washington Press.
Kan, S. (1989). *Symbolic immortality: The Tlingit potlatch of the nineteenth century*. Smithsonian Institution Press.
Kawagley, A. O. (1995). *A Yupiaq worldview: A pathway to ecology and Spirit*. Waveland Press.
Ḳi-ke-in (2013). "Hilth Hiitinkis—From the beach." In C. Townsend-Gault, Kramer, & Ḳi-ke-in (eds.), Native art of the northwest coast: A history of changing idea (pp. 26–30). : University of British Columbia Press.
Kohn, E. (2013). *How forests think: Toward an anthropology beyond the human*. University of California Press.
Krech, S. (1999). *The ecological Indian: Myth and reality*. W. W. Norton.
Krohn, E., & Segrest, V. (2010). *Feeding the people: Revitalizing northwest coastal Indian food culture*. Northwest Indian College.
Laugrand, F., & Oosten, J. (2015). *Hunters, predators, and prey: Inuit perceptions of animals*. Berghahn.
Legat, A. (2012). *Walking the land, feeding the fire: Knowledge and Stewardship among the Tłįchǫ Dene*. University of Arizona Press.
Lepofsky, D., Armstrong, C. G., Greening, S., Jackley, J., Carpenter, J., Guernsey, B., & Turner, N. J. (2017). Historical ecology of cultural keystone places of the northwest coast. *American Anthropologist, 119*, 448–463.
Lévi-Strauss, C. (1982). *The way of the masks*. Trans. by S. Modelski (Fr. orig. 1979). University of Washington Press.
Llanes Pasos, E. (1993). *Cuentos de cazadores*. Government of Quintana Roo.
Locke, J. (1924 [1690]). *Two treatises on government*. Everyman.
Lopez, K. L. (1992). Returning to fields. *American Indian Culture and Research Journal, 16*, 165–174.
Lowie, R. (1924). *Primitive religion*. Boni & Liveright.
Mandeville, F. (2009). *This is what they say*. Trans. by R. Scollon. Foreword by R. Bringhurst. University of Washington Press.
Manson, J. (2016). Place, indigeneity, and the social world: Lessons from the land, Lessons from the Bread-line. In *Paper, Society for Applied Anthropology, Annual Conference*.
Marshall, J. (1995). *On behalf of the wolf and the first peoples*. University of New Mexico Press.
Martin, C. (1978). *Keepers of the game*. University of California Press.

Martin, J. W. (2001). *The land looks after us: A history of native American religion*. Oxford University Press.
McClellan, C. (1987). *Part of the land, part of the water: A history of the Yukon Indians*. Douglas & MacIntyre.
McIlwraith, T. F. (1948). *The Bella Coola Indians* (Vol. 2). University of Toronto Press.
McIlwraith, T. F. (2007). *"But we are still native people": Talking about hunting and history in a northern Athapaskan Village*. Ph. D. dissertation, Department of Anthropology, University of New Mexico.
McIlwraith, T. F. (2012). *"We are still Didene:" stories of hunting and history from northern British Columbia*. University of Toronto Press.
Miller, J. (1988). *Shamanic odyssey: The Lushootseed Salish journey to the land of the dead in terms of death, potency, and cooperating shamans in North America*. Ballena Press.
Miller, J. (1997). *Tsimshian culture: A light through the ages*. University of Nebraska Press.
Miller, J. (1999). *Lushootseed culture and the shamanic odyssey: An anchored radiance*. University of Nebraska Press.
Miller, J. (2010). Jesus versus sweatlodge: Corpus Christi among the interior Salish on the Colville reservation of Washington state. *Journal of Northwest Anthropology, 44*, 45–51.
Miller, J. (2014). *Rescues, rants, and researches: A review of Jay Miller's writings on northwest Indian cultures*. Northwest Anthropology, Memoir 9.
Mishler, C., & Simeone, W. E. (2004). *Han: People of the river*. University of Alaska Press.
Nadasdy, P. (2003). *Hunters and bureaucrats*. University of British Columbia Press.
Nadasdy, P. (2007). The gift of the animals: The ontology of hunting and human-animal sociality. *American Ethnologist, 34*, 25–47.
Neihardt, J. (1932). *Black Elk Speaks*. William Morrow.
Nelson, R. K. (1973). *Hunters of the northern Forest*. University of Chicago Press.
Nelson, R. K. (1983). *Make prayers to the raven*. University of Chicago Press.
O'Brien, S. C. (2013). *Coming full circle: Spirituality and wellness among native communities in the Pacific northwest*. University of Nebraska Press.
Olson, R. L. (1936). *The Quinault Indians*. University of Washington Press, University of Washington Publications in Anthropology VI.
Olson, R. L. (1940). *The social Organization of the Haisla of British Columbia* (Vol. 2, p. 5). University of California, Anthropological Records.
Palmer, A. D. (2005). *Maps of experience: The anchoring of land to story in Secwepemc discourse*. University of Toronto Press.
Pelly, D. (2001). *Sacred hunt: A portrait of the relationship between seals and Inuit*. University of Washington Press.
Pentikäinen, J., Jaatinen, T., Lehtinen, I., & Saloniemi, M. (Eds.). (1999). *Shamans*. Tampere Museum Publication 45.
People of 'Ksan. (1980). *Gathering what the great nature provided*. University of Washington Press.
Pierotti, R. (2010). Sustainability of natural populations: Lessons from indigenous knowledge. *Human Dimensions of Wildlife, 15*, 274–287.
Pierotti, R. (2011). *Indigenous knowledge, ecology, and evolutionary biology*. Routledge.
Pierotti, R. (2020). Learning about extraordinary beings: Native stories and real birds. *Ethnobiology Letters, 11*, 253–260.
Pierotti, R., & Fogg, B. R. (2017). *The first domestication: How wolves and humans coevolved*. Yale University Press.
Pierotti, R., & Wildcat, D. (2000). Traditional ecological knowledge: The third alternative. *Ecological Applications, 10*, 1333–1340.
Pryce, P. (1999). *Keeping the lakes' way*. University of Toronto Press.
Ray, V. (1933). *The Sanpoil and Nespelem, Salishan peoples of Northeastern Washington* (Vol. V). University of Washington, Publications in Anthropology.
Redford, K. (1990). The ecologically Noble savage. *Orion Nature Quarterly, 9*, 25–29.

References

Reid, S. (1981). Four Kwakiutl themes on isolation. In D. Abbott (Ed.), *The world is as Sharp as a knife: An anthology in honour of Wilson Duff* (pp. 249–256). British Columbia Provincial Museum.

Ridington, R. (1981a). Technology, world view, and adaptive strategy in a northern hunting society. *Canadian Review of Sociology and Anthropology, 19*, 469–481.

Ridington, R. (1981b). Trails of meaning. In D. Abbott (Ed.), *The World is as sharp as a knife: An anthology in honour of Wilson Duff* (pp. 289–247). British Columbia Provincial Museum.

Ridington, R. (1988). *Trail to heaven: Knowledge and narrative in a northern native community*. University of Iowa Press.

Ridington, R. (1990). *Little bit know something, stories in a language of anthropology*. Douglas & McIntyre.

Ross, J. A. (2011). *The Spokan Indians*. Michael J. Ross.

Sapir, E. (2004). *The whaling Indians: West coast legends and stories, legendary hunters*. Canadian Museum of Civilization.

Schreiber, D. (2008). Glaciers listen: A review essay and response to Cole Harris. *BC Studies, 159*, 131–139.

Sharp, H. (1987). Giant fish, Giant otters, and dinosaurs: 'Apparently irrational beliefs' in a Chipewyan community. *American Ethnologist, 14*, 226–235.

Sharp, H. (2001). *Loon: Memory, meaning and reality in a northern Dene Community*. University of Nebraska Press.

Smith, D. M. (1999). An Athapaskan way of knowing: Chipewyan ontology. *American Ethnologist, 25*, 412–432.

Steller, G. (2003). *Steller's history of Kamchatka*. In M. Engel & K. Willmore (ed. and trans., German original 1774.). University of Alaska Press.

Stern, T. (1998). Cayuse, Umatilla, and Walla Walla. In D. E. Walker (Ed.), *Handbook of North American Indians: Plateau* (Vol. 12, pp. 395–419). Smithsonian Institution.

Storie, S. (1973). *Bella Coola Tales*. Indian Advisory Committee.

Sullivan, R. (2000). *A whale hunt: Two years on the Olympic peninsula with the Makah and their canoe*. Scribner.

Suttles, W. (1955). *Katzie ethnographic notes*. British Columbia Provincial Museum, Anthropology in British Columbia, Memoir 2.

Tanner, A. (1979). *Bringing home animals*. St. Martin's Press.

Teit, J. (1912). *Mythology of the Thompson Indians*. American Museum of Natural History, Memoir 12, Reports of the Jesup North Pacific Expedition, VIII.

Thomas, D. (2000). Face paintings from the Sapir collection. In A. L. Hoover (Ed.), *Nuu-chah-nulth voices, histories, objects and journeys* (pp. 172–209). Royal British Columbia Museum.

Thompson, M. T., & Egesdal, S. M. (2008). *Salish myths and legends: One people's stories*. University of Nebraska Press.

Trosper, R. L. (2009). *Resilience, reciprocity, and ecological economics: Northwest coast sustainability*. Routledge.

Turner, N. J. (2014). *Ancient pathways, ancestral knowledge: Ethnobotany and ecological wisdom of indigenous peoples of northwestern North America* (Vol. 2). McGill-Queen's University Press.

Viveiros de Castro, E. (2015). *The relative native: Essays on indigenous conceptual worlds*. HAU Books.

Wiget, A., & Balalaeva, O. (2011). *Khanty: People of the taiga*. University of Alaska Press.

Willerslev, R. (2007). *Soul Hunters: Hunting, animism, and personhood among the Siberian Yukaghirs*. University of California Press.

Chapter 9
Animism and Rationality: North vs "West"

Mystical spiritism and "irrational" vision questing outperformed economic rationality in managing resources in the Northwest. Indigenous people managed them sustainably for thousands of years; the settler society destroyed most of the resource base within 200. This is, obviously, a huge embarrassment for those who still hold that Western Man has some special pipeline to enlightened wisdom, although this is often rationalized in terms of "Western technology and economic growth are superior, therefore this is simply a price we pay. After all, the resource base exists so that we can exploit it, and only human life is worthy of consideration."

The reasons for the failure of rational scientific management are clear. The most obvious reason is the "steep discount slope," by which Western thought discounts the future. It could take care of itself. This is the attitude that has led to uncontrollable global climate change. In everyday economic thinking, only today, or next year, matters (see Kahneman, 2011 on such bad but natural habits of thought). This has been elaborated into various forms of corporate logic: shareholder payouts, government tax bases, effective demand, and the rest. The result is always the same: maximize profit now and forget the long term. A spirituality-based version of this can be seen in a discussion of how to deal with the alleged "religion of environmentalism" by a group that calls itself Capitol Ministries, where Clergyman Rolf Drollinger argues,

> To think that man can alter the Earth's ecosystem — when God remains omniscient, omnipresent and omnipotent in the current affairs of mankind — is to more than subtly espouse an ultra-hubristic, secular worldview relative to the supremacy and importance of man...God says He will continually renew the face of the earth until He forms a new heaven and a new earth in the end times (Rev. 21: 1). In the thousands of years of climate history since these words were recorded, the veracity of God's promises have [sic] proven to be reliable. So, who then should we trust? It follows that we can all rest assured and wholly rely on God's aforementioned promises pertaining to His ability and willingness to sustain our world's ecosystem (Pierotti, 2018: 299–300).

Alternatively, a American layperson of European descent responded to wildfires resulting from climate change in northern California as follows:

> It's "obvious"... "Look at the trees around you right now. The leaves are falling out of the trees when they shouldn't be. The environment is changing, and it's changing everywhere." But that doesn't mean he thinks it possible to do anything about it. "The good Lord has to fix it. We're not capable of it." (Pierotti, 2018: 300)

These two statements reveal what is called by Western philosophy a "faith-based" way of thinking. Adherents to this specific form of Christianity believe that humans cannot change the world, and only the actions of their creator matter. This absolves them from any responsibility. Runaway economic growth and its accompanying ecological impacts simply represent the will of this creator, because if the creator thought such actions were wrong, he (it is always he) would put a stop to it. Ironically, the layperson sees that climate change is real, but the learned helplessness imposed by his Christian beliefs keeps him from acting. Such attitudes clearly reveal one reason why the Western approach to management has failed, even though this approach led to massive accumulation for a while, supporting thousands of people.

Today, the landscapes in question are often ruined for the foreseeable future, and the short-term profits were usually not reinvested productively, even in terms of capitalist economic thinking. Over the medium or long term, Native management wins out. There are more people now in west Canada than in Aboriginal times, but that is because the current population gets its food primarily from elsewhere. Even the heavily farmed and productive parts of the northwestern USA imports considerable food.

The reasons for the success of spirit beliefs (e.g., "animism" in general anthropological parlance; Hunn & Selam, 1990: 230) are not always obvious. Leaving aside questions about the reality of sentient, agentive wills within bushes and clouds, there are at least four reasons. First, animism leads people to respect their environments, putting the survival and existence of other species on an equal footing with their own. Second, spirit vision questing validates individual knowledge and experience; people learn to take seriously their experience-based perceptions of the world because they are learning survival skills from interacting with it. Third, Northwest Coast religions are intensely social. The spirits are part of society. More to the point, life is based on interaction with a multitude of persons, not all of whom are human. One must be responsible toward them, including the children who will need fish and trees and roots 50 or 100 years from now (Trosper, 2009). Thus, Indigenous philosophy emphasizes making sure that seven or more generations into the future can have a quality of life equal, or even superior, to your own. Although, during much of the twentieth century, Americans were able to create conditions that allowed their children and grandchildren to have higher standards of living, it is clear that this achievement has ended, and, barring major changes, future generations will almost certainly have lower standards of living in a degraded environment. The capitalist experiment has failed the seven generations criterion. Fourth, respect for everything means that, when harvesting a living being for human use, everything should be used—it is disrespectful to ignore a being that is offering itself to you. Equally, if not more important, nothing must be overused; that too is a serious form of disrespect that damages the future for your descendants. As one of our colleagues has put it: "This belief that one's fellow animate beings are sensitive to any hint of disrespect on the part of a human highlights the Plateau social value placed on respect for the autonomy of the individual human" (Hunn & Selam, 1990: 236). Hunn was warned not to be vocally confident about finding a crooked stick for a root-digger; as James Selam says in the book in question, "if the trees hear you boast like that...All the bent branches will straighten up as you go by" (Hunn & Selam, 1990: 236). In short, animistic beliefs may not accord with modern science, but lead to conserving environments. Modern economic beliefs also fail to accord with science (or even common sense), and they lead to destroying environments. Which is more irrational?

Many readers of this book—if it reaches its intended audience—already see plants, salmon, and rocks as conscious agents who decide how to interact with their human friends. Not all those believing readers are Native American. We personally know anthropologists who converted to Native American spirit beliefs after experiences in the field. We have no quarrel with this conclusion, or with others that will appear below. What matters is that we understand that, as a result of their emphasis on connection, relatedness, and reciprocity, Native management was sustainable in important ways, while settler management has not been, and that this is a problem in need of exploration.

To Western biologists, the combination of intimate knowledge and "exotic beliefs" is often incomprehensible. Such a conflict in perceptions is the topic of Pierotti and Fogg (2020). Conversely, to First Nations people, the biologists' cold numbers and impersonal regard for animals as mere statistics are incomprehensible. A useful way to think about this is that Western science relies on data, whereas, in contrast, Indigenous philosophy and concepts rely upon stories, especially when they describe experiences (see Fogg et al., 2015; Pierotti & Fogg, 2020, Pierotti, 2015, 2016 for detailed examples). (This is not to say that bioscience never uses stories, nor do Indigenous people ignore data as such.) Paul Nadasdy, in particular, has described many cases of nonmeeting of minds. In management conflicts, the biologists generally win, with disastrous consequences not only for the Native people but for the animal populations. Nadasdy (2003) has more recently argued for adopting the Native viewpoint on spirits and personhood among animals, or at least a willing suspension of disbelief regarding it. Practical ways of achieving such a synergy have been discussed by Pierotti (2010, 2020a, b).

The best accounts of the thinking and understanding of Northwestern Indigenous Nations combine cognitive and phenomenological approaches with hard science (see, for example, Hunn & Selam, 1990; Thornton, 2008). These can be compared with such studies in other parts of the world as Steven Feld's *Sound and Sentiment* (1982), Keith Basso's *Wisdom Sits in Places* (1996), Deborah Rose's *Dingo Makes Us Human* (2000), Lynne Kelly's *Knowledge and Power in Prehistoric Societies* (2015), which discusses how landmarks are used to code stories and maps in Indigenous peoples around the world, and "Relationships between Indigenous

American Peoples and Wolves 1: Wolves as Teachers and Guides" (Fogg et al., 2015), which discusses relationships between US Indigenous Nations and wolves, and reveals the dynamics of a crucial social interaction between two important species. These works are thoroughly sophisticated in the use of ethnosemantic and psychological methodologies to investigate local vs. international-scientific views of the world.

First Nations people, at least those still living in rural or semirural situations, relate to the animals through lived experience, not through abstract book-learning. They experience the forest and its inhabitants directly, as a world of persons and spirits that watch and interact with them (Nelson, 1983; Ridington, 1988 provide superb descriptions). Their knowledge and teaching are experiential and procedural, not analytic and declarative. Being close to the animals in question makes avoiding the agency heuristic difficult or impossible. The animals are simply assumed to have human will and agency.

For these and other reasons, relationships with the animals are intensely emotional, and the emotions in question are complex and deep. Finally, this emotionality drives a relationship of mutual power and mutual empowerment; the animal persons have their power; humans have theirs. Biologists, on the other hand, assert a particular kind of power/knowledge (in Foucault's terms) and also represent the power of the State intruding on local societies (Nadasdy, 2003, produces an excellent, detailed Foucaultian account of this). In contrast, Creation stories from First Nations serve as heuristic tools. An effective way to evaluate creation stories is by understanding patterns which derive from nature itself, and determine if the pattern can be discerned in the story.

Stories instruct humans how to live effectively in a nonhuman world, and to humbly accept that this world was not constructed solely for humans, because such stories are based on behavior or habits of the animals, weather, rivers, or plants, and involve biological and geological aspects observed by many Indigenous peoples. Native people regard specific events they encounter as part of an ongoing pattern that emerges from ecological processes (Pierotti, 2011). When an individual animal is encountered it is recognized that the humans who are present experience the event as being in a state which is unfolding, and the humans must understand it as part of an emerging pattern. This can be thought of as requiring a story to examine the phenomenon, rather than a snapshot or a data set.

Traditional stories emphasize both the process and unpredictability of life and reveal that it is often necessary to change in response to changes in the environment (Ridington, 1988; Pierotti, 2011). The Western scientific term for such a process is, of course, *adaptation*, which reinforces the idea that Indigenous creator figures can be considered as forces which generate adaptive responses on the part of both human and nonhuman organisms. Creation stories involve apparent directional activity within an unpredictably varying environment, which helps resolve a major question in evolutionary biology, i.e., why there seems to be evidence of purpose and directed outcomes within a system resulting from apparently random events (Pierotti, 2020a).

In an exemplary account of one Northwest Coast society, the Kluane people of the British Columbia-Yukon border, Paul Nadasdy (2003: 112) maintains that anyone analyzing traditional worldviews cannot separate out the knowledge that outsiders see as "practical" from the knowledge that outsiders see as "mere cultural belief," i.e., the spiritual beliefs detailed above. There is some truth in this: within Kluane culture, there is no distinction. They do not see a separation between everyday observation and participation in spirit realms. One can take the clearly useful knowledge out of context and use it, but this does not represent the traditional view. It is, thus, inadequate for co-management or co-work purposes. Bureaucrats and outsiders continually try to separate the two realms, to the utter confusion of the Kluane.

Nadasdy sees no real hope of cooperation. The bureaucrats *must* assert power over knowledge; the Native people *must* lose. This is partly guaranteed, as he points out, by the fact that "numbers" and "scientific" evidence is all that is taken seriously by the Canadian government and its agencies; local knowledge is "mere anecdote" or "unverified." A better way of assessing local knowledge is desperately needed.

Since Nadasdy's account is particularly sensitive and insightful, and his field work eminently detailed, it is worth unpacking his analysis in some detail. In some ways, he adds to the confusion. Even when we sound a bit negative, it is precisely because his account is so perceptive and theoretically grounded that we discuss it. Nadasdy quotes one elder: "(traditional knowledge) is not really 'knowledge' at all; it's more a way of life" (Kluane First Nation member quoted in Nadasdy, 2003: 63). We understand the point Nadasdy is attempting to make, but he does little further analysis. The issue that Nadasdy and his Kluane colleagues emphasize is that to the Kluane people, hunting is a way of life, and hunting consists of the entire process, from the first thoughts about when to start, through the kill and preparation, on to ultimate allotment of the "meat" that is gathered as a part of the hunting process. Two different forms of knowledge are being discussed here: (1) patterns that have been shown to work over many generations which are the equivalents of premises or postulates in Western science (e.g., trophic dynamics) and (2) specific observations incorporated into these patterns (e.g., noting that one species preys upon or is preyed upon by another at certain times of the year).

Recall Nadasdy's account of Moose Jackson (above). Jackson was not operating in a "mystical" fashion" any more than is the field biologist. His empirical knowledge may be more complete. Nadasdy and some other recent anthropologists (such as Jean-Guy Goulet, 1998), in otherwise superb accounts of such knowledge, reify and essentialize culture, and sometimes seem to see experiential knowledge as confined to First Nations. They see it as showing "incommensurability" (Nadasdy, 2003: 62ff). This concept derives from Thomas Kuhn's *The Structure of Scientific Revolutions* (1962), in which Kuhn argues that paradigms within science are often so different that they cannot be directly compared; the very measurements involved in one paradigm are irrelevant to its alternatives. As Pierotti demonstrates, however, the two forms of knowledge are not "incommensurate." Moose Jackson is not employing "spirit vision"; he simply is not recording his observations by writing them down on paper, but by storing them away in his memory. Nadasdy seems to imply that spirit visions are incommensurable with biological measurements of growth hormones or bone density. However, when one deals with pragmatic knowledge such as the level of overhunting of a game resource, it is not the case, as Nadasdy admits (footnote, 2003: 277). The Native people obviously have an incredible pragmatic knowledge of their resources, fully commensurable with pragmatic bioscientific knowledge. What is incommensurable is the abstract ontology behind the different views: the Native spirit-based one and the biologists' lab-based reductionism.

Pierotti has more than 50 years of field experience working with seabirds, marine mammals, and dogs and wolves, and Anderson has similar experience with birds and coyotes. Only a fraction of the information we have acquired has ever been published, or even reported in papers given at national meetings. We can describe subtle movements in the head of an animal that communicate intentions. We can discuss fine differences in anatomy and behavior between dogs and wolves, as Pierotti has communicated as an expert witness in a courtroom, but which have never been written down or published. Moose Jackson has similar knowledge of the behavior and ecology of moose, which is how he earned his sobriquet. Jackson may believe in spirits, just as Pierotti has experienced phenomena inexplicable by contemporary science, but that does not make us "mystical" in our knowledge. Our knowledge is experiential and empirical. If we differ from Western science, it is because we do not depend primarily upon theory, which is, in many ways, more "mystical" than our experiences, having emerged from the thought processes of scientists.

Nadasdy appears to see a culture as a perfectly integrated, rather unchanging, whole, which greatly limits his ability to understand the phenomena he investigates. Anyone from any culture can learn practical skills and information. Consider baking: today's home breadmaking is not very different from that of the inventors of leavened bread thousands of years ago, but the worldview of those long-forgotten bakers—presumably animistic and spirit-haunted—is unimaginable now. Bread cares little about cosmovision, even though—as every baker knows—it is a living thing, thanks to its vibrantly alive yeast networks. It knows a sure hand from an incompetent one, but is not affected by the beliefs of the baker about spirits or yeast strains.

Cultures not only fail to be integrated, harmonious wholes; they may change rapidly. Anderson's elders in the Anglo-American midwest and south had many of the same attitudes as the Northwest Coast peoples. Children were taught to kill animals cleanly, not take too many, not let them suffer, and so on. We were taught to look down on indiscriminate shooters. We were taught to share our catches. Moreover, many, if not most, hunters and fishers believed that ultimately a catch was a gift from God, not just a matter of your own skill. You had to be grateful accordingly. Obviously, luck is involved in hunting and fishing, but so is skill, as Moose Jackson demonstrates. The difference is that an Indigenous hunter may view an animal as having "given itself to the hunter," which requires thanks and reciprocity through ceremonial practices, whereas a Western hunter may believe in his skill, and more importantly, his technology, and thank only his weapon rather than its target (Marshall, 1995: 43–64)—though that would be a very modest hunter. This way of thinking goes a fair way to bridging the gap between "Indigenous" and

"Western." Here as in many other cases, the opposition is between rural-experiential and urban-abstract as much as—if not more than—between Indigenous and "Western."

Worldviews do indeed influence the realities of counting sheep (Nadasdy, 2003), but do not prevent comparison of the results. Nadasdy's main example of a failure to communicate does not involve cosmovision at all, but was a simple disagreement over whether the local mountain sheep were overhunted or not. The more knowledgeable Native Americans thought so, and had all the good evidence. When the Kluane told the Ruby Range Steering Committee that large Dall Sheep rams should not be shot because they are the "teachers" of their society, they were simply citing a biological fact that they knew well; biologists have also found that older animals teach and lead younger ones, and that loss of them is devastating to younger animals trying to find their way on migration and feeding routes (see Festa-Bianchet and Côté for Northwest data). A Kluane elder went on: "Animals teach us things too … patience and respect and stuff like that … I always think of what we as a whole society can learn from animals … I mean we once lived with them" (Kluane First Nation elder quoted in Nadasdy, 2003: 101); this is neither mystical nor anthropomorphic. The biologists thought otherwise, but were more heavily influenced by economic issues after hearing from powerful trophy-hunter interests. One need not and should not go beyond common sense, of a sourly cynical sort, to understand this disagreement.

If people are trapped in narrow views, that is not incommensurability at work. It is just ordinary narrowness. In the case of settler society not understanding Indigenous society, it is all too often mere racism. Pierotti (2011: 33–34) points out that it follows logically from Indigenous thinking that if nonhumans are "persons," they must have cognitive abilities, which means they should recognize the danger of being hunted by humans. Thus, if a nonhuman is caught, it was assumed to involve some element of choice on their part (Anderson, 1996). This led to the concept of the prey "giving itself to you," which was the basis of the covenant that if you mistreat an entity, that entity may respond in a way that can cause harm to you and your people. This presumed gift required gratitude, as well as respectful treatment of the body of the nonhuman on the part of the human taking its life. Although the prey may not truly give up its life voluntarily, this assumption is an important guiding principle of the rituals that ensured that hunters and fishers treat their take with respect, so as not to offend the prey (Pierotti & Wildcat, 2000): "If we do not show respect to the bear when we kill him, he will not return" (traditional Mistassini Cree). As Darwin's defender Thomas Huxley said of the tendency toward selfishness in human social behavior, "Let us understand … that the ethical process of society depends, not on imitating the cosmic (natural) process, still less in running away from it, but in combating it" (Huxley, 1894). Pierotti would substitute "coming to terms with" for "combating," but otherwise Huxley's rule of dealing respectfully with others applies as much to nonhumans as to humans.

Not only Nadasdy, but also Thomas McIlwraith (2007), appears to find that biologists are so trapped in a mentality of lists and reductions that they cannot and therefore should not use "TEK" at all—they would only corrupt it. These two anthropologists seem to have been dealing with particularly clueless biologists and bureaucrats—urban youths with little or no bush experience or rural background. Such biologists are indeed fairly common, but they are far from universal. Their behavior does not fit the practice of a large number of biologists. Many of our friends and mentors in that profession, in fact, come from rural backgrounds where they were raised with understandings similar to those of Moose Jackson.

In the real world, there are reasons to pick out the most useful bits of traditional knowledge and bring them into wider practice. Of course, such cherry-picking might do violence to the holistic vision of culture, but so does all technical and medical practice. It is useful if scholars really try to understand the place of TEK in culture. It is easy to be intellectually bullied into thinking that all cultural information is literally true. However, it is also important to keep in mind, as Anderson has argued (1996), that all people and cultures make errors in logic, because of their assumptions (cf. Kahneman, 2011). The value comes from recognizing what is valuable in each culture and finding synergy in these areas of overlap. After all, does one treat malaria, or contemplate its holistic phenomenology? There is a place for the latter, and it may even be necessary, but direct treatment is what we usually wish. To cure malaria, however, we may need both ways of seeing and understanding.

In short, we are not dealing with totally incommensurable worldviews; we are dealing in large part with the difference between rural people who learn by experience and urban people who learn from books. This means, among other things, that we can and should all learn from each other. The extreme form of "cultural appropriation" that advocates never borrowing from other cultures because it would harm their integrity is based on an utterly false view of culture. Culture does not provide us with harmonious wholes protected by steel walls from other wholes. It provides us with a vast shifting set of adaptive strategies. The "world wide web" is not new: it is as old as the spread of language, stone tools, the bow and arrow, and every other shared innovation.

The Native view generates a particular morality, one of caring and nurturing; animals are protected and conserved out of "respect" or care, rather than because of mathematical models of populations (Nadasdy, 2003). Nadasdy points out that the Native people he knew tended to think of the mathematical models as disrespectful. Such models treat animals like things, instead of like the people—the "other-than-human persons"—that they are. This is another example of Nadasdy and the Kluane of not understanding alternative approaches. Indigenous stories are not that different from such models, because both function as metaphors that allow a philosophical framework to develop (Pierotti, 2011: 13). Neither stories nor models should be considered as literally representations of reality. Models of exponential and logistic population growth are not expected to literally represent the actual behavior of animal populations. If Nadasdy and the Kluane were encountering biologists who assumed that models represented reality, they should have learned about and examined the models to understand their limitations, so they could effectively argue against the biologists. This exemplifies a problem we regularly encounter in trying to establish the strengths of TEK: many Indigenous peoples criticize science without understanding how it functions and then argue that their own beliefs are literally true, which then creates a situation that is *really* incommensurable, as argued by Nadasdy (2003). Vine Deloria's *Red Earth, White Lies* (1995) and *Evolutionism, Creationism, and Other Modern Myths* (2002) are classic examples of Indigenous works that rather than trying to build bridges between cultural manifestations of knowledge simply tried to burn them down using arguments based on ignorance drawing evidence from the worst of Western attitudes (Pierotti, 2011, Chap. 9). Like mathematical models of population phenomena, traditional stories provide a context into which empirical observations can be placed and compared against the assumed state. TEK can be rebuilt and changed by each generation, in the same way that each new generation of graduates in ecology or evolutionary biology go out and make new discoveries that add on to the base of empirical knowledge that currently exists in those fields.

The problem identified by Nadasdy and the Kluane arises from the origins of modern Western science more than its current manifestations. This is because since the Enlightenment the traditional model of nonhuman organisms has been René Descartes' metaphor of the machine, which considers organisms to be made up of parts. This metaphor emerged from the materialist worldview and the links between economics and science that developed in Western Europe during the sixteenth and seventeenth centuries (Worster, 1994). Following Cartesian thinking, the most effective way to understand the organism is by understanding its constituent parts and how they fit together. The Cartesian approach creates a variety of conceptual problems in biology. For example:

> The task of evolutionary biology consists in providing a material, historically based explanation for current or extinct biological forms ... this task often involves the decomposition of organisms (or ecosystems for that matter) into simpler constituent parts, which can then be analyzed in greater detail. This approach, broadly known as *reductionism*, has proven enormously successful. Incautiously used, however, reductionism has led to a dangerously oversimplified shorthand in evolutionary biology—an interpretation in which organisms are decomposed into features and features are accounted for by single evolutionary causes ... a bird becomes no more than a set of individual features in search of explanation... While these explanations capture something crucial about the evolution of birds, they deliberately avoid what all biologists know to be true: Organisms are more than collections of independent features and features are rarely shaped by single causes. (Dorit, 2007: 234 as cited in Pierotti, 2011)

Combined with the machine metaphor, the use of averages as a stand-in for individuals leads most investigators in the Western scientific tradition to assume that any species under study can be represented by a typical individual, which is constructed of a combined set of adaptations, each of which solves a problem. To understand such thinking, it is helpful to consider the following question: Are dogs and cats in Anglo-American society mere numbers in a population graph, or are they individuals with complex lives and moral claims? How would you feel if your pet were considered to be a mere number, and could be culled if there are several other dogs in town? This, in fact, occurred in northeast Canada, when Inuit dogs were indiscriminately killed in the mid-twentieth century (the dogs were claimed to be disease-bearing and overnumerous), leaving the Inuit heartbroken.

Biologists talk of "harvesting" game, which is a metaphor derived from economic thinking that horrifies Indigenous peoples, because it illustrates how removed Western thought is from recognizing the value of individual lives. This is not to imply that wild animals are "pets" to the Native Americans; they are not. They are hunted for food, which in a sense is also "harvesting," but at a much smaller scale. Animals and plants are seen as having spirits that are powerful and potentially dangerous, if they are offended (Pierotti, 2010, 2011). The point is that other lifeforms are known to be complex, deep, individual beings, with intelligence and agency, not just numbers or soulless automatons. In this the Native people are closer to the truth than are "objective" scientists; animals may not have powerful guardian spirits, but they are *certainly* not automatons or creatures of instinct, let alone mere numbers. The religious views of Descartes were concerned about souls, not science. Long discredited in serious science, it remains basic to bureaucratic management and unfortunately to much of biomedical and physiological research. It is the bureaucratic view derived from Descartes' Renaissance Christian thinking that is the product of inaccurate religious-based belief, *not* the Indigenous one (Pierotti, 2011).

Proof that we are dealing with practical experience rather than essentialized culture can be found in eastern California. There, Kimberly Hedrick talked to ranchers who know the land and wildlife from experience. They often found it impossible to deal with, or communicate with, biologists who know it only from books (Hedrick, 2007). Even though their cattle are being raised only to be sold for slaughter, those cows are not mere numbers to the ranchers; they are living beings with personalities and with needs for care. Even though grass is merely a food for cows rather than a spirit being, the ranchers can tell if it is overgrazed or not. Hedrick found that many urban "experts" were clueless, and assumed the worst about rancher knowledge, thus ham-handedly interfering with ranching. She found serious disagreements among people who not only share an Anglo-American culture, but in some cases actually took more or less the same general science courses in college! Particularly ironic is that the range biologists in Hedrick's study took the same superior, unwilling-to-listen attitude to the ranchers that Nadasdy's bureaucrats did to the people of Yukon, though the ranchers were not only from the same culture but a good deal richer, and sometimes better educated, than the biologists! Status in these transactions does not depend solely on race or socioeconomic position. Hedrick's sober documentation makes all too believable the self-satisfied ignorance that Nadasdy, McIlwraith, and others describe (see also Pierotti, 2010, 2011, Chap. 4).

Nadasdy (2003) and many others have noted the double bind in which Indigenous people are often placed: if they change with the times, they are no longer practicing their "traditional, authentic" culture and thus are considered to no longer be truly "Indian," but if they do not change, they are "locked in blind tradition." We have also encountered this (Anderson, 1996; Pierotti, 2011). There was never a frozen "traditional, authentic" time. Cultures change all the time, often by borrowing good ideas from the neighbors. The problems arise when new ideas, such as those often coming along with firearms as opposed to bows and arrows, actually do damage culture and lead people away from respectful, reciprocity-based ways of living.

Nadasdy contrasts an extremely traditional and land-based group—the Kluane people—with an extremely urban bureaucratic world. The Anglo-Canadians who deal with the Kluane First Nation are not only ignorant of the bush; they are also trapped in Weberian bureaucratic rationalism. Wildlife biologists who know little about wildlife, except from books, try to deal with people who know everything about wildlife but have hardly read a book. These biologists may sometimes be unaware of how much actual biology has progressed, as compared to basic fisheries and wildlife management. Though the contrast he evokes is not all-pervasive or intrinsic, it is certainly real and extreme in northwest Canada (as any researcher in that area well knows). The result is predictable, and chaotic. Culture is real, and the fact that it is not necessarily destiny is easy to forget in the Yukon. But personal experience is real too, and surely the Kluane would be much more able to deal with Hedrick's ranchers than with biologists turned bureaucrats. The ranchers, some of whom work with Indigenous cowboys, would have a much easier time with the Kluane than the bureaucrats do.

Nadasdy contrasts Native and bureaucratic views of the land as "property." The Kluane were more or less nomadic. They used the land, managed it, and had a strong sense of it and of who belonged to what area, but did not have a concept of formal title and deed, either collective or individual. They, however, bore almost no resemblance to the mere aimless wanderers on a vast unowned space, as the Canadian government officially held them to be, until governmental attitudes changed very recently. Since "property" and "title," and even "rights" to land, are legal fictions rather than experience-derived knowledge, they are amenable to Nadasdy's cultural-idealist analysis. Among other things, he points out (2003: 233) that modern notions of "property" were developed partly in conscious opposition to the old (and mistaken) view of Native Americans as mere wanderers who owned and managed not at all. He is referring primarily to John Locke's famous, or infamous, line: "In the beginning, all the world was America" (Locke, 1924: 140).

More to the point, the idea that individuals can clearly own and control "private property" title is a relatively recent philosophical and economic concept, even in the Western tradition (Steinberg, 1995: 10–18). Until the Enlightenment and the establishment of the USA, land was generally held in trust, the way Indian reservations are in the USA today. Land was usually private in some sense, but subject to the will of the Crown or nobles or other authorities. Individuals (other than nobles, and even they had to answer to the king) did not usually own title to land, but were landholders, for specified periods of time. Until modern times, only in the USA did the idea of "private property" and the accompanying concept being able to do whatever you wished with land take hold as a dominant paradigm. Until the eighteenth century "property" did not necessarily include land (Steinberg, 1995: 12; it could refer to movable property as opposed to "real estate"). The USA can be considered as the first major nation to arise using Enlightenment concepts, and as such may be much less enlightened than it believes.

Much of the literature is far more extreme than Nadasdy's writings, or even Locke's. Stereotypes of "the Indian" range from "bloodthirsty savages" to Chateaubriand's and J. Fenimore Cooper's noble savages, and ultimately to Walt Disney's

Pocahontas. None of these reflect reality. Native Americans construct from their local knowledge a spiritual cosmovision that has certain logical features, as well as spirits of the animal and plant world. Even the spirits are logical in their way; humans everywhere notoriously ascribe agency to inanimate objects. Westerners had their own animistic beliefs until recently, and indeed still do in many quarters (perhaps especially when dealing with refractory cars and computers). The key to understanding this may lie in recognition that where metaphysics is concerned, Indigenous people recognize the significance of unusual or unique events, which the European tradition would characterize as anecdotes rather than data (Pierotti, 2011: 8). Where scientific knowledge is concerned, the European tradition assumes that a statistically "average" event carries more weight than the single "unusual" event. Unusual events are often attributed to the activities of "spirits" in Indigenous traditions. The English term is inadequate to translate the actual meaning of the concept (Marshall, 2005). In Indigenous knowledge traditions spirits refer to causes or events which are not seen as "supernatural." Thus, they are not outside the realm of what is real, but part of the natural order of things, readily subject to interpretation and understanding (Pierotti, 2011: 8).

Many of Hedrick's east California ranchers believe strongly in a Christianity that involves an intimate, personalistic relationship between God, humanity, and the land (Argandoña, 2012; Hedrick, 2007). Perhaps many urban Westerners have become "rational" in Weber's sense, but Westerners too have experiential and spiritual knowledge, though they may regard themselves as incapable of dealing with anything involving spirits, or even "the spirit."

Morality is constructed from experience, in all societies—indigenous or not (Bourdieu, 1977, 1990). Morals emerge and are negotiated within the family, the immediate peer group, and the local community—however far they may then be extended. Westerners usually confine morality largely to human persons, but do extend it at least to pets, and often to livestock, and even to some wild animals (at least "charismatic megafauna"). This situation arises because the most commonly used definition in contemporary American society is "a group of humans residing in the same locality and under the same government" (Pierotti, 2011: 26), which derives from Aristotelean concepts of ethics and personhood. In contrast, in cultural traditions of the Indigenous peoples of North America, distinction between social and ecological communities is not clearly delineated. Humans regularly have had social interactions and maintained social relationships with plants, animals, and features of the landscape. Thus, cultural essentialization and reification fail. Culture is not an arbitrary crust, a frozen crystalline array, or a unified, harmonious whole. It is working knowledge, shared to varying degrees, constructed into varying levels of abstraction, and—above all—moralized. It works from actual behavior to necessary working rules, and from there to abstract rules, covering both "is" and "ought."

Morality is, thus, not deduced from general principles or from its big, broad covering assumptions or laws. Quite the reverse: the latter are induced from daily practice. They are, to the extent they are general and all-covering, removed from reality. Reality tells us how to catch a fish, how to make a snare, and how to ride a horse. From this working knowledge we have to go farther and farther into speculation if we want grand generalizations. We induce or infer these from increasingly abstracted principles. The final claims—animals are persons, God is an angry god, science is objective—are the farthest from daily experience. This is where theory and models come into play, and these may feed back in practice, as in the miscounted Dall sheep, but even then the truth on the ground is more complex. Racism and economic concerns, which, although rarely recognized as such, are also theoretical concepts, had much more to do with the sheep count controversy than worldviews did.

References

Anderson, E. N. (1996). *Ecologies of the heart*. Oxford University Press.
Argandoña, M. (2012). *Every square inch: The fight for the California desert*. Ph.D. thesis. Department of Anthropology, University of California.
Basso, K. (1996). *Wisdom sits in places: Landscape and language among the Western Apache*. University of New Mexico Press.
Bourdieu, P. (1977). *Outline of a theory of practice*. Trans. R. Nice. Cambridge University Press.
Bourdieu, P. (1990). *The logic of practice*. Trans. R. Nice. Stanford University Press.
Deloria, V., Jr. (1995). *Red earth, white lies: Native Americans and the myth of scientific fact*. Harper & Row.
Deloria, V., Jr. (2002). *Evolution, creationism and other modern myths: A critical inquiry*. Fulcrum Press.
Dorit, R. (2007). Biological complexity. In R. M. May & A. McLean (Eds.), *Theoretical ecology* (3rd ed.). Oxford University Press.
Feld, S. (1982). *Sound and sentiment*. Princeton University Press.
Fogg, B. R., Howe, N., & Pierotti, R. (2015). Relationships between indigenous American peoples and wolves 1: Wolves as teachers and guides. *Journal of Ethnobiology, 3*, 262–285.
Goulet, J. A. (1998). *Ways of knowing: Experience, knowledge, and power among the Dene Tha*. University of Nebraska Press.
Hedrick, K. (2007). *Our way of life: Identity, landscape, and conflict*. Ph. D. thesis, Department of Anthropology, University of California, Riverside.
Hunn, E., & Selam, J. (1990). *Nch'i-Wana, the big river*. University of Washington Press.
Huxley, T. H. (1894). *Evolution and ethics*. Appleton & Co.
Kahneman, D. (2011). *Thinking, fast and slow*. Farrar, Straus and Giroux.
Kelly, L. (2015). *Knowledge and power in prehistoric societies: Orality, memory, and the transmission of culture*. Cambridge University Press.
Kuhn, T. (1962). *The structure of scientific revolutions*. University of Chicago Press.
Locke, J. (1924 [1690]). *Two treatises on government*. Everyman.
Marshall, J. (1995). *On behalf of the wolf and the first peoples*. University of New Mexico Press.
Marshall, J. (2005). *Walking with grandfather: Teachings from Lakota wisdom keepers*. Sounds True Press.
McIlwraith, T. F. (2007). *"But we are still native people": Talking about hunting and history in a northern Athapaskan Village*. Ph.D. dissertation, Department of Anthropology, University of New Mexico.
Nadasdy, P. (2003). *Hunters and bureaucrats*. University of British Columbia Press.
Nelson, R. K. (1983). *Make prayers to the raven*. University of Chicago Press.
Pierotti, R. (2010). Sustainability of natural populations: Lessons from indigenous knowledge. *Human Dimensions of Wildlife, 15*, 274–287.
Pierotti, R. (2011). *Indigenous knowledge, ecology, and evolutionary biology*. Routledge.

Pierotti, R. (2015). Indigenous concepts of 'Living Systems': Aristotelian 'Soul' meets Constructal Theory. *Ethnobiology Letters, 6*, 80–88.

Pierotti, R. (2016). The role of Myth in understanding nature. *Ethnobiology Letters, 7*, 6–13.

Pierotti, R. (2018). World views and the concept of traditional. *Ethnobiology Letters, 9*, 299–304.

Pierotti, R. (2020a). Historical links between ethnobiology and evolution: Conflicts and possible resolutions. *Studies in the History and Philosophy of the Biological and Biomedical Sciences, 81*, 101277. https://doi.org/10.1016/j.shpsc.2020.101277

Pierotti, R. (2020b). Learning about extraordinary beings: Native stories and real birds. *Ethnobiology Letters, 11*, 253–260.

Pierotti, R., & Fogg, B. R. (2020). Neocolonial thinking and respect for nature: Do indigenous people have different relationships with wildlife than Europeans? *Ethnobiology Letters, 11*, 48–57.

Pierotti, R., & Wildcat, D. (2000). Traditional ecological knowledge: The third alternative. *Ecological Applications, 10*, 1333–1340.

Ridington, R. (1988). *Trail to heaven: Knowledge and narrative in a northern native community*. University of Iowa Press.

Rose, D. B. (2000). *Dingo makes us human: Life and land in an aboriginal Australian culture*. Cambridge University Press.

Steinberg, T. (1995). *Slide mountain, or the folly of owning nature*. University of California Press.

Thornton, T. F. (2008). *Being and place among the Tlingit*. University of Washington Press.

Trosper, R. L. (2009). *Resilience, reciprocity, and ecological economics: Northwest coast sustainability*. Routledge.

Worster, D. (1994). *Nature's economy: A history of ecological ideas* (2nd ed.). Cambridge University Press.

Chapter 10
Respect and Its Corollaries

Respect as Basic

Respect, worldwide, is focally owed to people, based on their standing in society in relation to the respecter (e.g., Kan, 1989: 96–101), but in Native North America it is extended to all other sentient beings, and "sentient beings" can be mountains or rocks or the ocean. Above all, the Native peoples believe that one must show "respect" to animals and plants, or they will not offer themselves to be hunted or gathered successfully. The Secwépemc as described by James Teit are typical: "Respect was shown to animals to please them and to secure good luck in the chase" (Teit, 1909: 602). Bears were honored by songs and rituals. Trees, mountains, earth, and sky must also be respected. A common Tsimshian story tells of persons insulting the sky, which took revenge by causing heavy snows that failed to melt in spring; eventually the people escape to a place where summer had come (Barbeau & Beynon, 1987: 1: 255–261, three versions; Harris, 1974: 63–65, noted below). This story is widespread: in a Kathlamet version, a boy playing with and mocking his excrements is the cause, and he must be killed before the village can melt free (Boas, 1901: 216–220).

The most important management point of this truth is overhunting—taking more than one needs—or showing disrespect to the animals and the spiritual Keepers of the Game (Pierotti, 2010). Local disappearance or extirpation of game from overhunting is considered to be a consequence of disrespect. The animals are offended and leave or hide, which might well result in starvation of the disrespectful people. As we discussed in previous chapters, there is nothing mystical or magical in this; it is simply a result of cause and effect resulting from killing the wrong individual, or so many individuals that the local population on which you depend disappears. "The consequences are not the ridicule of one's peers, or the failure to get research grants, they are sickness, suffering, and death" (Alessa, 2009: 250). There is no "discount effect" in Indigenous economics because there is no profit in greed.

Even where the discourse of "respect" is eroding with assimilation and acculturation from colonizing practices, e.g., boarding schools, the practices of treating slain game animals politely and reverently still go on (McIlwraith, 2012). Disrespect to an animal can lead to its directly harming an individual, or kidnapping the individual for purposes ranging from teaching proper behavior to marriage, which involves both teaching and removal from human society.

The Koyukon, for instance, do not point at animals, or name powerful ones unnecessarily, or boast about success in hunting them. In addition, they do not usually keep them as pets, except for their sled dogs, which are given special, almost human, status, being descendants of wolves, the master predator among local animals (Nelson, 1983: 159). Like all other Indigenous groups treated in this book, they show respect to slain animals in many ways, first of which is by using all parts except those that must be spiritually or respectfully returned to the wild, such as bones (Nelson, 1983: 22–25). Every society has its own rules about this, but every society has important rules.

Menstruating women, particularly at first menses, are considered to be a threat to hunting success; they must keep away from game and be avoided by their husbands if hunting is planned. A girl who insisted on hunting and gathering during first menses turned into a monster and disappeared into the mountains (Kathlamet story; Boas, 1901: 221–224). Among the Koyukon, "women are completely forbidden to hunt, trap, or skin wolves...until they are too old to have children" (Nelson, 1983: 161). Avoidance of any sexual intercourse was necessary for hunters pursuing the highest risk game, such as whales. The Nuu-chah-nulth had a particularly complicated and difficult set of rituals associated with whaling, during which the wives of the whalers had to be inactive (Coté, 2010; Drucker, 1951; Jonaitis, 1999; Sapir, 2004).

It is particularly striking to find these beliefs still vividly and powerfully alive and motivating even among quite acculturated groups that have been subjected to Canadian colonial domination and coerced assimilation for a couple of centuries. As Thomas McIlwraith (2012) points out, these groups were supposedly assimilated more than 50 years ago, but they keep this ideology. McIlwraith cites a whole lineage of authors on this (McIlwraith, 2012: 25).

Marianne and Ron Ignace report from the Secwepemc: *x7ensqt,* "the land (and sky) will turn on you," if you don't make offerings and act respectful to all beings out there—"respect for places on the land imbued with spiritual power that derives from past events and experiences of ancestors.... The land communicates with people, and people communicate with it in song, prayer, story, and thought." Secwépemc law teaches respect, and violating it brings the *x7ensqt* down (Ignace & Ignace, 2020: 143).

Jean-Guy Goulet reports from the nearby, but linguistically unrelated, Dene Tha:

> signs of respect are faithfully given lest animals stop offering themselves to hunters.... A Dene Tha child learns this respect from his parents as they dispose properly of the bones of animals or the feathers of fowl and as they avoid talking negatively about animals (animals know how one talks about them and will not present themselves...to the hunter or trapper who speaks negatively about them (Goulet, 1998: 63).

Dene Tha must avoid eating animals related to their spirit-guardian animals (Goulet, 1998: 70), also a widespread practice. Also shared with all or almost all Northwest groups is reticence concerning revealing one's spirit animal (Goulet, 1998: 72–76).

It is easy to add to this list, e.g., from Haida sources (Anderson, 1996 and personal research; Boelscher, 1988: 176). The Dena (Alaskan Athabaskans) treated slain bears and other large predators with particular respect, and carefully cleaned up blood, hair, and the like from all kills (De Laguna, 1995: 59–61). As elsewhere, animals were once assumed to have been humanoid, and Raven or Traveler as transformer figures, established their proper behavior and the proper behavior of humans toward them. Slightly outside our region but sharing this cultural item, the Yup'ik of St. Lawrence Island thought that cutting up a not-quite-dead walrus caused the terrible winter of 1878, and disrespect of a polar bear by hunters led to the wreck of their snowmobile in more recent times (Jolles, 2002: 89, 183).

Giving away one's first kill (especially for men) or first basket of berries (for girls; to elders; Gahr, 2013: 68) is obligatory almost everywhere. There were ceremonies for the first salmon taken in the year (Gunther, 1926), and also first fruits and first roots ceremonies (Gahr, 2013: 68 for the Chinook, but every well-described group reports first salmon ceremonies or something equivalent).

Critically, respect involves not taking more than one needs for immediate consumption, including what is needed to supply family and needy neighbors. It also includes guests, especially for rituals involving necessary feasting. A large feast may mean a very large take. Still, taking too many animals is always bad and disrespectful and always subject to punishment. This clause frightens even the greedy (if they are at all traditional in their beliefs). It succeeds well in practice.

This wide concept of respect is found throughout the Northwest. It is a basic concept in Nuu-chah-nulth: ʔiisaak (Coté, 2022) or iisʔak̓ (Atleo, 2004), the word covering more than English "respect." It involves a sense of unity in difference, of interdependence, with all lives. It implies a need for mutual respect, care, awareness, and consideration. This broad concept is widely shared. The Gold Tungus of Siberia have the same basic attitudes (Arseniev, 1996), which are continuous right across Bering Strait. The Mongols share this; the word used in Mongolia is *shuteekh*, focally meaning respect for elders. *Khundelekh* is used among some Mongols (Marissa Smith, email to ENA, July 8, 2013), and in Inner Mongolia *bishirt* (Lulin Bao, pers. comm., 2014). Many other cultures, from Rumania to South America, have the same concept. The Chinese minority anthropologist Jianhua "Ayoe" Wang found respect to be the key concept for environmental management among his own people, the Akha of southern Yunnan (Wang, 2013).

The same basic stories about the person who killed excessively and was given stern warnings by the supernaturals are heard from Alaska to South America. Every Northwest Coast collection of tales records myths about this, and often actual stories of people punished for it, for example by having to give all their kill to relatives (Atleo, 2011: 50), or simply having accidents and bad luck (very widely). Overhunting was also punished by illness. "For the Heiltsuks disease was related to the larger moral state of the world.... For example, carelessness in hunting or mistreatment of game animals that resulted in their suffering inevitably led to illness and was considered a major cause thereof" (Harkin, 1997: 81). Any form of disrespect would do it; excessive take of game was one such, possibly the worst.

Sport hunting and trophy hunting are definitely out of favor, being clearly disrespectful. In fact, many indigenous peoples will not participate in recreational fishing and oppose it in general. Native people find that many fishing regulations make little sense in terms of the way they live their lives. Fishing regulations are designed to regulate greed on the part of fishermen, yet Indian people feel that the respect they hold for the fish means that they would not overexploit the resource, and are frustrated by regulations designed to solve problems for which they feel no responsibility. Catch-and-release fishing, considered a major tool for conservation by non-Indigenous sport fishermen, is regarded as "playing with the fish" because it shows no respect for the fish and the importance of its life (Nadasdy, 2003: 81–83; ENA has heard similar sentiments expressed). Deploring sport fishing as disrespectful "playing with fish" is so widely condemned that it has become the title of a book of "lessons from the north," by Robert Wolfe, who encountered this particular lesson among the Yup'ik of Alaska (Wolfe, 2006).

David Natcher et al. (2005) describe a situation in which Native people—Tutchone, in the Yukon—resisted advice from biologists to release caught fish that were small-sized or rare. The Native persons insisted that this would be disrespectful to the fish, whose spirits would be offended. Moreover, sport fishing was even more disrespectful of the fish—it was wanton destruction for fun. Needless to say, these beliefs made for major problems in communication between the Tutchone and the Anglo-Canadian biologists. McIlwraith (2007: 123) found the same opinion among the nearby Tahltan-speaking Dene of Iskut. Other conflicts between Native thinking and biologists' "scientific management" surface (Schreiber & Newell, 2006). Fisheries scientists now counsel against taking only large fish, since it selects against them, and they are usually relatively successful spawners. Taking only large fish leads to populations of small, frail fish. Once again, the Indigenous people have the successful approach to keeping the population persistent and functional because of their knowledge.

Indigenous people feel that if a fish is caught, that fish should be eaten; otherwise any suffering experienced by the fish during its capture is for no purpose. By showing respect for the sacrifice made by fish, Indigenous people reveal the true basis of their reciprocity, which, despite the arguments posed by Michael Harkin (2007) and others, is based on a specific ethical system that derives from Indigenous experience rather than the exploitative ideas that come from the Western conservationist tradition. This put many communities such as Iskut (Nadasdy, 2003) in another quandary: should they guide White sport hunters? Survival forced that issue in Iskut—they had to do it to get enough money to live—but it was a debate.

Respect includes positive valuing of other lives simply for being themselves. Nuu-chah-nulth scholar Charlotte Coté, who had a quite traditional upbringing, remembers gathering blackberries: "...we saw our grandma hunched over behind a fallen tree, picking berries.... On the other side of the tree stump was a black bear eating berries from the same vine. Trying to stay calm we inched closer to Grandma, trying not to startle the bear. When we were near enough, we said to her softly,

'Grandma, there' a bear on the other side of the stump.' My grandma replied, 'I know. I see her. Leave her alone. She loves the berries too.' And my grandma just kept on picking" (Coté, 2022: 26).

Spirit Guardians

The management rules were often enforced by Masters of the Game: magical animals, sometimes part humanoid and sometimes large and splendid specimens of their species (see also Anderson & Medina Tzuc, 2005; Martin, 1978; Pierotti, 2010). The Katzie tell of a girl shamed by her brothers for eating part of a deer's heart. She left them, became a huge deer with a human head, and remains with us as what Old Pierre called the "queen deer" (Old Pierre in Jenness, 1955: 32)—a mistress of the game. She allows hunters to kill only one or two deer at a time before punishing them, but can also give hunting power.

"Some Yukon Indians thought that there was a kind of powerful master spirit or headman for each kind of animal" (McClellan, 1987: 260). A few thought there was a Mountain Man who ruled animals and the weather (McClellan, 1987: 268). Indigenous tradition teaches that productivity in animal populations is determined and regulated by specific high-status individuals that are variously referred to as "keepers of the game," "animal masters," or those "who cannot be killed." These "keepers" are individuals that "decide" whether their species will continue to be available for exploitation by humans. The relationship between these animal populations and human hunters depends upon the maintenance of respectful relationships between humans and these "keepers." If humans are greedy and hunt to excess, the "keepers" will withdraw their species from accessibility to humans (Pierotti, 2011).

"Keepers" are typically considered by Western anthropologists to be spirits, somewhat akin to Platonic "ideals," or to be mythological constructs that serve a symbolic purpose. In contrast, we have the example of Kluane First Nations people expressing "concern over the practice of restricting hunters to shooting only full curl rams (mature rams eight years old or more). They argued that these older animals are especially important to the overall sheep population because of their role as teachers; it is from these mature rams that younger animals learn proper mating behavior ... (and) more general survival strategies" (Nadasdy, 2003: 127).

Pierotti has been told by non-Native reviewers that it would be offensive to traditional Native Americans to think of "keepers of the game" as being anything other than spirits. The quote from Nadasdy shows that, in fact, this concept appears to be based on real social dynamics, albeit ones that are lacking from traditional wildlife science. The Kluane people contend that, "Killing full curl rams has an impact on the population far in excess of the number of animals actually killed by sport hunters" (Nadasdy, 2003: 127). The same belief in the reality of leaders of the game exists among the Maya of Mexico (Anderson & Medina Tzuc, 2005). Most Indigenous Americans we have asked about this disagree with the idea that "keepers of the game" are spirits, which makes us suspect that non-Natives are very protective of the images they create for Natives, even when these images may be the result of their own imaginations.

Pierotti (2010, 2011) points out that in every wild population of animals some individuals are especially powerful or especially successful at reproducing, and this seems to lie behind the belief. Research into variation in individual quality, and reproductive performance of animals, led Pierotti to examine whether the "keepers of the game" concept might be based upon actual experience with real animals, with the stories serving as heuristic devices to teach hunters to practice restraint and caution, and to avoid taking individuals that might be responsible for a large percentage of the reproductive output of a population (Pierotti, 2011: 82–83; this also fits with Maya views).

The Nuxalk River Guardian (a title granted by a higher being at the beginning of time) enforces respect in fishing. He must make sure that no one throws refuse into the rivers, that salmon guts are deposited in the woods (as elsewhere, the bones must be returned intact to the water, so the fish will be reborn), and that other rules are followed (McIlwraith, 1948: I: 263, 664). Salmon must be kept entire, with heads and tails (McIlwraith, 1948: I: 664). When Raven learned to return the bones to the river and to keep newly caught salmon entire, he married a salmon woman, but became stingy and eventually broke the entire-head rule, whereupon his salmon wife revived all the dried salmon in the house and went off with them into the river, never to return (McIlwraith, 1948: II: 416–419, 472–481)).

James Teit records for the Tahltan:

> The Meat-Mother watches her children, the game, and also the people. When people do not follow the taboos, and do not treat animals rightly, the latter tell their mother; and she punishes the people by taking the game away for a while, or by making it wild, and then the people starve...the Moose children are the most apt to tell their mother of any disrespect shown them: therefore people have to be very careful of to how they treat moose (Teit, 1919: 231–232, as cited in McIlwraith, 2012: 68–69).

Respect must also be given to the supernatural powers (as in all cultures). In Haida Gwaii, every stream has a Creek Woman, a female spirit being who guards and protects that creek and its salmon. A man once swore at the Creek Woman of Pallant Creek, because he kept losing his hat (a hat he was not entitled to wear in any case). The woman destroyed the whole town (Enrico, 1995: 160–169). This prohibition on disrespect for spirit guardians of waters carries right across the Bering Strait, being important to the Mongols (Roux, 1984: 132ff.). Insulting supernaturals draws collective punishment, as did disrespect for God in ancient Israel.

Yup'ik (Yup'ik) people are observed by a high god, *Ellam Yua*, "person of universal awareness." In modern Alaska, belief in *Ellam Yua* has fused to some extent with worship of the Christian God, but evidently an Indigenous high being existed before contact. "*Ella* can mean 'outside,' 'weather,' 'sky, 'universe,' or 'awareness'" (Fienup-Riordan, 1994: 262). The *ellam iinga*, all-watching eye, is shown as a small circle with a dot in the center, a universal motif in Yup'ik art (p. 254 ff.), which is also shared in Siberia. The Siberian Khanty, who at least until recently practiced sustainable hunting and careful reindeer herding, have it (Wiget &

Balalaeva, 2011). So do the Koryak (Jochelson, 1908). Even the Yukaghir, who have lost much of their old beliefs and are now apparently less conservation oriented than Northwest Coast peoples, still have something of this idea (Jochelson, 1926: 144–145; Willerslev, 2007). Individual animals have their own spirit guardians among the Yukaghir (Jochelson, 1926: 145–146).

The same idea survives among the Maya (Anderson & Medina Tzuc, 2005; Llanes Pasos, 1993). Evert Thomas (pers. comm. to ENA at Society of Ethnobiology meeting, 2009) recounts that among the Bolivian Quechua, *susto* (fear sickness) in children can come from overhunting by their parents; the masters of the game punish the hunter by making his family sick.

The game masters have sometimes been replaced by higher gods, thanks to missionary teaching. Thomas Thornton's Tlingit consultant, Herman Kitka, recalls "Uncle noticed how we enjoyed the [halibut] fishing; he told us that to fish for the halibut for fun we would be wasting our food supply—to do so would offend the Holy Spirit and cause us to lose our blessing and go hungry" (Thornton, 2008: 124). He lets others use the resources at Deep Bay, of which he was designated monitor: "'I don't own it,' I tell them, 'I'm just taking care of it'" (Thornton, 2008: 168). Thornton comments on this: "This typifies the Tlingit attitude towards conservation: it is not a matter of 'resource management' as much as a matter of taking care of places." He also describes Kitka noting that when a weir blocks the river totally, the salmon leave because they are "insulted," not because they are physically stopped: "Those little sockeye get offended if you don't leave them a hole in your [fish] weir; they won't come back…" (Thornton, 2008: 173–174). Tlingit often tell guests, in formal contexts, *Tleil dagák' ahwateeni yík*, "Don't leave insulted like those little sockeyes" (Thornton, 2008: 173). We should note that this last story is, in bioscientific terms, the polar opposite of the first one. Fish avoiding an area because of habitat damage is basic ecological response. To the Tlingit, however, there is no opposition here.

Weirs themselves also need respect. They are ritually constructed and often consecrated, and thus they become persons. The great fish weir at Kepel on the Klamath River, reconstructed with major rituals every year, was an important person that had to be treated with respect (Waterman & Kroeber, 1938). Weirs need proper attention, and were given wives—at least among the Iroquois and Algonquian, in early times, though not on the Northwest Coast (Miller, 2014: 135–142).

The countless intricate observances hunters must observe before hunting are intended to show—among other things—respect for the animals. If respected properly, they will give themselves to the hunter. Stories (see Chap. 10) occur everywhere of a boy or man who mistreated or misused a piece of salmon and was abducted by the salmon people. Even criticizing a bit of mold on a dried salmon can earn this fate (Swanton, 1905: 7–14, Swanton, 1909: 300–320—Haida and Tlingit variants of the same story; Langdon, 2007; Trosper, 2009: 40–41 have commented on it). They teach him proper behavior and gratitude toward the fish, which sacrificed its life so he could be fed. Many rules of the Tsimshianic groups are incorporated in stories (e.g., Barbeau & Beynon, 1987: 1: 109–112; Miller, 1997: 24–29). In one Tsimshian story, it was the boy's uncle who mistreated the salmon;

the boy is taken and instructed, and marries the beautiful salmon girl, but eventually cheats on her and is destroyed. In one Nuxalk story, it is a girl that is disrespectful of black cod (sablefish), and is taken below and instructed (McIlwraith, 1948: 1: 351–355).

Tsimshian learning includes the point that heads and tails can be cut off, but only by using a mussel-shell knife (Miller, 1997: 29). When a man on the Nass harmed a small salmon gratuitously, his whole village was destroyed by fire (Barbeau, 1950: 77). Salmon remains are usually returned to the rivers, insuring good nutrition for the young fish, but among the Gitksan (and many other groups) they are burned, and failure to burn every leftover causes trouble (Barbeau, 1950: 165–176; People of 'Ksan, 1980: 30). The Quinault burned the heart of the salmon, since eating it would stop the runs; burying a fish or its heart would also stop runs. The Quinault replaced the bones in the river, as did most groups (Olson, 1936: 34–35).

One common, widespread story tells of a man who retained or lost one of the small skull bones of a salmon. He was confronted by a chief whose human-appearing son had terrible head pains and disability, because of lack of the relevant bone. The unaware eater returns the bone, and the child is healed. Old Pierre of the Katzie told a particularly dramatic version of this story to Diamond Jenness (1955: 19). This is combined with the bone-burning story in the Skidegate Haida version; the child is healed when inadvertently overlooked parts of his salmon self are burned (Swanton, 1905: 7–9).

In another version, poor management of fisheries leads to a chief's son being taken by the salmon to the depths of the sea, where they are humanoid and live in carved houses, each species with its own house. He is taught proper behavior: eat all that is edible, burn the rest carefully, drink fresh water after eating, and do not close the whole river with nets (Boas, 1916: 192–206, part of a long and complex myth; comparisons with other stories on p. 770). Other versions of this widespread story tell of a boy insulting salmon (Cruikshank, 1979: 104–110 for northern Athapaskans) or wasting salmon meat.

William Beynon provided several versions of a long, brilliant, tightly organized version of this story (Barbeau, 1953: 338–367). Salmon that are caught must be eaten soon, and among the Gitksan anything not eaten must be burned (as opposed to the usual returning to the water). Respectful treatment causes the salmon to be reincarnated. Violation of this rule leads to famine; successful following of it leads to the young man in question being taken undersea to the village where the salmon put off their disguise and appear as human, or humanoid. He learns of the way they willingly sacrifice themselves for humans who treat them well (Cove, 1987: 53–64). This is a universal theme on the northern Northwest Coast (e.g., Old Pierre's stories in Jenness, 1955: 19–20). It extends far into Siberia (e.g., an east Siberian story of respecting salmon in Dolitsky, 2002: 26–27).

More common and interesting is the instruction to return all bones (and often other waste) to the water (e.g., Adamson, 2009: 90 for Salish; Boas, 1998: 76 and McIlwraith, 1948: 2: 416–419, 472–481 for Bella Coola; Boas & Hunt, 1905: 390–392 for Kwakwaka'wakw, but several groups, including the Tillamook and Chinook, did *not* do this). As noted above, it restores necessary nutrients that are

Spirit Guardians

211

generated by adult carcasses decomposing in spawning streams. The nutrients not only enter the streams, but are often spread by bears or flooding many meters into the surrounding forest, helping generate a suitable habitat (Bilby et al., 1996; Bartz & Naiman, 2005) The Tututni had a rite in which:

> [a] religious leader offered thanks to the fish for returning once again...the villagers sent out their private thoughts of gratitude. This giving of respect to the salmon—and in other settings to deer and elk, roots and berries, and other gifts from the lands and waters...was not some romantic construct. The Tututnis saw themselves as...citizens along with the plants and animals, and it was proper to show appreciation (Wilkinson, 2010: 22–23).

The Katzie were taught by one of their creator-transformer heroes, whose daughter was the ancestor of the sockeye salmon,

> At a certain time of the year all her relatives shall visit you. You may eat them, but of the first ones you catch you must throw back into the water the bones, the skin, and the intestines. Then their souls will return hither and take on new bodies (Old Pierre in Jenness, 1955: 35).

South Wind, the Tillamook Salishan trickster and transformer hero, learned his rules the hard way. He caught no fish, day after day, till it was explained to him that he would have to cover any fish he caught with green branches and leaves and then burn the bones and skin in a fire that is to be extinguished. Water from washing up must go on that fire, not in the bush (Deur & Thompson, 2008: 38). Similar stories come from the Chinook (Boyd, 2013).

Most (probably all) groups gave distinctive ritual treatment of the first salmon caught in the season. George Hunt recorded a particularly detailed rite among the "Kwakiutl" (Kwakwaka'wakw) for the first dog salmon. The fisherman's wife is the ritual keeper. She prays: "O Supernatural Ones! O, Swimmers! I thank you that you are willing to come to us. Don't let your coming be bad, for you come to be food for us. Therefore, I beg you to protect me and the one who takes mercy on me, that we may not die without cause, Swimmers!" After the salmon is consumed, the leftovers are returned to the water (Boas, 1921: 609). Another prayer to the first salmon (species unspecified) ran:

> Welcome, Swimmer: I thank you, because I am still alive at this season when you come back to our good place; for the reason why you come is that we may play together with my fishing tackle, Swimmer. Now, go home and tell your friends that you had good luck on account of your coming here, and that they shall come with their wealth bringer, that I may get some of your wealth, Swimmer; and also take away my sickness, friend, supernatural one, Swimmer (Boas, 1921: 1319).

Fishermen all had their own variations of these prayers. Similar prayers (not recorded) were given over other first fish. A long series of prayers for catching halibut involves prayers to the halibut hook—it had to be large, and possess its own power—as well as to the halibut (Boas, 1921: 1320–1327). Once again, thanks to the meticulousness of Boas and Hunt, we have detailed records of women's beliefs and behaviors; this is only one of many charms recorded from them. The record is unique among early ethnographies. When ENA was a student, Boas was regularly excoriated in the literature for wasting time recording and publishing such material. It is a measure of the progress of the field that women's knowledge is now valued along with men's. Such was not always the case.

Respecting salmon, for the Twana of Puget Sound, involves keeping the river clean for them. From late summer, before the main fall run, "no rubbish, food scraps or the like, might be thrown in the river; canoes were not bailed out in the river; and no women swam in the river during menstrual seclusion....the 'salmon people' were beings with supernatural powers who had to be treated with respect and according to correct procedure" (Elmendorf, 1960: 62–63). First salmon rites showed respect for the fish, and could be complex. The first salmon of the year went to the children. A young man had to catch a large salmon early in the season, and eat it. Bones were returned to the water; similarly, the first elk of the year had to be eaten entirely (Elmendorf, 1960).

Hunting was more strictly sex-segregated, with women barred, and men barred if they had had sexual intercourse or any contact with a menstruating woman; they also had to be physically clean, not overly well fed (in other cultures, actually fasting), and in touch with guardian spirits via spirit dancing. A woman had to remain quiet while her husband was hunting (Elmendorf, 1960: 85). The man had to have spirit power for hunting, and the spirits of the animals had to cooperate by voluntarily giving themselves up. It is always useful to recall that in pre-gun times the Indigenous hunters got as close to the animal as they could, since the bows had limited range, and visibility was often limited (Elmendorf, 1960: 92). This made stealth, quiet, and care essential. Moreover, at all times, people had to be relatively calm and respectful around game animals, to prevent them from becoming "wild."

Insulting fish cost *Wegyet*, the Gitksan trickster figure, his beautiful hair (Kitanmax School of Northwest Coast Indian Art, 1977: 22). Insulting the lowly "bullhead" (sculpin) led to destruction of the villains by a whirlpool (Boas, 1916: 285–292). Respect for the bullhead continues today (Anderson, 1996: 62). It is treated with respect by the Nuxalk, who free sculpins from nets, and it is reported to return the favor by saving the lives of drowning fishermen, expanding its size from minute to huge for the purpose (McIlwraith, 1948: 1: 74, noting also that toads can help humans).

Rules regulating menstruating and pregnant women were often complex, and related to respect for animals as well as to health and good fortune of the women. Gitksan women in such condition were not allowed to have fresh meat; eating it would spoil the hunting (People of 'Ksan, 1980: 45–46). Similar rules are universal in most of the Northwest, and there is no need to list them all; suffice it to say that they all involve forbidding some protein foods. Fresh meats can spoil and produce bacteria that cause miscarriage, so there is probably a functional subtext here, but the main motives were to teach young girls respect (in their first menstrual seclusion) and to preserve luck for everyone. Nuxalk women are not allowed to bathe in the river during their periods, "lest a speck of blood should blind the fish.... Occasionally an olachen with red eyes is caught; this is assumed to be caused by some heedless woman" (McIlwraith, 1948: I: 263).

Respect is also a fundamental value basic to Plateau societies. It "included not only other human beings but also food, air, water, the plants and animals whose lives were taken, and all other aspects of nature.... The very acts of making or teaching art were highly regarded" and artists were respected for their special skills (Loeb & Lavadour, 1998: 79).

Rodney Frey, writing about the educational function of Native American folktales, uses as examples the moral injunctions to hunters: "'Nothing will they throw away,' 'you must not be proud,' 'you must not kill too many of any kind of animal'" (Frey, 1995: 173). These are his prime examples not of Native conservation, but of Native moral discourse; he is not singling out game management to talk about. It was simply what came first to hand. Note that *hybris* is condemned as well as waste. Respect means proper humility before one's animal spirit beings, as well as intelligent management. For one thing, it means not mentioning the name of whatever one is hunting or fishing for, or otherwise being overconfident about one's expedition (see, e.g., Palmer, 2005). It also means killing the animal cleanly, with one shot (an important mark of respect for the game among Mexican Maya also).

"Respect," however, goes much farther than this. One does not speak ill of animals, or insult them. Saying anything disparaging about a fresh kill is particularly serious (see, e.g., McIlwraith, 2007: 115). Animals must be killed as cleanly and painlessly as possible. McIlwraith was told quite graphically, to be sure a fish was dead before gutting it: "'How would you like to be gutted if you were still alive?'" (McIlwraith, 2007: 117; ENA remembers being told similar things by Anglo-American fishermen in his childhood). As noted above, fresh water is offered to a newly killed animal (especially sea mammals, along the coast). Various ceremonies, some of them extremely elaborate, must be performed. A large animal may be decorated before being brought home. This practice extends right across the Bering Strait into Siberia, where it is widespread. Permission must be asked, and thanks given, even to small bushes and herbs for use of their valued parts (e.g., Turner, 2014–2: 316–320, and for ceremonial recognition 326–328, 337–342).

One is also not supposed to boast of success. Hunters and fishers return saying they have taken nothing much. (This also is shared widely; Willerslev (2007: 37) reports it for the Yukaghir, and ENA can testify that it is a rule in Finland.) Boasting brings ridicule and censure, but also can actually drive the animals away, since it is disrespectful of them. Anyone with much experience in the field will understand these and related concepts. Most individuals with field experience in these matters have seen more than one proud White hunter or field naturalist humbled, by failing to find the animal they were "sure" to get, to the loud amusement of their friends. One need not believe in spirits to see the value of humility in the bush. Among other things, humbler hunters may not be as obvious, and can approach more closely without being noticed.

The rigors of hunting must be undergone, partly to show the animals that the hunter really wants and needs them. Sometimes a mountain can withhold them. Near Iskut is a hill called Stingy Mountain, because the mountain goats there are notoriously good at escaping hunters. An outsider would probably find that the mountain affords good lookout and listening posts for the wary goats, but the local people say the mountain is deliberately refusing to let them be hunted (McIlwraith, 2007: 116). Different personal guardian spirits can give very different abilities in hunting. At least among the Yukaghir, one's own protective spirit can be stingy (Willerslev, 2007: 43).

Frey (2001: 9) tells of a Coeur d'Alene woman who resisted a pest control worker's call to get rid of spiders. As she said, "'...that Spider might have had a message for me, something to say to me. Why would I want to kill him?'" Two pages later, we read of a man who wanted to rob a muskrat's store of food, but was prevented by his wife (Frey, 2001: 11). The Coeur d'Alene pray extensively before hunting or gathering wild foods, and also give thanks, often by leaving some tobacco where they have gathered (Frey, 2001: 180–182). He quotes a wonderful oration about respect for deer who have given their lives to the hunters as food (p. 205).

An extreme (and possibly a bit self-serving) bit of rhetoric on the Northwest Coast is a Nuu-chah-nulth speech to a just-harpooned whale:

> Whale, I have given you what you wish to get—my good harpoon. And now you have it. Please hold it with your strong hands. Do not let go. Whale, turn towards the fine beach...and you will be proud to see the young men come down...to see you; and the young men will say to one another: What a great whale he is! What a fat whale he is! What a strong whale he is! And you, whale, will be proud of all that you hear them say of your greatness (Coté, 2010: 34, quoting text recorded and translated by T. T. Waterman).

Significantly, the Nuu-chah-nulth abstained totally from the very active whale fishery initiated by Anglo-Canadian settlers on Vancouver Island (Coté, 2010: 64). Apparently, they saw that the whales were not treated with respect, the meat, for instance, being wasted.

Similar beliefs extend to Mexico. A Nahuatl fish-trapping charm recorded around 1600 says, in part:

> I have come to lay for you your oriole arbor, your oriole fence, within which you will be happy, within which you will rejoice, within which you will seek all kinds of food, the flowery food...My elder sister the Lady of Our Sustenance...spread out for you your oriole mat, your oriole seat...she is awaiting you with her atole drink... (Ruiz de Alarcón, 1982: 166–167, modified).

The oriole arbor is a fish trap, and this is a charm to lure fish into it. Apparently, the idea of promising much to a flighty water creature is widespread. (Orioles, stunningly lovely birds, are a standard trope for beauty among the Nahuatl.)

The Yup'ik of southwest Alaska have enormously elaborate rules (Fienup-Riordan, 1994, 2005) for paying respects to mammals and fish, even the small, poor-quality freshwater blackfish (valuable because once a winter staple for dogs as well as a backup food for humans). Fish see traps and nets differently according to the behavior of the fishers; the fish want to sacrifice themselves to those that have treated their previous incarnations with respect (see Fienup-Riordan, 1994: 123). For instance, if the blackfish are taken with an ordinary dip net, instead of a special device reserved for them, "the blackfish will become extinct and be depleted" (Clara Aagartak, quoted Fienup-Riordan, 1994: 119). Most of this long book consists of similar rules. The great Greenlander ethnographer Knut Rasmussen found equally elaborate rules in the central Arctic (Rasmussen, 1931, 1932; Bown, 2015).

Seals, like salmon, are sometimes thought to live in a giant ceremonial men's house under the sea. Here they arrange themselves by social rank, as do humans. They discuss "whether to give themselves to human hunters" (Fienup-Riordan, 2005: xvi), giving themselves only to those that show respect. When seals are caught and shared out, the seals' *yuit* (persons; sing. *yua*) go into their bladders, so these are saved and restored to the sea at a major festival at the start of winter (the sealing season and the ceremonial season; the Yup'ik have an extremely rich and evocative ceremonial life, beautifully described by Fienup-Riordan). Here too, bones had to be ritually disposed of; seal bones had to go into fresh water (1994: 107), but salmon bones were not returned to rivers. "Yup'ik cosmology is a perpetual cycling between birth and rebirth, humans and animals, and the living and the dead" (1994: 355), and between land and sea as well. Seals, like other animals and major fish, have immortal souls that keep reincarnating in bodies available to grateful, respectful humans. In other groups, hunters who disrespected seals might be taken to the undersea home of the seals and taught better manners (e.g., Andrade, 1969: 155–165 for the Quileute; also a tradition with gray seals (selkies) in the British Isles).

Generosity to other humans is rewarded by generosity from nature; generous hunters are successful hunters (Fienup-Riordan, 2005: 38; a belief that seems to be region-wide, e.g., noted among Interior Salish by M. B. Ignace, 1998: 208, with stinginess causing bad luck). Other social rules too lead to good fortune if followed and bad fortune if violated.

In general, there was a strong separation between sea creatures and land ones, but they were interlocked by trails and processes. As an example, scientists in Alaska could not understand why belukha whales, *Delphinapterus leucas*, did not enter certain river systems while they freely entered other rivers (Huntington & Myrmin, 1996). Local Indigenous hunters attributed the avoidance of certain river systems by belukha to the presence of beavers, *Castor canadensis*. When the Western scientists expressed confusion at this interpretation, the Yupiaq explained that beavers built dams, which prevent salmon from reaching spawning areas in those rivers, whereas in rivers where beavers were absent, salmon were abundant. The belukha were drawn to salmon, not to rivers. As a consequence, they avoid rivers where beavers and their dams are present (Pierotti, 2011: 76–77). This is a classic example of how the Indigenous Yup'ik saw the world in terms of relationships, and readily understood the links between these three species. This is also a classic example of the compartmentalization of Western scientific knowledge (Nadasdy, 2003: 123), and where the Indigenous way of seeing the world had a ready explanation based on connections for a phenomenon that perplexed the specialized thinking of Western science.

Sea creatures wanted land amenities (hence the seal bones in fresh water); land creatures wanted sea goods (see 1994: 140ff.). This fits well with Claude Lévi-Strauss' analysis of the separation of land and sea, and its mediation by travel and myth, among the nearby Tlingit and other groups (Lévi-Strauss, 1958, 1995).

As in most of the Arctic and much of Asia, there were elaborate rules for dividing up large animals and giving particular parts to particular categories of relatives. Any violation of rules led to the animals going away. The Tłı̨chǫ Dene, far to the northeast of the Northwest Coast, have very similar concepts. Blood of a slain animal must be cleaned up, butchering must be done properly, and there are many rules for such matters. "The Tłįchǫ elders attribute the disappearance of wildlife to industrial development and to a lack of respect, which inevitably causes destruction" (Legat, 2012: 30). On one hunting trip, caribou did not show themselves, so the hunt leader made the people eat all the meat they had brought and made Dr. Legat send away her leather backpack—whereupon the caribou immediately appeared. Since there was meat and hide in the camp, the caribou had not felt *needed* (Legat, 2012: 84).

Among the Spokan, as reported by John Alan Ross: "After killing a deer, the hunter exercised considerable care in butchering...and skinning...for fear that an act of disrespect would impair his future hunting success, or worse, keep the offended species away from the hunting range of his group" (Ross, 2011: 256). Ross also records several other respectful practices, including slitting the throat of a slain game animal to release its life force (p. 257). He also watched as a youngster borrowed his great-grandfather's traps and went out to trap coyotes without thinking to carry out the necessary ritual; the elder expected both of them would now be kept awake at night by the laughter of the coyotes, who, of course, would laugh at the trapper rather than approaching the traps (Ross, 2011: 290).

A touching example of respect and care is the universal custom of leaving something for rodents when taking resources they use. People often find and steal the caches of seeds, nuts, and roots that rodents, e.g., white-footed mice (*Peromyscus*) and wood rats (*Neotoma*), make, but they always leave enough to keep the animals from starving. Ross (2011: 249) reports that, also, cattail leaves were left for squirrels for nesting material. This has some self-interest to recommend it—without the mice, rats, or squirrels, no more caches—but less practical is the widespread tendency to leave something valuable to humans also. The Tanana leave bits of cloth (see accounts, Guédon, 1974: 29). The Itelmen of Kamchatka also do this, as described by Georg Steller in the eighteenth century. They do not kill the mouse, they leave some roots for it, and they supply some "old rags, broken needles, fireweed, cow parsnips, pine nuts..." in trade (Steller, 2003: 65).

Respecting Plants

Respect also extends to plants, including realizing that plants may produce berries for people and feel disrespected if these are not picked, although one suspects that bears and birds might address this issue, even if humans do not. Susie Sampson Peter, superb storyteller and tradition-keeper of the Skagit people, said: "I *wonder what* the berries are thinking.... 'Is somebody going to come and pick me? Or am I just going to just fall? Am I just going to get ripe for nothing, and nobody's going to come and enjoy me'...I'm ripe. No one, none of you, are going to come and pick me?'" Ms. Peter picked berries until she was very elderly and blind (Peter, 1995: xiv).

Plants also need respect, including forbearance by the picker on eating berries until the picker has returned to share them with others (McIlwraith, 1948: I: 691). Annie Miner Peterson, last speaker of the Miluk Coos language of the Oregon coast, after commenting on the need for a girl to share her first berries, and a boy his first game, without eating them themselves, continued with advice: "Annie said that picking flowers and gathering shells and rocks were also prohibited. 'It might start to rain or bring bad luck.' She said, 'Things were just to be there, not to be picked.' It was also bad luck to catch wild things" (Youst, 1997: 68). This appears to be a widespread belief. Elsewhere in Oregon, the Tillamook told a story of two Younger Wild Women making the trees and berry bushes what they are today, with their appearances and uses. They asked the plants if they (the wild women) looked good in their face paint, and rewarded those who said yes, punished those who said no—a nice reversal, with the plants disrespecting the humans (or humanoids) rather than the other way round (Jacobs & Jacobs, 1959: 148–150; the original Wild Woman was a major character in earliest mythic times; the younger ones were transformers from a later mythic era).

From the Spokan, once more: when berrying, shouting "to a distant partner...was considered as being disrespectful to the plants," and would cause them to withhold berries. Recent widows could not pick (Ross, 2011: 352). Ceremonies, especially for the very valuable huckleberries, paid respect to the berry plants. Roots were the subject of ceremonies, including one that recognized a young girl when she got her first digging stick and another when she dug her first root—which was, of course, given away (Ross, 2011: 335; or the root might be worn as a charm). Eugene Hunn (Hunn & Selam, 1990) reports comparable Sahaptin rituals and practices. In a much more local—apparently strictly Spokan—tradition, wild rose was ritually very important, used to drive away evil spirits and influences, cleanse after a death, and the like; brushes of the plants were used, and water in which flowers or hips were soaked (Ross, 2011: 351 and passim).

Trees also warrant respect. Traditional people still pray to them, and apologize for hurting them by taking bark. Several such prayers were recorded more than a century ago from the Kwakwaka'wakw by George Hunt, again preserving women's lore. A woman taking bark from a cedar would say:

> Look at me, friend! I come to ask for your dress, for you have come to take pity on us; for there is nothing to which you cannot be used, because it is your way that there is nothing for which we cannot use you, for you are really willing to give us your dress. I come to beg you for this, long-life maker, for I am going to make a basket for lily roots out of you. I pray you, friend, not to feel angry with me on account of what I am going to do to you; and I beg you, friend, to tell our friends about what I ask of you. Take care, friend! Keep sickness away from me, so that I may not be killed by sickness or in war, O friend! (Boas, 1921: 619).

Another such request was an apology for burning berry bushes to make them yield well:

> I have come Supernatural Ones, you, Long-Life-Makers, that I may take you, for that is the reason why you have come, brought by your creator, that you may come and satisfy me; you Supernatural Ones; and this, that you do not blame me for what I do to you when I set fire to you the way it is done by my root (ancestor) who set fire to you in his manner when you get

old on the ground that you may bear much fruit... (here as cited in Turner & Peacock, 2005: 127).

A male counterpart comes when a canoe-maker cutting a cedar would throw four chips behind it, saying to the first, "O supernatural one! Now follow your supernatural power!" Then, with another chip, he would say "O, friend! Now you see your leader, who says that you shall turn your head and fall there also." Similar charms go to the next two. The tree will then fall where the canoe-maker wants it to fall—after careful cutting, of course (Boas, 1921: 617–618). (Anyone long in the rural Northwest Coast learns how to make a tree fall where it is supposed to.) This passage is followed by a long account by George Hunt concerning weather magic, and then of life-cycle rites, which often involve ceremonial relationships with cedar, salmon, and other powers.

Even stones and rocks get due respect, especially when made into tools. Arrow and spear points, adzes, and other tools become persons. As usual, this belief extends widely. The K'iche' Maya of Guatemala do not allow men to touch the metates that the women use to grind maize, because for a man to touch it "shows the stone no respect" (Searcy, 2011: 93). Mongol respect for stones has been noted above. Small cairns of piled stones are made at passes and other key points to protect travelers, a belief that is absolutely worldwide, from the Northwest to Peru, traditional Hawai'i, Celtic Europe, and, again, Mongolia, where these *ovoo* or *obo* continue to be made and decorated with animal skulls and sacred sky-blue cloths.

Conflicts with Respect

However, respect can be anti-conservation. Respect includes taking an animal that offers itself to you. This is another belief shared all around the subarctic, from the Cree (Brightman, 1993) to the Yukaghir (Willerslev, 2007: 35). Recall Nadasdy's observation (2003) that local Indigenous people ran afoul of conservation authorities by keeping undersized fish because throwing back a fish that offered itself would be "disrespectful." Similarly, Old Pierre, the Katzie Salish authority who gives us many of the best accounts of traditional conservation ethics, records that advice to keep only the biggest fish from a huge haul was countered by pointing out that the master of salmon would be offended by throwing back what he had given.

Such beliefs go down to South America. In Paraguay, Mario Blaser studied a situation in which a Native American game management program ran into heavy water because the local people, the Yshiro, were willing to conserve, but had to take peccaries that offered themselves, thus severely discomfiting the biologists (Blaser, 2009; there was special concern for the rare and local Flat-headed Peccary, *Coregonus* sp.).

If the animals' souls go on to reincarnate, there is little recognition of the finite quality of game. A Cree from eastern Canada told Harvey Feit (2006): "The animals are still there, they just do not want to be caught." Long experience with hunters and fishers of various races teaches the authors of this book that almost all of them say this, whether or not they believe in reincarnation of animal souls. Fishermen in particular almost never admit the fish are actually getting scarce. The fish are always out there, "just not biting today." Sometimes they secretly know the truth, and will admit that over a drink, but usually are genuinely in denial; such at least is the experience of many who study fisheries development (see, e.g., McEvoy, 1986).

The offering-and-reincarnating belief as charter for overhunting has occurred among the Yukaghir. They once had the same basic beliefs about animals as the Northwest Coast peoples, but are now Russianized enough to overhunt game seriously (Willerslev, 2007: 30–35). The rapid decline in game numbers is met with the worldwide refrain: "They've just gone elsewhere for a while." This is in some ways the assumed response of an insulted species or population, i.e., that the Keeper would remove the species from human access for some undefined period. This is what in fact has happened, revealing again that First Nations knew what would happen, regardless of who it was that committed the insult, or showed the lack of respect. Even the Yukaghir have traces of a respect-by-not-overhunting belief. Willerslev expresses very well a tension clearly found all over the Siberian and Native American world: "In fact, it would be fair to say that we find two polar tendencies in Yukaghir subsistence practices: One in the direction of overpredation..., the other in the direction of limiting one's killings to an absolute minimum..." (Willerslev, 2007: 49). The Yukaghir have rather different reasons for this from Native Americans, but the basic ideas are the same. The Yukaghir's neighbors, including the Mongols, Tuvans (Kenin-Lopsan, 1997), Altaians, and Tungus, are far more conservationist in their ideology and behavior, but face some of the same tension because all have been impacted by Russia and its Westernized way of thinking. This was the basis of a classic paper by Pierotti and Daniel Wildcat, "The Third Alternative," which argued that even though Marxist and Capitalist concepts were in conflict, both assumed they were in charge of the natural world and could exploit it as much as they wanted (Pierotti & Wildcat, 1999). This is also the experience of ENA and his Chinese minority student Jianhua Wang in China (Wang, 2013).

The Yukaghir must have once had a stronger restraint on hunting. They still show the pattern of proper respect for a slain game animal, for instance (Jochelson, 1926: 147, and see below, Chap. 9). They would long ago have hunted themselves into suicidal starvation without it. Apparently, Soviet cultural oppression (chronicled in harrowing detail by Willerslev) ended it, in a similar fashion to the way cultural traditions were crushed by colonial ignorance, and the assumption that European knowledge was automatically superior, a situation we still faced in North America in 2021. Willerslev cites Soviet ethnography claiming that massive overhunting is aboriginal and traditional in the Arctic, but this is so obviously tendentious as to be ridiculous. Like Americans, European Russians are quite capable of dishonesty and racist attitudes. The Siberian Arctic has been settled for centuries by Russians, and their devastating cultural impact on hunting beliefs was felt by the early eighteenth century. Georg Steller's observations in Kamchatka in the 1740s record massive overhunting by Russians, including the rapid extinction of Steller's sea cow,

Hydrodaailis gigas, a species that had probably existed for millions of years in the Sea of Okhotsk, and was wiped out within two decades of its discovery by Europeans, to the sorrow of local Itel'men (Steller, 1988, 2003).

The Northwest Coast First Nations of old preferred to err on the side of caution. The literature suggests that overhunting seems much more common in Siberia and east Canada. Again, this is probably due to outside influence, but there is a difference of opinion on this. Problems with overhunting, justified by reincarnation, are reported for eastern Canada by Robert Brightman (1993), Shepard Krech (1999), and Calvin Martin (1978). Of this troika, Brightman and Krech appear to the present authors to be cultural supremacists. Calvin Martin tried to argue that irresponsible behavior by Europeans devastated Indigenous cultures, through introduction of both disease and capitalist approaches to fur-bearing animals and resource management, and the Indigenous people lost their faith in the environment and thus their drive to conserve it. Not surprisingly, of the three, Martin had the least successful career in academia; he was simply ahead of his time. Krech built his career on a volume that was critical of Martin's way of thinking (Krech, 1981). In contrast, Martin continued to engage Indigenous scholars and publish on Indigenous ways of understanding (Martin, 1993, 1999), before abandoning academia. (For detailed critiques of Krech's best seller, *The Ecological Indian*, see Pierotti, 2011, Chap. 8; Anderson, 2000; Trosper, 2009: 25–39.)

Most sources seem to maintain that eastern Canadian hunters maintain respect and hunt with some care (Barsh, 2000; Berkes, 2008; Tanner, 1979), but unquestionably there has been overhunting in the past, especially for the fur trade, where the America constructed by beavers, over many millennia, was systematically destroyed through excessive trapping to supply the fashion industry in Europe (Martin, 1978). Overhunting by Native people does occur in the Northwest, but experience of many researchers suggests that overhunting is far less common than it would be otherwise, and definitely less common than among local whites. The Coast Salish traditions, for instance, hold that a family could take only one grouse per year (Krohn & Segrest, 2010: 97), a tight limit.

If an animal must be taken when it offers itself, problems also arise with moose conservation (the same problem appears with the Yukaghir). McIlwraith (2007) reports that the people of Iskut refrain from killing cow moose when they possibly can, especially if the cow is likely to be pregnant; so, if a cow just stands there when a hunter comes up on her, the hunter is in a terrible quandary. The ban on killing female game animals in breeding season (and moose pregnancies last almost a year) is widespread if not often emphasized in the literature (see, e.g., Miller, 1999: 98 for the Salish), and is another part of the "respect" rule that says "don't waste."

This dilemma has caused confusion for both Indigenous people and anthropologists. The indigenous people must always decide whether respect in the form of not taking too much, or respect in the form of taking what is offered, prevails in a given situation. Anthropologists argue about whether the "natives conserve" or not, depending largely on whether they have observed people take what is offered or go with the conserving option, or whether they assume that First Nations should act as the functional equivalents of white environmentalists (Krech, 1999; Pierotti,

2011). Individual hunters differ greatly in this regard. The general pattern is likely to be what Anderson found both in the Northwest Coast and among the Maya: greedier people go with taking what is offered, the more thoughtful ones go with conservation, and all pay some attention to how much they need for food. This is certainly true of Indigenous fishing in the contemporary Northwest. A few Indigenous people overfish mercilessly, while most maintain traditional self-limits (our personal research).

Critically, however, respect also involves the point mentioned above: not taking more than one needs for immediate consumption includes community needs. It also includes guests, especially for rituals involving necessary feasting. A large feast may mean a very large take. Still, taking too many animals is always bad and disrespectful and always is subject to punishment. This clause frightens even the greedy (if they are at all traditional in their beliefs).

Significantly, only wild game is thought of this way. Domestic animals get less respect because their lives are not their own, but under human control (Pierotti, 2011, Chap. 2). Teaching stories turn on the issue that some people are greedier than others. The whole point of the stories is that greedy people who overhunt endanger their entire communities and cultural traditions, which were designed to prevent such actions. This is a perfectly hardheaded observation. It is maintained by belief in animal spirits to whom humans are connected and related, but it is founded on fact. We are not speaking of Krech's "ecologically noble savage," which was always a fragment of European imaginations, but of people who are quite aware that personality and situation will make some people greedy or desperate.

References

Adamson, T. (2009). *Folk-tales of the Coast Salish*. (New edn. with introduction by W. R. Seaburg and L. Sercombe; original, New York: American Folklore Society, Memoir 27, 1934). University of Nebraska Press.
Alessa, L. (2009). What is truth? Where Western science and traditional knowledge converge. In M. Williams (Ed.), *The Alaska native reader: History, culture, politics* (pp. 246–251). Duke University Press.
Anderson, E. N. (1996). *Ecologies of the heart*. Oxford University Press.
Anderson, E. N. (2000). Review of *The ecological Indian* by Shepard Krech. *Journal of Ethnobiology, 20*, 37–42.
Anderson, E. N., & Medina Tzuc, F. (2005). *Animals and the Maya in Southeast Mexico*. University of Arizona Press.
Andrade, M. (1969). *Quileute texts*. (Orig. Columbia University Press, 1931). AMS Press.
Arseniev, V. K. (1996). *Dersu the Trapper (Dersu Uzala)*. Trans. Malcolm Burr. (Russian original, early 20th century). McPherson & Company
Atleo, E. R. (2004). *Tsawalk: A Nuu-chah-nulth worldview*. University of British Columbia Press.
Atleo, E. R. (2011). *Principles of Tsawalk: An indigenous approach to global crisis*. University of British Columbia Press.
Barbeau, M. (1950). *Totem poles*. National Museum of Canada.
Barbeau, M. (1953). *Haida myths illustrated with Argillite carvings*. National Museum of Canada, Bulletin 127.

Barbeau, M., & Beynon, W. (1987). *Tsimshian narratives*. Canadian Museum of Civilization, Mercury Series, Paper 3.

Barsh, R. L. (2000). Taking indigenous science seriously. In S. A. Bocking (Ed.), *Biodiversity in Canada: Ecology, ideas, and action* (pp. 152–173). Broadview Press.

Bartz, K. K., & Naiman, R. J. (2005). Effects of Salmon-borne nutrients on riparian soils and vegetation in Southwest Alaska. *Ecosystems, 8*, 529–545.

Berkes, F. (2008). *Sacred ecology* (2nd ed.). Routledge.

Bilby, R. E., Fransen, B. R., & Bisson, B. A. (1996). Incorporation of nitrogen and carbon from spawning Coho Salmon into the trophic system of small streams: Evidence from stable isotopes. *Canadian Journal of Fisheries and Aquatic Sciences, 53*, 164–173.

Blaser, M. (2009). The threat of the Yrmo: The political ontology of a sustainable hunting program. *American Anthropologist, 111*, 10–20.

Boas, F. (1901). *Kathlamet texts*. Smithsonian Institution, Bureau of American Ethnology, Bulletin 26.

Boas, F. (1916). *Tsimshian mythology* (pp. 27–1037). Smithsonian Institution, Bureau of American Ethnology, Annual Report for 1910.

Boas, F. (1921). *Ethnology of the Kwakiutl* (Vol. 2). United States Government, Bureau of American Ethnology, annual report for 1913-1914.

Boas, F. (1998). *Franz Boas with the Inuit of Baffin Island, 1883-1884: Journals and letters*. Trans. by W. Barr. University of Toronto Press.

Boas, F., & Hunt, G. (1905). *Kwakiutl texts, part 2*. American Museum of Natural History, Memoir V.

Boelscher, M. (1988). *The curtain within: Haida social and mythical discourse*. University of British Columbia Press.

Bown, S. R. (2015). *White Eskimo: Knud Rasmussen's fearless journey into the heart of the Arctic*. Da Capo Press.

Boyd, R. T. (2013). Lower Columbia Chinookan ceremonialism. In R. T. Boyd, K. M. Ames, & T. A. Johnson (Eds.), *Chinookan peoples of the lower Columbia* (pp. 181–198). University of Washington Press.

Brightman, R. (1993). *Grateful prey: Rock Cree human-animal relationships*. University of California Press.

Coté, C. (2010). *Spirits of our whaling ancestors: Revitalizing Makah and Nuu-chah-nulth traditions*. University of Washington Press.

Coté, C. (2022). *A drum in one hand, a Sockeye in the other: Stories of indigenous food sovereignty from the northwest coast*. University of Washington Press.

Cove, J. (1987). *Shattered images: Dialogues and meditations on Tsimshian narratives*. Carleton University Press.

Cruikshank, J. (1979). *Athapaskan women: Lives and legends*. National Museums of Canada, National Museum of Man, Mercury Series, No. 57.

De Laguna, F. (1995). *Tales from the Dena: Indian stories from the Tanana, Koyukuk, and Yukon Rivers*. University of Washington Press.

Deur, D., & Thompson, M. T. (2008). South Wind's journeys: A Tillamook epic reconstructed from several sources. In M. T. Thompson & S. Egesdal (Eds.), *Salish myths and legends: One people's stories* (pp. 2–59). University of Nebraska Press.

Dolitsky, A. B. (2002). *Ancient tales of Kamchatka*. Trans. by Henry N. Michzael. Alaska-Siberia Research Center.

Drucker, P. (1951). *The northern and central Nootkan tribes*. Smithsonian Institution, Bureau of American Ethnology, Bulletin 144.

Elmendorf, W. W. (1960). *The structure of Twana culture, with comparative notes on the structure of Yurok culture [by] A. L. Kroeber*. Research Studies, Washington State University, Monographic Supplement 2, part 1.

Enrico, J. (1995). *Skidegate Haida myths and histories*. Queen Charlotte Islands Museum.

References

Feit, H. (2006). "The animals are still there, they just do not want to be caught": Exploring nonmodern conservation and caring in James Bay Cree spirituality and hunting. In *Paper, American Anthropological Association, Annual Conference.*

Fienup-Riordan, A. (1994). *Boundaries and passages: Rule and ritual in Yup'ik Eskimo Oral tradition.* University of Oklahoma Press.

Fienup-Riordan, A. (2005). *Wise words of the Yup'ik people: We talk to you because we love you.* University of Nebraska Press.

Frey, R. (1995). *Stories that make the world: Oral literature of the Indian peoples of the inland northwest.* University of Oklahoma Press.

Frey, R. (2001). *Landscape traveled by Coyote and Crane: The world of the Schitsu'umsh (Coeur d'Alene) Indians.* University of Washington Press.

Gahr, D. A. T. (2013). Ethnobiology: Nonfishing subsistence and production. In R. T. Boyd, K. M. Ames, & T. A. Johnson (Eds.), *Chinookan peoples of the lower Columbia* (pp. 63–79). University of Washington Press.

Goulet, J. A. (1998). *Ways of knowing: Experience, knowledge, and power among the Dene Tha.* University of Nebraska Press.

Guédon, M.-F. (1974). *People of Tetlin, why are you singing?* National Museum of Man, Mercury Series, Ethnology Division, #9.

Gunther, E. (1926). An analysis of the first Salmon ceremony. *American Anthropologist, 28*, 605–617.

Harkin, M. E. (1997). *The Heiltsuks: Dialogues of culture and history on the Northwest Coast.* University of Nebraska Press.

Harkin, M. E. (2007). Swallowing wealth: Northwest coast beliefs and ecological practices. In M. E. Harkin & D. R. Lewis (Eds.), *Native Americans and the environment: Perspectives on the ecological Indian* (pp. 211–232). University of Nebraska Press.

Harris, K. B. (1974). *Visitors who never left.* University of British Columbia Press.

Hunn, E., & Selam, J. (1990). *Nch'i-Wana, the big river.* University of Washington Press.

Huntington, H. P., & Myrmin, N. I. (1996). Traditional ecological knowledge of beluga whales: An indigenous knowledge pilot project in the Chukchi and northern Bering seas. In *Paper, Inuit Circumpolar Conference.*

Ignace, M. B. (1998). Shuswap. In D. E. Walker (Ed.), *Handbook of North American Indians: Plateau* (Vol. 12, pp. 203–219). Smithsonian Institution.

Ignace, M., & Ignace, R. (2020). A place called Pípsell: An indigenous cultural keystone place, mining, and Secwépemc law. In N. J. Turner (Ed.), *Plants, people and places: The roles of ethnobotany and ethnoecology in indigenous peoples' land rights in Canada and beyond* (pp. 131–150). McGill-Queen's University Press.

Jacobs, E. D., & Jacobs, M. (1959). *Nehalem Tillamook Tales.* University of Oregon Books.

Jenness, D. (1955). *The faith of a coast Salish Indian.* British Columbia Provincial Museum [now Royal British Columbia Museum], Memoir 3.

Jochelson, W. (1908). *The Koryak.* Memoir of the American Museum of Natural History, X, Reports of the Jesup North Pacific Expedition, VI.

Jochelson, W. (1926). *The Yukaghir and the Yukaghirized Tungus.* Memoir of the American Museum of Natural History, XIII, Reports of the Jesup North Pacific Expedition, IX.

Jolles, C. Z. (2002). *Faith, food and family in a Yup'ik whaling community.* University of Washington Press.

Jonaitis, A. (1999). *The Yuquot whalers' shrine.* University of Washington Press.

Kan, S. (1989). *Symbolic immortality: The Tlingit potlatch of the nineteenth century.* Smithsonian Institution Press.

Kenin-Lopsan, M. B. (1997). *Shamanic songs and myths of Tuva.* Akadémiai Kiadó.

Kitanmax School of Northwest Coast Indian Art. (1977). *We-gyet wanders on: Legends of the northwest.* Hancock House.

Krech, S., III. (1981). *Indians, animals, and the fur trade: A critique of keepers of the game.* University of Georgia Press.

Krech, S. (1999). *The ecological Indian: Myth and reality*. W. W. Norton.
Krohn, E., & Segrest, V. (2010). *Feeding the people: Revitalizing northwest coastal Indian food culture*. Northwest Indian College.
Langdon, S. (2007). Sustaining a relationship: Inquiry into the emergence of a logic of engagement with Salmon among the Southern Tlingits. In M. E. Markina & D. R. Lewis (Eds.), *Native Americans and the environment: Perspectives on the ecological Indian* (pp. 233–273). University of Nebraska Press.
Legat, A. (2012). *Walking the land, feeding the fire: Knowledge and Stewardship among the Tłįchǫ Dene*. University of Arizona Press.
Lévi-Strauss, C. (1958). *Anthropologie structurale*. Plon.
Lévi-Strauss, C. (1995). *The story of Lynx*. Trans. by C. Tihanyi. University of Chicago Press.
Llanes Pasos, E. (1993). *Cuentos de cazadores*. Government of Quintana Roo.
Loeb, B., & Lavadour, M. W. O. (1998). Transmontane beading: A statement of respect. In S. E. Harless (Ed.), *Native arts of the Columbia plateau: The Doris Swayze bounds collection* (pp. 71–86). High Desert Museum and University of Washington Press.
Martin, C. (1978). *Keepers of the game*. University of California Press.
Martin, C. (1993). *The American Indian and the problem of history*. Oxford University Press.
Martin, C. (1999). *The way of the human being*. Yale University Press.
McClellan, C. (1987). *Part of the land, part of the water: A history of the Yukon Indians*. Douglas & MacIntyre.
McEvoy, A. F. (1986). *The Fisherman's problem: Ecology and law in the California fisheries, 1850-1980*. Cambridge University Press.
McIlwraith, T. F. (1948). *The Bella Coola Indians* (Vol. 2). University of Toronto Press.
McIlwraith, T. F. (2007). *"But we are still native people": Talking about hunting and history in a northern Athapaskan Village*. Ph.D. dissertation, Department of Anthropology, University of New Mexico.
McIlwraith, T. F. (2012). *"We are still Didene:" stories of hunting and history from northern British Columbia*. University of Toronto Press.
Miller, J. (1997). *Tsimshian culture: A light through the ages*. University of Nebraska Press.
Miller, J. (1999). *Lushootseed culture and the shamanic odyssey: An anchored radiance*. University of Nebraska Press.
Miller, J. (2014). Rescues, rants, and researches: A review of Jay Miller's writings on Northwest Indien cultures. In D. C. Stapp & K. N. Powers (Eds.). Northwest Anthropology, Memoir 9.
Nadasdy, P. (2003). *Hunters and bureaucrats*. University of British Columbia Press.
Natcher, D. C., Davis, S., & Hickey, C. G. (2005). Co-management: Managing relationships, not resources. *Human Organization, 64*, 240–250.
Nelson, R. K. (1983). *Make prayers to the raven*. University of Chicago Press.
Olson, R. L. (1936). *The Quinault Indians*. University of Washington Press. University of Washington Publications in Anthropology VI:I.
Palmer, A. D. (2005). *Maps of experience: The anchoring of land to story in Secwepemc discourse*. University of Toronto Press.
People of 'Ksan. (1980). *Gathering what the great nature provided*. University of Washington Press.
Peter, S. S. (1995). *Gwəqwulćaʔ*. Lushootseed Press.
Pierotti, R. (2010). Sustainability of natural populations: Lessons from indigenous knowledge. *Human Dimensions of Wildlife, 15*, 274–287.
Pierotti, R. (2011). *Indigenous knowledge, ecology, and evolutionary biology*. Routledge.
Pierotti, R., & Wildcat, D. R. (1999). Traditional knowledge, culturally-based world-views and Western science. In D. Posey (Ed.), *Cultural and spiritual values of biodiversity* (pp. 192–199). United Nations Environment Programme.
Rasmussen, K. (1931). *The Netsilik Eskimos* (Vol. 8). Gyldendal Bog Handel, Nordisk Forlag. Reports of the Fifth Thule Expedition.

References

Rasmussen, K. (1932). *Intellectual culture of the copper Eskimos* (p. 9). Gyldendal Boghandel, Nordisk Forlag, Reports of the Fifth Thule Expedition, vol.

Ross, J. A. (2011). *The Spokan Indians*. Michael J. Ross.

Roux, J.-P. (1984). *Religion des Turcs et des Mongoles*. Payot.

Ruiz de Alarcón, H. (1982). *Aztec sorcerers in seventeenth century Mexico: The treatise on superstitions*. State University of New York, Institute for Mesoamerican Studies, Publication 7.

Sapir, E. (2004). *The whaling Indians: West coast legends and stories, legendary hunters*. Canadian Museum of Civilization.

Schreiber, D., & Newell, D. (2006). Negotiating TEK in BC Salmon farming: Learning from each other or managing tradition and eliminating contention? *BC Studies, 150*, 79–102.

Searcy, M. T. (2011). *The life-giving stone: Ethnoarchaeology of Maya Metates*. University of Arizona Press.

Steller, G. (1988). *Journal of a voyage with Bering, 1741-1742*. (M. A. Engel & O. W. Frost, ed. and trans.). Stanford University Press.

Steller, G. (2003). *Steller's history of Kamchatka*. M. Engel & K. Willmore (ed. and trans. German original 1774). University of Alaska Press.

Swanton, J. R. (1905). *Haida texts and myths: Skidegate dialect*. Bureau of American Ethnology, Bulletin 29.

Swanton, J. R. (1909). *Tlingit myths and texts*. Bureau of American Ethnology, Bulletin 39.

Tanner, A. (1979). *Bringing home animals*. St. Martin's Press.

Teit, J. (1909). *The Shuswap*. Memoir of the American Museum of Natural History, The Jesup North acific Expedition, Vol. II, part VII.

Teit, J. (1919). Tahltan tales. *Journal of American Folk-Lore, 32*, 198–250.

Thornton, T. F. (2008). *Being and place among the Tlingit*. U niversity of Washington Press.

Trosper, R. L. (2009). *Resilience, reciprocity, and ecological economics: Northwest coast sustainability*. Routledge.

Turner, N. J. (2014). *Ancient pathways, ancestral knowledge: Ethnobotany and ecological wisdom of indigenous peoples of northwestern North America* (Vol. 2). McGill-Queen's University Press.

Turner, N., & Peacock, S. (2005). Solving the perennial paradox: Ethnobotanical evidence for plant resource management on the northwest coast. In D. Deur & N. Turner (Eds.), *Keeping it living: Traditions of plant use and cultivation on the northwest coast of North America* (pp. 101–150). University of Washington Press.

Wang, J. (2013). *Sacred and contested landscapes*. Ph.D. dissertation, Department of Anthropology, University of California.

Waterman, T. T., & Kroeber, A. L. (1938). *The Kepel fish dam* (Vol. 35, p. 6). University of California Publications in American Archaeology and Ethnology.

Wiget, A., & Balalaeva, O. (2011). *Khanty: People of the taiga*. University of Alaska Press.

Wilkinson, C. (2010). *The people are dancing again: The history of the Siletz tribe of Western Oregon*. University of Washington Press.

Willerslev, R. (2007). *Soul hunters: Hunting, animism, and personhood among the Siberian Yukaghirs*. University of California Press.

Wolfe, R. (2006). *Playing with fish and other lessons from the north*. University of Arizona Press.

Youst, L. (1997). *She's tricky like coyote: Annie Miner Peterson, an Oregon coast Indian woman*. University of Oklahoma Press.

Chapter 11
Teachings and Stories

Stories in Contexts

In the Northwest, teaching and discussing was most usually done via stories—highly personal, specific, geographically grounded narratives (Attla, 1983, 1989, 1990; Cove, 1987; Cruikshank, 1979, 1983, 1998, 2005; Frey, 1995; Goulet, 1998; Legat, 2012; Mandeville, 2009; Pettitt, 1950). In fact, this is a worldwide phenomenon. Talking about anything so grounded as hunting and practical everyday conservation in personal experience and local knowledge simply cannot be reasonably done any other way.

On the Northwest Coast, as we have seen, stories were owned by families and descent groups. They provided origin records and histories of these kinship affiliations. Isaiah Wilner (2018) has provided a moving account of Franz Boas' discovery of this link—discovering that stories were not just folktales (as in Germany) but were intimately involved in the personhood and social lives of people. They were danced with masks, sung, acted out, and, of course, narrated.

Education was gender- and kin-structured. Among the matrilineal Gitksan, for instance, much education was expected to come from the father's side. This was partly due to gender roles (a father would teach his son hunting), but partly to the desire to keep the father's side involved in the ordinary and ritual life of the child. Many privileges, ritual, and ceremony came from the mother's side, since the child would succeed to those, not to his or her father's privileges. Conversely, in patrilineal societies, a mother and her kin would teach the child both women's work and their own family stories and privileges.

Other education processes involve meetings of the whole community, which may include long, ritualized speeches and ceremonial rhetoric in general, as well as storytelling. The Okanagan anthropologist Jeannette Armstrong records her nation's political process known as En'owkin (Armstrong, 2005a), basically a council meeting of a type familiar over most of Native North America. Everyone gets together and talks out issues, with elders—not simply older people, but rather citizens with special knowledge or political ability—addressing the issues. Sharing and community are addressed as major values in such meetings.

Native American oral literature includes enormous amounts of material that is great verbal art by any standards. Northwest Coast collections of myths and tales, often recorded in the original languages, fill dozens of large books. Unfortunately, what we have is only a fraction of what was known. Most of the treasure trove was never recorded, and most of the rest molders unread and unappreciated in ancient volumes and manuscripts stored in academic libraries. This material was recorded at a time when it was fast disappearing, and ethnographers rushed to get it down before it was forgotten. They had no time for analyses or for literary renderings. Popular renderings of traditional narratives are too often bowdlerized, shortened, and badly translated.

Fortunately, a number of recent books provide easier and better-explained access to some of this material. A brief period of serious attention to worldwide Indigenous oral literature appeared in the 1970s, marked by the short but intense life of the journal *Alcheringa*, edited by Jerome Rothenberg. The writings of Dell Hymes (1981, 2003), Robin Ridington (1988), and others echo that time of serious poetic rendering and explanation of texts. The tradition continues, sporadically, with the work of scholars such as Crisca Bierwart, Rodney Frey, and Thomas Thornton. Nora and Richard Dauenhauer have been especially productive, with superbly written and analyzed Tlingit stories (Nora Dauenhauer is herself Tlingit; Dauenhauer & Dauenhauer, 1987, 1990, Dauenhauer et al., 2008). John Enrico (1995) has translated or retranslated material collected by John Swanton over a century ago and made them available in Haida Gwaii; this sort of reanalysis and republication needs to be done much more often. The time is long overdue for lovers of great literature, and especially for Native American people, to republish—hopefully in better-analyzed form, as recommended by Hymes—serious collections of this material. A famous analysis of Northwest Coast literature, *The Content and Style of an Oral Literature* by Jacobs & Jacobs (1959), has been supplemented by later analyses by Dell Hymes and William Seaburg (Hymes, 1981, 2003; Hymes & Seaburg, 2013). Claude Lévi-Strauss (1982, 1995) also analyzed Northwest Coast myths. Such analyses and reanalysis are very rare.

Northwest Coast storytellers are famous (Frey & Hymes, 1998; Miller, 2014). Folklore scholar Rodney Frey reports that he observed Mari Watters, Nez Perce storyteller, hear a twenty-minute dramatic Sioux presentation that was new to her, and promptly reproduce it with her own added touches (Frey, 1995: 152). In the more complex coastal societies, experts on tales were not just any elder person; stories were associated with high status and with particular descent groups. One had to have permission of the group to tell them. An unwritten but very real copyright law existed, and still exists today. The Tsimshian ethnographer William Beynon was a high chief (Halpin, 1978; Roth, 2008: 78), and this gave him power and possession over many stories (MacDonald & Cove, 1987).

Nancy Turner has compiled dozens of stories relating to plant management (2014: 2: 231–296). Crisca Bierwert, a story collector herself, has analyzed sensitively the problems of collecting and publishing Native traditions, and the variations, multiple versions, and sometimes downright corruptions that occur (Bierwert, 1999). Even the long and detailed stories collected in the late nineteenth and early twentieth centuries were condensed and sometimes partially forgotten, and early recorders were not perfect (Cove, 1987).

Lacking are analyses of the conservation and resource management stories that are scattered thickly throughout Native American oral literatures. Many recent scholars, notably Ronald and Marianne Ignace and Nancy Turner, have drawn on this material, but there is no comprehensive study of it. Failing a complete study and analysis, the following materials are presented.

Individuals might know many books worth of stories. Beynon, his uncle Henry Tate, the adopted-Kwakiutl (in Boas' term), George Hunt, and the Nuu-chah-nulth Native ethnographer Tom Sayach'apis (working with Edward Sapir) all dictated or wrote book-length materials. Chief Kenneth Harris of the Gitksan put together the origin stories of his people into a single epic narrative (Harris, 1974). A long book of superb stories of animals and divinities has come from notes taken by early ethnographers on the narrations of Coquelle Thompson, an Oregon Coast Athabaskan, and this is only part of his record; more is unpublished (Seaburg, 2007). John Swanton's records of Haida tales include long stories, adding up to book-length materials from several consultants. The Nez Perce anthropologist Archie Phinney (1934) recorded a long book of superb stories, many running to several pages, from his mother; this is a unique collection in that neither language nor culture were barriers. Phinney could simply write down his mother's dictation in their native speech, just as told. This saves us from the distortions of mishearing, uncomfortable tale-tellers, and above all after-the-fact editing, which is massive in many of the collections.

Sadly, few early scholars cared about Native stories, and fewer still had the talents of Beynon, Tate, Hunt, Phinney, Sapir, and Swanton as recorders. Much more work has been done recently, but the stories are now shorter, simpler, and less pervaded with wild and strange magic and supernatural powers.

Songs also abound, but are very poorly documented. Many of these taught both environmental knowledge and the moral and aesthetic sensitivity that people felt in the presence of the wide world. A detailed study of one Tlingit song reveals its situation in place, its preservation by the family of the singer, and its association with old-time and subsequent interactions with the place of its composition in the early twentieth century (Thornton et al., 2019); unfortunately, it does not come with a recording. Countless songs on the Northwest Coast have similar histories.

One Tsimshian storyteller told Beynon that "what he had just told would normally take a full day to narrate" (Cove, 1987: 44). The quite long and detailed version Beynon recorded was nowhere close to that. Sadly, we have no records of these day-long sessions, which must have rivaled *The Iliad* in length and detail. The problems of transcribing texts in the Indigenous languages, and getting them translated, kept stories short—to our cost. Dell Hymes pointed out that early recorders missed or lost much of the poetry and narrative art (Hymes, 1981, 2003), but modern ethnographers like Rodney Frey (1995, 2001) are highly sensitive to these aesthetic concerns. Films can now capture some of the vocal styles, gestures, and other elaborations that are essential to Native American storytelling but are not transcribed in texts.

These stories get around—sometimes around the world. Franz Boas (2006 [1895]: 660) traced many elements all over North America and even to Asia. Some, like the Flood, Swan Maiden, and Orpheus stories (see Chap. 15), are almost worldwide, which is unsurprising. Death and fears connected with it are universal. Floods occur everywhere; stories may go back to the melting of continental ice sheets and subsequent flooding and sea level rise (see Nunn, 2018, for Australian examples), and countless local floods have occurred since.

Less connected with experience, but strangely and marvelously evocative everywhere, is the Swan Maiden story. It or similar transformation tales occur almost everywhere. A Haida version of the Swan Maiden story collected by John Swanton from Walter McGregor (Swanton, 1905: 264–268), in which a man marries two geese, is strikingly similar to the Irish tale of Oengus (see Jackson, 1971: 93–97 for a particularly early and moving version of that story), with the difference that instead of humanizing the geese, the man joins the geese and becomes one himself. An interesting touch is that in this and other Haida stories, a mouse—really the powerful Mouse Woman—is helped by the hero and later helps him at critical times; this remains a standard teaching about respecting animals in Haida culture and throughout the Northwest. The great poet Gary Snyder (1979) wrote a study of McGregor's story, with comparisons to world literature, in his early days as an anthropology student.

Swanton collected a very similar version from Tlingit sources (Swanton, 1909: 55–57), with the geese specifically identified as brants. In a marvelous collection of tales from the Thompson Indians of south-central British Columbia (Hanna & Henry, 1995), Annie York, a Nlaka'pamux elder and major authority on Interior traditions, told a highly creative version of the Swan Maiden myth that harmoniously blends European and Native elements of this story. This is one of many reminders in the collections that stories do not stand still. They change, incorporate new materials, drive new morals, and adapt to particular situations and audiences.

Local stories capture truths. Recent analysis showed that myths of a vast serpent under the earth coincide perfectly with the location of the giant east-west fault that endangers Seattle (Krajick, 2005). The local Salish had obviously observed enough earthquakes to know where to expect them, and they shared the worldwide, and thoroughly common-sense, belief that earthquakes are caused by a monster in the earth shaking himself. This belief extends to the Athapaskan tribes of the interior (Sharp, 1987). It also exists in China (ENA, personal research), where people know that dramatic scarp faces are earthquake-prone locations, and hold that a dragon in the earth is responsible. It is worth noting that the scientific explanation for earthquakes has emerged only since the 1960s. Before that, the dragon was still as good an explanation as anything dominant in academic geology.

All Northwest Coast groups that have been studied report teachings about conservation and wise management. These include cautionary tales, outright instruction, prayers, and other oral literature (see, e.g., Ignace & Ignace, 2017: 203–210). Indigenous tale-tellers would often include applications and explanations of their stories in their narration, especially if the stories were in older and arcane language, and ethnographers sometimes included these in their records. The very early recordings—those by Boas, Swanton, Teit, and other early investigators—contain many more conservation and respect stories than later collections. It appears that such stories are easily lost, which probably explains their scarcity in Oregon and western Washington records, since recording was later and settler influence earlier there.

The number of conservation stories in a group's legacy, however, is subject to great variation. The southern groups—people south of central Washington—appear less conservation-minded than the northern ones, but conservation comes back into the fore in California and probably in southern Oregon (where our knowledge is tragically poor, thanks to the Rogue River War and other genocidal acts). Archie Phinney's comprehensive collection of his mother's Nez Perce stories contains nothing specific about conservation, except a short version of the Coyote-breaking-the-dam tale (Phinney, 1934: 26–29, 380–381), though most of Ms. Phinney's narratives turn on morality toward and between animals in some way. The same is true for Deward Walker's Nez Perce collection (1994). Early collections from southwest Washington and western Oregon are also rather short of conservation. Coquelle Thompson's southwest Oregon repertoire had almost nothing, for instance (Ramsey, 1977; Seaburg, 2007). This, however, may be due to cultural loss. Thompson was narrating of a world long destroyed. At the extreme other end, dozens of explicitly conservationist stories are recorded from the Tsimshianic groups.

Translation and heavy revision have cost us much. As Pierotti has written in another context, discussing the Colville/Okanagan author Christine Quintasket, writing under her Indigenous name of Mourning Dove or Cogawea:

> Translation is a particular issue in the case of Mourning Dove, who is known to be one of the most heavily edited of Native American authors... Her works were changed considerably by her editor, L. V. McWhorter, ... "at times scholars will not know if they are reading McWhorter or Mourning Dove...unless they are very familiar with the primary documents" ...McWhorter enlisted a close friend, Yakima newspaperman Heister Dean Guie (1896-1978), to help with the shaping of Mourning Dove's traditional stories. Guie's wife, Geraldine (1897-1994), happened to be one of the first graduates of the University of Washington's anthropology program, and her methods may have influenced many of the book's editorial decisions. Because Guie envisioned the collection as a series of children's bedtime stories, all mentions of sex and violence were eliminated, and most of the legends were simplified and shortened. Mourning Dove and McWhorter, who remained active in the editing process, removed moral points, "superstitions," and creation stories that might bring ridicule from a white audience. When *Coyote Stories* was published in 1933, it included editing credits to Guie and McWhorter...Many... stories as published were unrecognizable to the Colville-Okanagan elders who originally told them (Nisbet & Nisbet, 2010; Pierotti & Fogg, 2017: 168–169).

Overhunting

As the previous chapter showed, throughout the Northwest Coast, stories tell of people who were disrespectful to fish, mountain goats, and other animals, often by overhunting them. The prey animals move away, leaving people to starve, which puts teeth into the obligations required of ceremonies and rituals based on showing respect and care. Native American storytellers and ethnographers often stress these stories, and non-Native anthropologists have learned to recognize their importance as moral conservation teachings. "A common theme in Heiltsuk mythology is that of the (potential) chief who goes out from a starving village to encounter a supernatural creature. The creature gives him the power that enables him to obtain a wealth of food with which to feed his people" (Harkin, 1997: 164).

This theme is not confined to the Heiltsuk. Almost always, the starvation results from disrespecting animals, and the potential chief is a child who has resisted the temptation to do that. Ice, the uncouth slob of Nehalem Tillamook storytellers, often disrespects other lives—usually by being too greedy—and suffers for it, leaving his people to be rescued by Raven or by young people who learn better (Jacobs & Jacobs, 1959: 3–42). The morals are clear.

Many geological features remind children of such matters. Typical is a story McIlwraith (2007: 146) brings us from Iskut: A hunter protested against a goat refusing to be caught, and the hunter and his dog were promptly turned to stone. The rock is evidently one of those vaguely human-shaped rocks that people everywhere love to tell stories about. It is pointed out to children as a cautionary tale. A Tahltan version of the story, recorded by James Teit, involves a hunter turned to stone with his dog for hunting too many mountain goats (Teit, 1919: 241–242). The same story is told by the Siberian Yukaghir, where the goat is a different species (Jochelson, 1926: 156). There are similar cautionary rocks everywhere in the Northwest, and for that matter throughout the world. Alternatively, a mountain may heave up to frighten people out of overhunting goats (McIlwraith, 1948: 1: 93).

Franz Boas' Tsimshian recorder Henry Tate sent him a mythic story of a man who overhunted porcupines to the point of local extirpation. The porcupines caught him, and sang to him that he had to say their leader's name or be struck by their quilled tails. He failed to guess the name and his face was thoroughly struck by the exceedingly painful and hard-to-remove quills, till Mouse Woman (ever the small helper on the northern Northwest Coast) told him the right name. After that, the porcupines removed the quills with the unpleasant aid of porcupine excrement and urine, warned him never to overhunt, and sent him home (Boas, 1916: 108–110; Tate, 1993: 13–21). This story is evidently related to one from the Tagish of a man who insulted a porcupine and was trapped in a porcupine lair for two months, to be saved—once again—by the mouse (McClellan, 1975: 153). In a connected Tsimshian story, overhunting by humans caused the animals to declare permanent winter, which may be a memory of the last glaciation event. The porcupine allowed a compromise by creating the four seasons (Boas, 1916: 106–108).

The porcupine story is followed in Tate's book by one of a man so overcommitted to hunting raccoons that his wife turned herself into a beaver, or allowed herself to become one (Tate, 1993: 22–29). After this comes the widely known story of the Bear Mother. The latter features a girl out berrying who steps accidentally in bear dung, angrily insults bears, and is captured by them and made the wife of one. Various adventures then occur, some tragic. Every large collection of northern Northwest Coast stories contains a version of this, and often other versions substituting other animals. (e.g., Barbeau, 1953: 108–146; Boas, 2006: 470–472, 581–589; Tate, 1993: 30–41, from Boas, 1916: 278–284; Cruikshank, 1979: 118–122; Davis & Saunders, 1980: 155–157; Harris, 1974: 83–95; McClellan, 1987: 261–264; McIlwraith, 1948: I: 678–682; Naziel & Naziel, 1978: 14; Storie, 1973: 30–35; Swanton, 1905: 336, 1909: 126 and 252, followed by a variant involving an octopus, 130–133, and later ones with other creatures, including a snail, 175; a frog, 236–237; a small fish, 237–238; and even a fire, 239–240).

Tate climaxes his set with his version of what can only be called the national epic of the Tsimshianic groups, also told by the neighboring Wetsu'weten: the story of the goats of Dumlakam (also spelled Dimlahamid, Temlaxam, or Temlaham) who punished a village of greedy hunters, and established self-conscious conservation as a firm rule (Boas, 1916: 131–135; Glavin, 1998; Harris, 1974; McClellan, 1970). The village was destroyed by an avalanche, except for one man (in this version) who had been more respectful of the animals. A huge one-horned goat appears in many versions of the story.

Tate's nephew William Beynon's Temlaxam story concerns a young hunter who killed all of a herd of mountain goats except one, painted it red, and turned it loose. Soon after, "two men came to the village, wearing white blankets"—mountain goat fur, made by magic to look as if spun. They lured the village to a feast, where a huge goat created a rockslide that killed them all, except the hunter, who was saved by the goat he had released (Barbeau & Beynon, 1987, vol. 1: 246–247). Other versions (ibid.: 248–253) name the village site, which is still known, and the rockslide is well known also, pointed out to visitors with the instruction that Temlaxam lies under it to remind people not to kill too much. This story is still very much alive among all Tsimshianic groups, and taken very seriously as a charter for proper relations with the nonhuman world.

The climax of developing this story, and of recorded conservation stories in general, is Chief Kenneth Harris' book-length version (1974) of the Gitksan creation cycle. Many classic Gitksan stories are grouped together. Playing with a ball made from a bear stomach causes disasters to the village (pp. 26–37). Overhunting goats later leads to the main punishment. One hunter refrains and paints a small goat's face red to recognize it. He is later taken into the realm of the mountain goats. A one-horned goat appears and dances. A strange young man with painted face turns out to be the kid who had been painted and spared, who then guides the young hunter back toward the human world, gives him goat powers, and tells him the village has been destroyed in punishment (pp. 45–51). This story is followed by a flood myth, and then another punishment tale: the village is punished for playing with trout by making headdresses of them. The *medeek*, a huge bearlike monster, appears and kills them (pp. 56–60), giving later Gitksan both an emblem and a warning. Later, people mock the sky, which takes revenge by snowing them in; eventually the few who were less insulting escape from a permanent snowfall to a world where summer still exists (63–65; recall earlier note on this tale). Starving, they are saved by the one-horned goat, appearing now as savior (66–68). The last story involving animal respect is the familiar Bear Mother story (83–95).

A happier outcome is found among the Nlaka'pamux (Thompson): the hunter listens to advice. A hunter of mountain goats was taken in by them in their humanoid form. (The reasons they took him in are unclear in this version, but presumably the full form would have the plot of the hunter who has helped a goat and been rewarded.) He learned their ways, including being able to have sex only during their rutting season. He marries and has a son. He returned to teach his people to respect them: "When you kill goats, treat their bodies respectfully, for they are people. Do not shoot the female goats, for they are your wives and will bear your children. Do not kill kids, for they may be your offspring. Only shoot your brothers-in-law, the male goats. Do not be sorry when you kill them, for they do not die but return home. The flesh and skin (the goat part) remain in your possession; but their real selves (the human part) lives just as before, when it was covered with goat's flesh and skin" (Teit, 1912: 258–262, quoted Lévi-Strauss, 1995: 70; more or less the same story comes from the Secwepemc; Boas, 2006: 76–77). Claude Lévi-Strauss (1995) discusses this myth at length, showing its relationship to other myths of kinship to the animal people.

Related are Kwakwaka'wakw stories of men who got special hunting charms or powers, abused them (evidently by overhunting goats), and were in one case turned into a grizzly bear and in another into a wolverine (Boas & Hunt, 1905: 7–25, 36–45). In another a mountain falls on the goat hunter and kills him (Boas, 2006: 373–374). A slightly happier outcome involved overhunting goats; however, the youngest of five brothers was more moral, and survived to get supernatural powers (Boas, 1910: 5–15). Nlaka'pamux related stories include Coyote's overuse of magical means to get food, and consequent loss (Hanna & Henry, 1995: 49–55).

The Tlingit share the general plot; in one story a village has been killing too many animals and disrespecting their remains, so a man is captured by the mountain sheep (probably goats are meant; they are described as bearded) and instructed in proper behavior. He tells his village and it survives (Swanton, 1909: 58–61).

Another Nlaka'pamux version of this story concerns deer, not mountain goats. A man overhunts deer, is seduced by a deer woman, marries her, and learns that if he hunts deer but uses every edible part of them, puts all the bones into the water, and otherwise treats them with respect, the deer will immediately be reborn. Or he can burn the bones; then the deer "really die, but they will not find fault with you." He is told that other less respectful treatment will cause the deer not to revive, explaining failures of hunting. He moves back and forth between deer and human worlds, eventually stabilizing as a deer (Boas, 1917: 40–43, from James Teit's recording).

Often in such stories, there is one villager who does not overhunt, and is saved from disaster (the famine caused by overhunting) by the animals, and taught better behavior. In a Tsimshian story (Boas, 1902: 211–216), a boy refrains from wanton killing of squirrels, is told to burn their skins and dried meat so that they can be reincarnated, and rewarded with shamanic powers. Another version, or related story, recounts that mistreating a squirrel led to a horrific back-and-forth war in which squirrels and humans died, leaving only one of each in the group (Boas, 1916: 345, 2006: 595). Versions of this story were told to Boas by Mathias and by Henry Tate. It appears that the Tsimshianic peoples are particularly aware of conservation, as documented by the recent work of the Tsimshian anthropologist Charles Menzies (2006, 2016). The Kwakwaka'wakw have many of the same stories (Boas, 1910: 9–15; see also Frey & Hymes, 1998, esp.: 585–587). These stories blend into a vast number of stories and story cycles of individuals acquiring power, from hunting success to shamanic ability. There are dozens of such stories in the great collections (e.g., Barbeau & Beynon, 1987; Elmendorf, 1960).

Many stories include incidental comments about saving a few animals to reproduce and maintain the population (e.g., Mandeville, 2009: 214). Sometimes these few are referred to as "seeds." The general motif occurs with strikingly little variance from Alaska to the Popoluca of far southern Mexico, who tell it about a man who hunted armadillos (Foster, 1945).

Rodney Frey (1995) gives several versions of stories about overhunting and its punishment. Since his book is about Native storytelling, not about animals or conservation, this represents a fair sampling of how important the story is. The purest form he provides is a Wishram story, originally published by Edward Sapir and beautifully re-edited by Frey, of a man who got elk power, took too many elk, and was punished by death (Frey, 1995: 179–182 from Sapir, 1990: 283–285).

A story from the Klickitat (southern Washington):

"Coyote would go about hunting;
    he would shoot and kill all sorts of things.
And now then he shot and killed a deer.
Then he butchered it.
There were two young ones in its belly.
He threw them away and left them there,
    at that place there.
The two young deer lay there.
And rain and snow came upon them.
And the Deer who he had killed felt very badly at heart about them.
Then after that Coyote went all over
    and shot and killed nothing."

(For the obvious reason. He is instructed to sweat in the sweat lodge five times—the Klickitat sacred number—and then never do such a thing again. He leaves this instruction for human beings, who are soon to come into the world:)

"'All right then....
And this is the way the people will do whenever they shoot
    and kill game.
They will carry all of it back home.
Nothing will they throw away.
And always will they sweat for the purpose of the hunt...."

(Joe Hunt, collected and translated by Melville Jacobs in the 1920s, ed. and publ. Frey, 1995: 49–52).

Wasting anything killed gets punishment: no more game. The slain deer survives in spirit form and is able to bring this about. Killing a pregnant doe is already bad, and made much worse by any lack of respect shown to the fawns.

As usual, there are Siberian equivalents; Vladimir Jochelson recorded several from the Yukaghir (Jochelson, 1926: 147–150). In one case, people took too many reindeer, and mistreated one, so "the keeper of wild reindeer...was insulted by the cruelty and ridicule inflicted and led his reindeer away..." (Jochelson, 1926: 149; see also Pierotti, 2010). The spirits of game "are friendly to the hunter as long as he observes certain regulations and kills only what he actually needs for his livelihood... these religious conceptions...correspond to and compare perhaps favorably with our hunting laws and game preserves" (Jochelson, 1926: 150).

Returning to the conservation-conscious Tsimshianic peoples, young people "would disregard the many taboos and also never heeded the warnings to respect the many animals of the hills but killed these needlessly, and having no need of the flesh at the time, would leave it to rot in the woods. They began to do the same with fish...." This can be considered as a pre-flood story, because a deluge followed, exterminating all but one girl who had been respectful and observant. She escaped, saved by a supernatural being, and founded a noble lineage among the Nisga'a (Barbeau & Beynon, 1987, vol. 1: 340–342; A Lakota version of this story, narrated by Joseph Marshall, can be watched at https://www.youtube.com/watch?v=TZSdJQFiljw).

An actual recent event—not a folktale—was related by Lawrence Aripa (quoted in Frey, 1995: 177–179). There was an individual named Cosechin who was a doctor and prophet of the nineteenth century. However, this story probably does not apply to him, but to another Cosechin, or to some generic bad actor who has, in the story, picked up the name. In any case, it is considered historically true. The ellipses represent Mr. Aripa's pauses in the story, not my leaving out material; italics represent spoken emphasis.

> A person named Cosechin.
> "was just *cruel*...to animals.
> He would knock down trees
>   and not use them.
> He would grab leaves from the trees in the springtime
>   and scatter them *to* try to *kill* the trees.
> And you know it's the custom of the Indian that when they're going to use something from a tree
>   or...even fish or hunt,
>   they...ask permission first...
>   'Mr. Tree may I use some part of you
>   or I need it...for warmth for my children,'"

And so on. Cosechin treated humans just as badly, but notice the animals and plants come first.

> Cosechin was told to reform, and threatened. He shaped up for a while, but
>   "he went back to his old ways like.
> And then...all of a sudden...he disappeared...."

The obvious implication is that someone did away with him, and no one was about to inquire into the matter.

Now and then someone overhunts and escapes, but so frightened that he warns his people never to overhunt again, as in the case of a Tsimshian Wolf clan member who overhunted and was chased by wolves (Barbeau & Beynon, 1987: 281–283).

Respect Shown to Individual Animals

The critical value of respect is the subject of countless stories. Tsimshian ethnographer Charles Menzies repeats from Beynon a story (mentioned above) of a salmon woman who married Raven, the creator-trickster; of course, he inevitably disrespected her (as Raven is wont to do), and she and the salmon left him to starve. Other stories tell of disrespectful behavior toward salmon that leads to the same result (Menzies, 2016: 87–90). Menzies concludes:

> If respected, they will reward the harvester. History has taught us that catching too much salmon...will result in a marked decline or total extirpation...The same history has also shown that not taking enough seems to have a similar effect [because of overspawning, as well as ease of spreading disease—ENA]. Thus, oral histories provide guidelines... (Menzies, 2016: 90; see also Barbeau & Beynon, 1987: 109–112 for the classic story of the boy who mistreated a piece of salmon and was carried down to the undersea salmon home and taught better).

The Coos of Oregon told a story of children who caught and roasted a raccoon, laughing at its shrinking up and looking funny while roasting; all died except one who did not laugh. This was universally known, and taught to keep children from laughing at animals. However, the irrepressible Annie Miner Peterson, narrator of this story to ethnographer Melville Jacobs (1940: 146–147), said she and other children sometimes laughed at this story and at other animals anyway, and survived. (On her unforgettable character, see the delightful and moving biography by Lionel Youst, 1997. The title, *She's Tricky like Coyote*, translates her Coos name, and shows how she was regarded by her people; it is not a criticism, but a recognition of her cleverness and survival ability.)

Peterson also narrated a story of a man who was pitied by the supernaturals for his poverty; he was given a strange fish for good luck; his wife made him ask for too much, and he lost the luck (Jacobs, 1940: 133–135). This is a version of a story almost universally known from ancient Greece to China to modern America, where variants of it survive in current European "three wishes" stories of people who wished for too much and lost all. The Nlaka'pamux (Thompson) of British Columbia are among groups who have tales of Coyote making this mistake—in their tales, he helps other animals who give him magical powers to get meat, but he is as greedy as ever, presumes too much on their grateful gifts, and loses all (Hanna & Henry, 1995: 49–55). East Siberian stories often turn on the same theme of respect for animals being rewarded (e.g., Dolitsky, 2002: 209–211) and the reward being lost from greed. As in the Northwest, humans and animals communicate easily and sometimes shape-shift into each other, giving point to animal fables that carry moral teachings. Siberians—both hunters and herders—butchering animals may apologize, or pray, or even say "I had no idea to kill, I killed you by mistake" (Stépanoff et al., 2017: 58).

Humans who disrespect animals suffer for it. In addition to the Bear Mother story, the Haida have a tale of a man who insulted black bears for robbing his fish trap. They assumed human form, captured him, and enslaved him, and he had a hard time escaping. When he did, the animal people united to make war on humans, and had to be fought down (Swanton, 1908: 518–523). In a Bella Coola story, a man violated the rule that an animal killed by a hunter should leave it to anyone who had been stalking it seriously at the time, by killing and taking a seal that was being hunted by a pack of wolves. The wolves hunted him down and killed him (Davis & Saunders, 1980: 173–184; see also related story, 185–191). Even in the modern rather cursory form, this story is well constructed and moving; the full old-time form must have been dramatic, and even frightening, to hear around a fire in midwinter, and would certainly have taught children this bit of politeness. One can imagine the narrator putting full drama into the story, while the firelight flickered and the children's eyes opened wide.

In short, if X (whether animal, human, or supernatural) cheats, insults, tricks, or maliciously harms Y, who is usually assumed to be weak or defenseless, Y always gets revenge (Boas, 1916: 463–464, 749–750). X's various deeds may all be seen as disrespecting as well as materially harmful. Bad treatment is punished. Related to the Bear Mother tale is one of a man who mistreated a frog, was carried off by the frog people, and threatened with death if she died and ordered to marry her if she survived. She lived, and they had a good marriage (Swanton, 1908: 557–560). In other versions, it is a girl who features in this story (tale by famous Nuxalk storyteller and authority Margaret Siwallace, Storie, 1973: 23–24; Siwallace was so helpful, and her knowledge so valuable, to so many scholars that she received a degree from the University of Victoria). In a Tlingit version recorded by Swanton (1909: 53–54), the girl is recovered but soon dies; in another, she is saved (Swanton, 1909: 236–237). "Frog" is used here for convenience, because all the sources call it that, but in fact there are no frogs in Haida Gwaii or anywhere on the British Columbia coast; the animal in question is the Western Toad (*Bufo boreas*).

Another Haida story (collected by Swanton, back-translated into Haida and published by John Enrico, 1995: 163) tells of people torturing a frog by putting it in the fire to watch it explode; of course, it destroys them. The Haida also tell of people who tortured frogs by burning them, the result being that the frogs burned their village. They also have an important story of insulting water, burning a frog, and paying accordingly (Swanton, 1905: 316–317). (For this and other versions of destructive frogs, see Barbeau, 1950: 65–75, 1953: 28–37; Boas, 1916: 260–270, 2006: 608–609, a very long and complex version; Miller, 1997: 73–77; many other versions exist). Frogs could also do the insulting; frog women insulted several animals so badly that Beaver called rain on them and killed them, driving them out of one Kwakwaka'wakw area (Boas & Hunt, 1905: 318–321).

Frogs are widely connected with illness and health; they are associated by the Heiltsuk with pregnancy and feminine fertility (Harkin, 1997: 84). Like many animals that cross the boundary between land and water, frogs are powerful. George Emmons records an actual case in 1899 of a chief falling ill from fear after offending a frog (Emmons, 1991: 359). Here, we may be speaking of a real frog, since the Wood Frog *Rana sylvatica* reaches the Alaska coast.

A similar Tlingit story, which adds mocking a loon to human vicissitudes, is recounted by Emmons (1991: 103, from an actual occurrence in 1881). Tlingit respect for animals is well known and taught by many stories (De Laguna, 1972; Newton & Moss, 1983: 28–30; Thornton, 2008). Mocking or ineptly treating animals could lead to mental damage. Shamans had to be particularly careful in their searches for power from animals (Emmons, 1991: 360). One Tlingit story combines insulting a halibut and insulting ducks, with horrific punishments following (Swanton, 1909: 38–40).

The story of insulting a frog is told also of a dog. In the commonest form, the Dog Mother tale, a woman who insulted a dog was lured away by the dog, in human form, and married him, eventually having several puppies; these grew, and she found they could shed their dog skins and appear as humans, so she burned the skins. The story typically ends tragically. Julie Cruikshank (1983: 15; see her recording, 1979: 114–118) mentions that James Teit recorded this story from 20 Northwest Coast groups (see Teit, 1898: 27, 1912: 354–355). Some of the variants are given in Boas (1917: 30); the story is known much more widely in the continent. Some versions end with more charity (e.g., a Tlingit version, Swanton, 1909: 100–101, in which the pups become heroes). In a Nuxalk version, the woman and pups survive, including a girl pup with the wonderful name of *Nuskwuplxmx*, "one who knows how to track and can seize the right opportunity to act" (Boas, 2006: 532–535)—a name surely to be recommended to modern parents and dog owners. Thomas MacIlwraith later recorded two other Nuxalk versions (McIlwraith, 1948: 1: 6432–645), but they are fragmentary. Many other Dog Mother stories are told, from north (e.g., Tsimshian, Barbeau & Beynon, 1987: 1: 89–92; Tlingit, Swanton, 1909: 99–106) to south (for Oregon, Boas, 1901: 155–165; Frachtenberg, 1913: 167–171; Jacobs & Jacobs, 1959: 22–24; Jacobs, 1940: 159–162, 1945: 174). Boas noted versions as far afield as Greenland (Boas, 2006: 656). Sometimes it is a man that marries a female dog (Frachtenberg, 1920: 125–149, for the Alsea).

Mistreated dogs also cause disasters such as earthquakes (Boas, 1935: 118–124), but treating dogs well leads to their protection, even reviving the dead (Boas, 1935: 143–147). Urinating on a small fish caused lightning to strike the offending woman, in a Comox story (Boas, 2006: 161–162). Mismanaging fish remains causes the mistreater to be taken below the sea and instructed (Boas, 2006: 578–580; McIlwraith, 1948: 52; apparently every British Columbia coastal tribe has a version of this story).

A Kwakwaka'wakw trickster figure once insulted geese by telling them they ate sand and seaweed; they carried him off and instructed him better, showing him all the wonderful roots they actually eat, thus teaching root use to humans. They gave him magical vessels that produced these foods, but a malevolent rival told people how the trickster acquired them; the ban on telling the source of one's supernatural powers led to loss of the vessels (Boas, 1910: 161–165).

Chipmunk insulted Owl, who tore his back with talons, causing the stripes on the chipmunk, which now remind us not to insult others (Teit, 1909: 654; several other versions exist, e.g., Hanna & Henry, 1995: 81). Steelhead insulted King Salmon and a noble woman, and was deprived of fat, which is why steelhead are usually lean today, perhaps also why they do not die after spawning (Hilbert, 1985: 127–129; Peter, 1995: 235–237). In a "Coast Salish" version of this story, Trout rescues Steelhead and makes some parts of his body from vine-maple, hence his thinness; "Steelhead…has some of the purity of the vine-maple in his body—its pure leaves and blossoms. Its limbs are his bones. Some of its oil is in his flesh. It would seem that vine-maple was his relative" (Jonas Secena in Adamson, 2009: 73–74). It is possible to see almost all Northwest Coast myths and folktales as somehow involving respect or disrespect.

Many of the stories in Swanton's Tlingit material turn on gratitude for respect and punishment for lack of it; even chunks of floating ice return respect (Swanton, 1909: 52). The Haida and other groups share belief in the power of mockery. Swanton recorded a Haida story in which laughing at a sea otter caused a flood (Swanton, 1908: 400–407), and a Tlingit one in which Raven hit his wife with a piece of dried salmon, and she—being Fog—disappeared, along with all the salmon she had prepared (Swanton, 1909: 168). Recall from above the Tsimshian trickster who married a salmon woman, and thus caught vast quantities of salmon, but then insulted his wife, and she left, followed by all the stored and dried salmon in the house, which returned to life and went off with her (Barbeau & Beynon, 1987: 32; Menzies, 2016). There are many variants of this story all along the coast, as far south as the Coos of the Oregon Coast (Frachtenberg, 1913: 21–38).

Themes and Variations

Julie Cruikshank, a formidably impressive and insightful collector and interpreter of stories, began her career by recording several of these stories of insulting fish, marmots, and other life forms, and being taken and instructed by them (Cruikshank, 1979: 104–114). She notes these stories are typically about boys or young men, who do the hunting and are at the age to seek and get supernatural power from animals and also to learn the lesson of what it means to screw up. Women are more apt to be simply stolen by the animals, often Lynx, who in biological reality is a notoriously sly, isolated, but powerful predator. They must also fear land otters (see Cruikshank, 1983: 90–95), who, here as elsewhere in the northern coast, take humans and transform them into otter people. Such entities are portrayed as villains in Haisla author Eden Robinson's fascinating *Trickster Trilogy* (Robinson, 2017, 2018, 2021), especially the first volume *Son of a Trickster*.

Cruikshank's collection of tales about stolen or wandering women makes many points about these differences. "By far the largest number of male-centred stories are 'helper' stories.... All have a common pattern. In some, a starving man meets an animal helper who brings him vast amounts of food, often enough to save many other people. Other times, a man befriends an animal, who later repays him by becoming his helper. In even more dramatic instances, men who have been mistreated or abandoned to certain death are rescued by an animal who not only helps the protagonist to get home, but also helps him to seek revenge... (Cruikshank, 1983: 14)." Very often the man gets shamanic powers. McIlwraith (1948: 1: 619–653) provides 20 such stories, including variants of the widely known wolf-marriage and Dog Mother tales, as well as several related stories about shamans (pp. 661–678). Supernatural guardians can also pity the deserving. The daughter of the spirit guardian of fish, for instance, might marry a forsaken boy and provide him with rich stocks of fish, as in a Wishram story (Sapir, 1990: 169).

Such stories are worldwide. Exactly the same outline—the poor, abandoned, and/or mocked young person who is saved by a being with special powers or by a supernatural—is the standard plot in a large collection of Chinese folk stories about medicinal plants and animals (Miao, 2008). Apparently, according to legend, the standard Chinese medicines were all shown to humans this way. Similar stories occur worldwide.

Some stories turn on successful respect. Among the many stories John Swanton recorded from the Haida is a tale of a Haida man who hunted grizzlies too successfully for their comfort, so they took his daughter and married her to one of their own. He sought and found her, but treated her and her grizzly husband and family well, and they rewarded him by giving him meat and helping him hunt successfully (Swanton, 1908: 508–512). The story provides a model of respectful treatment of animal persons and of in-laws in general. Another Haida story tells of a girl who fed a raven, who then steered more and more food to her and her community (Swanton, 1905: 48–51).

In an odd Nuxalk variant, a boy abandoned by his people and left to die catches four "bull-heads" (sculpins) and throws them back in disgust, only to be reprimanded by a beautiful supernatural woman, who says "Why did you throw away my children?... They showed kindly feeling toward you in allowing themselves to be caught, but you angrily threw them away. None the less I have...come to help you" (McIlwraith, 1948: 1: 522–524, quote from p. 523). We have already met this moral teaching of keeping even small fish that offer themselves.

A widely known story tells of a boy who treated wolves with respect, and was able to marry a wolf and live happily with them (Boas, 2006: 362–363), or get a magical hunting plume from them that enabled him to feed his people (Swanton, 1909: 33–36). Alternatively, a wastrel could be pitied by wolves and helped by them (Barbeau & Beynon, 1987: 1: 275–277). There is even a Tsimshian story of a man who seriously overhunted game and was pursued by wolves for it, but eventually escaped (Barbeau & Beynon, 1987: 1: 281–283), a rare case of getting away with it, though the story is scary enough to deter any reasonable hunter. (Recall the story above of the man who cheated the wolves and died for it.) A related story tells of a boy taking out a bone stuck in a wolf's throat, and being rewarded by the helpful wolf (Barbeau & Beynon, 1987: 1: 290–291); this seems completely independent of

Aesop's story of Androcles and the lion, but there may be very distant connections. Many stories of girls marrying wolves teach aspects of good behavior toward these fellow social predators. A particularly fine one explains the origin of the Tsimshian wolf crest (Barbeau & Beynon, 1987: 295–304). Eagles can also star in such stories. In one, a boy treats them well and gets fed, but not married (Boas, 2006: 589–594; compare Swanton, 1905: 356–357).

In various Kwakwaka'wakw stories, a child received medicine power from animals; good behavior toward dogs was rewarded; and mistreating them was punished (Boas, 1935: 79–85, 118–147). A Tsimshian family had major trouble with bears, but finally the youngest brother was good to two tiny dogs, who became huge and saved them from the bears; he taught his people, "You must respect bears and address them as your equals and kill only what you need for your own welfare.... Treat them with respect. And the same with all animals..." (Barbeau & Beynon, 1987, vol. 1: 216).

Animals must respect each other, too. Many myths turn on disrespect in the time when animals were humanoid. Deer disrespected the wolves in a Nuxalk story, with the inevitable conclusion: wolves still prey on deer (McIlwraith, 1948: 425–427). Coyote got angry and disrespectful to drying salmon when his hair was caught in the gills of one, so all the dried salmon and their oil and roe came back to life, rushed back to the river, and left Coyote hungry (Teit, 1909: 637; compare stories, above, of this happening to humans). These are only two particularly clear cases; hundreds of the animal stories of the Northwest Coast involve some sort of offense by one animal to another, and all teach respect.

Related to stories of respectful treatment are the "bungling host" stories, universal in our area (but beyond the scope of this book): the trickster is treated by a series of beings to food they magically produce; he tries to do the same for them, showing proper reciprocity, but burns or injures himself trying to imitate their magic. The moral is that reciprocity is proper, but each being has his or her own way of treating others respectfully. Magical feeding is the animal powers equivalent of showing one's own skills and regional delicacies in the human world. Show your strengths, not your imitations.

Transformers

Origin stories of the Northwest Coast all tell of a Transformer who traveled the world, making it what it is today. He (he is always male) turns a formless, amoral cosmos into a formed and moral one, where humans can exist and follow the principles (Trosper, 2009). Sometimes he is Raven, or Steller's Jay in coastal Washington from the Quinault south. In other places, he is Mink. On the Plateau and nearby areas (including Oregon and California), as throughout most of the west, he is most often Coyote. Some groups, especially on the Plateau, have most or all of these, each with his own cycle. Sometimes these tricksters are usually human (as opposed to humanized animals), but can be animals at will. Sometimes one is wholly supernatural. Often, supernatural beings opposed him, and he changed them into what they are today. Most widespread of these stories is that the ancestral deer was a humanoid, found sharpening two stone knives to kill the Transformer. The latter jammed the knives onto his forehead, turned them into ears, and made him a deer, running off to be food for people henceforth. Every group in Washington and neighboring parts of Oregon, British Columbia, and Idaho seems to have this story.

Transformer cycles can merge into poetic celebrations of the glory and diversity of this wonderful world. One such is an Alsea text recorded by Leo Frachtenberg far back in 1910 (Frachtenberg, 1920: 35–55). These may be compared with the amazing epic poem created in California by the Kato elder Bill Ray for Pliny Goddard in 1906 (Goddard, 1909).

Such stories deserve recovery, and may actually want and need it. On the Northwest Coast, stories, like trees and mountains, are living things themselves, and can desire, become lost, return, and otherwise act as full agents in society. Bruno Latour's actor-network theory (2005) is nowhere more necessary to an analyst than in Native North America. Latour held that inanimate things become active agents in society when we think or act as if they do—even if it is merely personalizing one's computer or car with a human name, or cursing at an inanimate object we trip over in the dark. Indigenous peoples of America and Siberia anticipated this theory by thousands of years, but take it more literally: the "inanimate" things in their stories really have agency and action. One cannot understand their societies without taking that into account.

Local to the Nimpkish Kwakwaka'wakw was a powerful spirit who generously helped the Transformer, and was rewarded by being given the privilege of choosing what to be. He chose to become the Nimpkish River (on Vancouver Island; versions in Barbeau, 1950: 150; Boas, 2006 [1895]: 306). That is why this beautiful stream has all the kinds of salmon in it—or had before the logging companies massacred it, to the heartbreak of the Nimpkish people.

Old Pierre, the Katzie elder who gave Diamond Jenness a moving account of the history of the world, provided a particularly detailed account of the travels of transformer heroes, especially Khaals, the main transformer of the lower Fraser River area (Jenness, 1955: 21–33). Pierre reported that the Katzie transformer Khaals punished several unhospitable and unfriendly people, by turning them into rocks or animals, but gave special powers to those who had some good to offer. A man who loved to wander alone in the forest became, with his family, the ancestral wolves, which have power and can help people hunt (Jenness, 1955: 22). A one-legged fisherman was pleasant enough to be pitied for his condition and turned into the master of the salmon (Jenness, 1955: 24–25).

Pointing Morals

Myths, as elsewhere, convey the basic moral and personal lessons of the societies that tell them. Richard Atleo (2011) gives an extended account of their uses in Indigenous education (see Armstrong, 2005b; Basso, 1996, for other Indigenous accounts of traditional education). Long epics of human/nonhuman transformation and interchange are universal. Famous examples include the Bear Mother story and the abovementioned stories of humans and salmon. All these reinforce the general message of respect for animals and concern for the entire living world.

Even Coyote himself was vital to creation. He made the world what it is, but he was always a zany, greedy, irresponsible, delightfully silly rogue (Ramsey, 1977). The idea of the Creator as a wild, wise, poetic fool not only makes beautiful poetry; it solves the "problem of evil"; evil was created by mistake and through foolish, self-indulgent actions. Raven, along the coyote-less northern coast, has a similar role, but is smarter, greedier, and rather more sinister. He attracts the same mix of respect and scorn. Thus, Native Americans were saved from the Eurasian paradox of a good deity creating evil, or of an evil one allowing good.

One of the finest Coyote narrations ever recorded is Lawrence Aripa's telling of the Coeur d'Alene story of "Coyote and the Green Spot" (Frey, 2001: 134–144). It was originally told in English, allowing us to share the wonderful style that is otherwise lost in translation. In it, Coyote's long-suffering wife Mole reflects:

> "This worthless Coyote,
>   he's no good,
>   he's mischievous,
>   he does things that are bad...
>   But he's a Coyote....
>   and we have to have Coyotes on this earth."

She reflects on this several times as she revives him from the dead—a wife's role in many a Coyote tale. Coyote's tricks always fail, but he is always revived—because, as she says, "Well we do need a Coyote." He promises to reform when he revives, but of course is soon back to his tricks, and Aripa ends the story:

> "But he is still the Coyote,
>   and he will always remain the Coyote."

This sense that all beings are necessary to the world, no matter how inconsequential or annoying they seem, is absolutely basic to Native North American religion, and has been underemphasized in the past. We need it now.

A sad indication of how much was lost by poor recording or transcribing comes from a Salishan tale of a man who went off to join the seals. (Or sea lions—they are not distinguished in Boas' or other older collections.) Franz Boas recorded this story in the 1890s (Boas, 2006: 221–224), but without a moral, leaving it merely a story of a man who unaccountably went off to become a seal. Around a hundred years later, Honoré Watanabe recorded a version from Sliammon narrator Mary George. In

George's version, the man is actively captured or persuaded away by the seal folk, and becomes an ancestor of the Sliammon. She added a long moral (extracted here):

> "You always see seals.
> > If you see them, you appreciate them.
> > Talk nicely to them....
> They are a part of us.
> That's why we really respect seals today....
> > Seals are our people." (George & Watanabe, 2008: 128–129).

This story reminds us of the were-seal (selchie, sealkie) legends of Scotland and Ireland; as so often, northwest Europe and Northwest North America run parallel. (See Thomson, 1954, and the classic ballad *The Great Silkie of Sule Skerry*.) A number of other Sliammon moral and Transformer tales are found in Kennedy and Bouchard (1983).

A related story was told by Annie Miner Peterson of the Coos (Jacobs, 1940: 149–150), and a long and complex seal narrative filled with ideals of respect for animals was recorded from Sauk-Suiattle Salish elder Martha Lamont (Bierwert, 1996: 230–309). Presumably most other early collectors failed to hear or record morals in cases like this. We are deprived of many applications that we now need to know.

Sometimes conservation is stressed indirectly. The Kutenai culture hero Yaukekam (in simplified spelling) had to get the components of arrows from their animal keepers. Ducks had the feathers, and he gave them the power to grow a new crop of beautiful feathers every year, freeing up the old ones for fletching, during the annual molt (Boas, 1918). Ducks do not make the greatest arrow feathers, but we assume geese (more beneficial to fletchers) are included here.

Plants too can be over-collected. From Columbia River people comes a story of a mouse who stole too many roots, which loaded her down so much she drowned crossing a creek; mice have been thieves ever since (Ackerman, 1996: 21–22).

Finally, in the mid-nineteenth century, Chief Seattle (Sealth, Siatl) gave a famous speech defending his people's land and land rights. This speech was, unfortunately, heavily "edited" by a Baptist group in the twentieth century; much of the militant defense of land was taken out, and some syrupy Christian-style rhetoric substituted. Alas, the syrupy additions tend to be the parts most often quoted. This has given rise to a myth that the speech is purely fraudulent, leading to the bizarre conclusion that its phoniness discredits all claims of Native American sustainable use (see Trosper, 2009: 25–26). This is not the case. Seattle's famous and passionate defense of Indian rights and of the hunting-gathering way of life against white theft and ruin was real. However, no full version of it appeared for about 30 years, so the one that finally emerged is rather suspect. Shorn of the purple English added in late versions, it still rings true to the general spirit of northwestern American rhetoric of the time. The whole story has been told by Albert Furtwangler (1997), and by Rudolf Kaiser (1987), who salvaged what can be found of the original.

References

Ackerman, L. A. (Ed.). (1996). *A song to the creator: Traditional arts of native American women of the plateau.* University of Oklahoma Press.

Adamson, T. (2009). *Folk-tales of the Coast Salish.* (New edn. with introduction by W. R. Seaburg and L. Sercombe; original, New York: American Folklore Society, Memoir 27, 1934). University of Nebraska Press.

Armstrong, J. (2005a). En'owkin: Decision-making as if sustainability mattered. In M. K. Stone & Z. Barlow (Eds.), *Ecological literacy: Educating our children for a sustainable world* (pp. 11–17). Sierra Club Books.

Armstrong, J. (2005b). Okanagan education for sustainable living: As natural as learning to walk or talk. In M. K. Stone & Z. Barlow (Eds.), *Ecological literacy: Educating our children for a sustainable world* (pp. 80–84). Sierra Club Books.

Atleo, E. R. (2011). *Principles of Tsawalk: An indigenous approach to global crisis.* University of British Columbia Press.

Attla, C. (1983). *Sitsiy Yugh Noholnik Ts'oin': As my grandfather told it: Traditional stories from the Koyukuk.* Yukon-Koyukon School Board and Alaska Native Language Center.

Attla, C. (1989). *Bakk'aatugh Ts'uhuniy: Stories we live by.* Yukon-Koyukon School Board and Alaska Native Language Center.

Attla, C. (1990). *K'etetaalkkaanee: The one who paddled among the people and animals.* Yukon-Koyukon School Board and Alaska Native Language Center.

Barbeau, M. (1950). *Totem poles.* National Museum of Canada.

Barbeau, M. (1953). *Haida myths illustrated with argillite carvings.* National Museum of Canada, Bulletin 127.

Barbeau, M., & Beynon, W. (1987). *Tsimshian narratives.* Canadian Museum of Civilization, Mercury Series, Paper 3.

Basso, K. (1996). *Wisdom sits in places: Landscape and language among the Western Apache.* University of New Mexico Press.

Bierwert, C. (Ed.). (1996). *Lushootseed texts: An introduction to Pugeet Salish narrative aesthetics.* University of Nebraska Press.

Bierwert, C. (1999). *Brushed by cedar, living by the river: Coast Salish figures of power.* University of Arizona Press.

Boas, F. (1901). *Kathlamet texts.* Smithsonian Institution, Bureau of American Ethnology, Bulletin 26.

Boas, F. (1902). *Tsimshian texts.* Smithsonian Institution, Bureau of American Ethnology, Bulletin 27.

Boas, F. (1910). *Kwakiutl tales. Columbia University contributions to anthropology* (Vol. 2). Columbia University Press.

Boas, F. (1916). *Tsimshian mythology* (pp. 27–1037). Smithsonian Institution, Bureau of American Ethnology, Annual Report for 1910.

Boas, F. (1917). *Folk-Tales of Salish and Sahaptin tribes.* American Folk-Lore Society.

Boas, F. (1918). *Kutenai tales.* United States Government, Bureau of American Ethnology, Bulletin 59.

Boas, F. (1935). *Kwakiutl Tales: New series* (Vol. 1). Columbia University Press.

Boas, F. (2006). *Indian myths and legends from the North Pacific coast of America.* [German original 1895.] Trans. D. Bertz. Ed. R. Bouchard & D. Kennedy. Talon.

Boas, F., & Hunt, G. (1905). *Kwakiutl texts, part 2.* American Museum of Natural History, Memoir V, Part 2.

Cove, J. (1987). *Shattered images: Dialogues and meditations on Tsimshian narratives.* Carleton University Press.

Cruikshank, J. (1979). *Athapaskan women: Lives and legends.* National Museums of Canada, National Museum of Man, Mercury Series No. 57.

References

Cruikshank, J. (1983). *The stolen woman: Female journeys in Tagish and Tutchone.* National Museums of Canada, National Museum of Man, Mercury Series No. 87.

Cruikshank, J. (1998). *The social life of stories: Narrative and knowledge in the Yukon territory.* University of Nebraska Press.

Cruikshank, J. (2005). *Do glaciers listen? Local knowledge, colonial encounters, and social imagination.* University of British Columbia Press.

Dauenhauer, N. M., & Dauenhauer, R. (1987). *Haa Shuká, our ancestors: Tlingit Oral narratives.* University of Washington.

Dauenhauer, N. M., & Dauenhauer, R. (1990). *Haa Tuwunáagu Yís, for healing our Spirit: Tlingit oratory.* University of Washington.

Dauenhauer, N. M., Dauenhauer, R., & Black, L. (2008). *Anooshi Lingit Ani Ka/Russians in Tlingit America.* University of Washington Press.

Davis, P. W., & Saunders, R. (1980). *Bella Coola texts.* British Columbia Provincial Museum. Heritage Record No. 10.

De Laguna, F. (1972). *Under mount Saint Elias: The history and culture of the Yakutat Tlingit.* Smithsonian Institution. Smithsonian Contributions to Anthropology 7.

Dolitsky, A. B. (2002). *Ancient tales of Kamchatka.* Trans. by Henry N. Michzael. Alaska-Siberia Research Center.

Elmendorf, W. W. (1960). *The structure of Twana culture, with comparative notes on the structure of Yurok culture [by] A. L. Kroeber.* Pullman, WA: Research studies, Washington State University, monographic Supplement 2, part 1.

Emmons, G. T. (1991). *The Tlingit Indians.* Edited with Additions by F. de Laguna. American Museum of Natural History.

Enrico, J. (1995). *Skidegate Haida myths and histories.* Queen Charlotte Islands Museum.

Foster, G. (1945). *Sierra Popoluca Folklore and beliefs.* University of California Press.

Frachtenberg, L. J. (1913). *Coos texts* (Vol. 1). Columbia University Press, Columbia University Contributions to Anthropology.

Frachtenberg, L. J. (1920). *Alsea texts and myths.* Government Printing Office, Bureau of American Ethnology, Bulletin 57.

Frey, R. (1995). *Stories that make the world: Oral literature of the Indian peoples of the inland northwest.* University of Oklahoma Press.

Frey, R. (2001). *Landscape traveled by coyote and crane: The world of the Schitsu'umsh (Coeur d'Alene) Indians.* University of Washington Press.

Frey, R., & Hymes, D. (1998). Mythology. In D. E. Walker (Ed.), *Handbook of North American Indians. Vol. 12, plateau* (pp. 584–599). Smithsonian Institution.

Furtwangler, A. (1997). *Answering chief Seattle.* University of Washington Press.

George, M., & Watanabe, H. (2008). The seal, a traditional Sliammon-Comox story. In M. T. Thompson & S. Egesdal (Eds.), *Salish myths and legends: One people's stories* (pp. 121–129). University of Nebraska Press.

Glavin, T. (1998). *A death feast in Dimlahamid.* New Star Books.

Goddard, P. E. (1909). *Kato texts* (Vol. 5, pp. 65–238). University of California Publications in American Archaeology and Ethnology.

Goulet, J. A. (1998). *Ways of knowing: Experience, knowledge, and power among the Dene Tha.* University of Nebraska Press.

Halpin, M. (1978). William Beynon, ethnographer, Tsimshian, 1888-1958. In M. Liberty (Ed.), *American Indian intellectuals* (pp. 140–156). West Publishing.

Hanna, D., & Henry, M. (1995). *Our tellings.* University of British Columbia Press.

Harkin, M. E. (1997). *The Heiltsuks: Dialogues of culture and history on the northwest coast.* University of Nebraska Press.

Harris, K. B. (1974). *Visitors who never left.* University of British Columbia Press.

Hilbert, V. (1985). *Haboo: Native American stories from Puget Sound.* University of Washington Press.

Hymes, D. (1981). *"In vain I tried to tell you": Essays in native American Ethnopoetics.* University of Pennsylvania Press.
Hymes, D. (2003). *Now I know only so far: Essays in Ethnopoetics.* University of Nebraska Press.
Hymes, D., & Seaburg, W. (2013). Chinookan oral literature. In B. J. Colombi & J. F. Brooks (Eds.), *Keystone nations: Indigenous peoples and Salmon across the North Pacific* (pp. 163–180). School of American Research Press.
Ignace, M., & Ignace, R. (2017). *Secwépemc people, land and Laws.* McGill-Queen's University Press.
Jackson, K. H. (1971). *A Celtic Miscellany.* Penguin.
Jacobs, M. (1940). *Coos myth texts* (Vol. 8, pp. 127–260). University of Washington, Publications in Anthropology.
Jacobs, M. (1945). *Kalapuya texts.* University of Washington. Publications in Anthropology 11.
Jacobs, E. D., & Jacobs, M. (1959). *Nehalem Tillamook Tales.* University of Oregon Books.
Jenness, D. (1955). *The faith of a coast Salish Indian.* British Columbia Provincial Museum [now Royal British Columbia Museum], Memoir 3.
Jochelson, W. (1926). *The Yukaghir and the Yukaghirized Tungus.* Memoir of the American Museum of Natural History, XIII, Reports of the Jesup North Pacific Expedition, IX.
Kaiser, R. (1987). Chief Seattle's speech(es): American origins and European reception. In B. Swann & A. Krupat (Eds.), *Recovering the word; essays on native American literature* (pp. 497–536). University of California Press.
Kennedy, D., & Bouchard, R. (1983). *Sliammon life, Sliammon lands.* Talonbooks.
Krajick, K. (2005). Tracking myth to geological reality. *Science, 310,* 762–764.
Latour, B. (2005). *Reassembling the social: An introduction to actor-network-theory.* Oxford University Press.
Legat, A. (2012). *Walking the land, feeding the fire: Knowledge and Stewardship among the Tłı̨chǫ Dene.* University of Arizona Press.
Lévi-Strauss, C. (1982). *The way of the masks.* Trans. by S. Modelski (Fr. orig. 1979). University of Washington Press.
Lévi-Strauss, C. (1995). *The story of lynx.* Trans. by C. Tihanyi. University of Chicago Press.
MacDonald, G. F., & Cove, J. (eds.). (1987). *Tsimshian narratives,* 2 vols. Canadian Museum of Civilization. Directorate Paper 3.
Mandeville, F. (2009). *This is what they say.* Trans. by R. Scollon. Foreword by R. Bringhurst. University of Washington Press.
McClellan, C. (1970). *The girl who married the bear.* National Museum of Man, Publications in Ethnology 2.
McClellan, C. (1975). *My old people say.* National Museum of Man, Publications in Ethnology 6.
McClellan, C. (1987). *Part of the land, part of the water: A history of the Yukon Indians.* Douglas & MacIntyre.
McIlwraith, T. F. (1948). *The Bella Coola Indians* (Vol. 2). University of Toronto Press.
McIlwraith, T. F. (2007). *"But we are still native people": Talking about hunting and history in a northern Athapaskan Village.* Ph.D. dissertation, Department of Anthropology, University of New Mexico.
Menzies, C. M. (Ed.). (2006). *Traditional ecological knowledge and natural resource management.* University of Nebraska Press.
Menzies, C. R. (2016). *People of the saltwater: An ethnography of Git lax m'oon.* University of Nebraska Press.
Miao, W. (2008). *Herbal pearls: Traditional Chinese folk wisdom.* Trans. by C. Yue, edited and annotated by S. Foster. (Chinese orig. 1981.) Ozark Beneficial Plant Project in association with Boian Books.
Miller, J. (1997). *Tsimshian culture: A light through the ages.* University of Nebraska Press.
Miller, J. (2014). *Rescues, rants, and researches: A review of Jay Miller's writings on northwest Indian cultures.* Northwest Anthropology, Memoir 9.

References

Naziel, C., & Naziel, R. (1978). *Stories of the Moricetown carrier Indians of northwestern B.C.* Moricetown Indian Band Council.

Newton, R., & Moss, M. (1983). *The subsistence lifeway of the Tlingit people: Excerpts from Oral interviews.* United States Forest Service, Alaska Region, Report No. 179.

Nisbet, J., & Nisbet, C. (2010). *Mourning Dove (Christine Quintasket).* Retrieved from http://www.historylink.org/index.cfm?DisplayPage=output.cfm&file_id=9512

Nunn, P. (2018). *The edge of memory: Ancient stories, oral tradition, and the post-glacial world.* Bloomsbury.

Peter, S. S. (1995). *Gwəqwulćəʔ*. Lushootseed Press.

Pettitt, G. (1950). *The Quileute of La Push, 1775-1945* (p. 1). University of California, Anthropological Papers 14.

Phinney, A. (1934). *Nez Percé texts* (Vol. XXV, pp. 1–497). Columbia University Contributions to Anthropology.

Pierotti, R. (2010). Sustainability of natural populations: Lessons from indigenous knowledge. *Human Dimensions of Wildlife, 15*, 274–287.

Pierotti, R., & Fogg, B. R. (2017). *The first domestication: How wolves and humans coevolved.* Yale University Press.

Ramsey, J. (1977). *Coyote was going there: Indian literature of the Oregon country.* University of Washington Press.

Ridington, R. (1988). *Trail to heaven: Knowledge and narrative in a northern native community.* University of Iowa Press.

Robinson, E. (2017). *Son of a Trickster.* Knopf.

Robinson, E. (2018). *Trickster drift.* Knopf.

Robinson, E. (2021). *Return of the Trickster.* Knopf.

Roth, C. F. (2008). *Becoming Tsimshian: The social life of names.* University of Washington Press.

Sapir, E. (1990). *The collected works of Edward Sapir. VII. Wishram texts and ethnography.* Mouton de Gruyter.

Seaburg, W. R. (Ed.). (2007). *Pitch woman and other stories: The Oral traditions of Coquelle Thompson, Upper Coquille Athabaskan Indian.* University of Nebraska Press.

Sharp, H. (1987). Giant fish, Giant otters, and dinosaurs: 'Apparently irrational beliefs' in a Chipewyan community. *American Ethnologist, 14*, 226–235.

Snyder, G. (1979). *He who hunted birds in his Father's village: The dimensions of a Haida myth.* Grey Fox Press.

Stépanoff, C., Marchina, C., Fossier, C., & Bureau, N. (2017). Animal autonomy and intermittent coexistences: North Asian modes of herding. *Current Anthropology, 58*, 57–81.

Storie, S. (1973). *Bella Coola Tales.* Indian Advisory Committee.

Swanton, J. R. (1905). *Haida texts and myths: Skidegate dialect.* Bureau of American Ethnology, Bulletin 29.

Swanton, J. R. (1908). *Haida texts, Masset dialect. Memoirs of the American Museum of Natural History, X, Jesup North Pacific expedition reports, II.* Brill.

Swanton, J. R. (1909). *Tlingit myths and texts.* Bureau of American Ethnology, Bulletin 39.

Tate, H. (1993). *The porcupine hunter and other stories: The original Tsimshian texts of Henry Tate.* Talonbooks.

Teit, J. (1898). *Traditions of the Thompson River Indians of British Columbia.* Houghton Miffling & Co. for American Folkore Society.

Teit, J. (1909). *The Shuswap* (Vol. II). Memoir of the American Museum of Natural History, The Jesup North Pacific Expedition.

Teit, J. (1912). *Mythology of the Thompson Indians.* American Museum of Natural History, Memoir 12, Reports of the Jesup North Pacific Expedition, VIII.

Teit, J. (1919). Tahltan tales. *Journal of American Folk-Lore, 32*, 198–250.

Thomson, D. (1954). *The people of the sea: A journey in search of the Seal Legend.* Turnstile Press.

Thornton, T. F. (2008). *Being and place among the Tlingit.* University of Washington Press.

Thornton, T. F., Rudolph, M., Geiger, W., & Starbard, A. (2019). A song remembered in place: Tlingit composer Mary Sheakley (Lyook) and Huna Tlingits in Glacier Bay National Park, Alaska. *Journal of Ethnobiology, 39*, 392–408.

Trosper, R. L. (2009). *Resilience, reciprocity, and ecological economics: Northwest coast sustainability*. Routledge.

Turner, N. J. (2014). *Ancient pathways, ancestral knowledge: Ethnobotany and ecological wisdom of indigenous peoples of northwestern North America*. McGill-Queen's University Press.

Walker, D. E., Jr., with Matthews, D. N. (1994). *Blood of the monster: The Nez Perce coyote cycle*. High Plains Publishing Co.

Wilner, I. L. (2018). Transformation masks: Recollecting the indigenous origins of global consciousness. In N. Blackhawk & I. L. Wilner (Eds.), *Indigenous visions: Rediscovering the world of Franz Boas* (pp. 3–41). Yale University Press.

Youst, L. (1997). *She's tricky like coyote: Annie Miner Peterson, an Oregon coast Indian woman*. University of Oklahoma Press.

Chapter 12
The Visual Art

This section of our review is confined to the coast from northwest Washington to Alaska. The interior and southerly groups have their own art forms. These are not only very different from the classic "Northwest Coast" art; they are also less directly concerned with animals and the spirit world. They are usually geometric and ornamental, rather than visual statements of a whole philosophy.

The classic analyses by Franz Boas that gave rise to structuralism in anthropological analysis of the arts were largely devoted to analyzing the structural systems and pictorial language of the Northwest (Boas, 1955, 1995). There has recently appeared a superb and encyclopedic collection of new studies, *Native Art of the Northwest Coast*, edited by Charlotte Townsend-Gault, Jennifer Kramer, and Ḳi-ḳe-in (2013; for a similar but less overwhelming work on the interior parts of the region, see Ackerman, 1996; see also Bunn-Marcuse & Jonaitis, 2020).

This contrasts with ideology and verbal art, where the commonalities overwhelmingly outweigh the differences. The obvious practical reason for the difference is the more stable, settled life on the coast and the consequent greater opportunity to amass whole arrays of tools as well as gigantic works of art, e.g., totem poles. The Plateau tribes and the Oregon and California coastal ones had large permanent houses and sophisticated tool arrays. However, there is more to it than that. There are clearly some aesthetic and philosophical differences that so far defy analysis, but that probably have an important effect.

It was briefly thought in the early twentieth century that the spectacular art of the Northwest was largely a post-contact phenomenon, which is strange considering that the Western tradition has never produced art like that found in this region. Since that speculation, archaeology has found plenty of examples of fine art in the Northwest going back more than 2000 years (Duff, 1975). The climate is not kind to wood or even to soft stone, so little remains except hard-stone work. However, the stonework is revealing. Also, we can compare the ancient ivories from Alaska (Wardwell, 1986), dating back as far as 2000 years, which reflect a style rather close to Northwest Coast art—considerably closer to it than are contemporary Yupiq and

Inupiaq arts. (Compare the plates in Wardwell, 1986: 71–91 with the stonework in Duff, 1975 and with the earliest historic northern Northwest Coast art.) The sheer volume of art appears to have increased after contact; however, this may simply be a result of the introduction of a literary tradition which documented this work, because the style and content were developed before.

Art on the Coast was intensely social. House fronts were painted. Totem poles stood in front of the major houses. Carving ranged from house posts to small personal charms. Most impressive and important were the masked dances. Chosen individuals, often chiefly in status, carried out dramatic and athletically demanding dances with heavy masks, rattles, and other gear. These dances were associated largely with winter ceremonies, and especially with potlatches. The main central coast groups had whole cycles of winter dances, and spent much of the winter engaged in such activities. Darkness and constant rain kept people largely indoors around the fires, where storytelling, ceremony, and dance made the winter a time of intense social reaffirmation.

Most of the art was in wood. Early wood carvings were allowed to return to the landscape, which means that most early work went unrecognized, or more likely, unseen. Small ivory, bone, and stone pieces existed and were better preserved, but large pieces were almost all made of cedar or similar wood. Textiles made of yellow cedar bark, mountain goat wool, and, locally, dog wool in Salish communities and among the Makah displayed crests as well as simple geometric designs.

Sheer aesthetic wonder at the bounty of fish and other resources must lie behind the Northwest Coast attitude to nature and art (Lam & Gonzalez-Plaza, 2007). It is, however, not the major drive. The main focus is on spirits, and the unity of cosmos and local community that exists on the spiritual level and that is maintained by spiritual exchanges, gifts, and actions.

The highest purpose of this art is to display kinship affiliations and stories and associated hereditary privileges within the kinship groups. Dance costumes including huge masks, blankets, frontlets, and other goods were designed specifically for displaying in ritual dances—usually kin group-owned—at potlatches and cemeteries.

"Totem poles" have nothing to do with totemism; they display important events in the ancestry of the chiefly descent group that set them up. The great ones are usually house poles; smaller ones are memorial poles, so similar in purpose to gravestones that many groups have switched to the latter without changing the associated ceremonies. Still others are for art's sake, especially in the contemporary world, for museums, or parks. Whatever the purpose, the poles always commemorate the chiefly ancestry of the carvers or their people. They are visual cues for stories, often long and detailed ones, from the histories of the families mounting the poles. (For many stories, see Barbeau, 1929, 1950; Garfield & Forrest, 1961; Halpin, 1981b).

Aesthetic power shows spiritual power. An artist whose work has a powerful emotional effect on the viewer has received power from the appropriate spirit. The Haida have a Master Carpenter (or Carver) as one of their most important supernaturals. The evocative quality of the art is greatly enhanced for its intended audience by the social connections, family histories, and personal connections.

Indigenous lawyer and leader Douglas White, with his sister, visited the American Museum of Natural History. They came upon a screen that had belonged to their Nuu-chah-nulth great-grandfather.

> My sister and I were stunned and deeply moved. After standing in front of the screen for five minutes...we realized we had to do something. I told my sister that we should go back to the screen, and I would sing one of our family's potlatch songs, and she should dance." [Which they did], "for a full five minutes—tears streaming down our faces at the emotional and spiritual impact of an unexpected reconnection with such a critical part of our family's existence.... No one disturbed us, not even the security guard, who came close but did not intervene...my sister and I were transformed that day by the experience of singing and dancing in front of our screen (White, 2013: 642).

Thus, gifts of carved, painted, or woven art works are the highest of gifts, and are appropriate prestations to new title-holders at the most important events. The best Northwest Coast art still usually stays in this highly charged, emotional world, and rarely gets out to the museums, at least until everyone has seen it and viewed it appropriately. Exceptions occur when a museum or public body can directly commission a work of art for an appropriate public ceremonial purpose. It is within this personal universe that craftsmanship becomes both a social and a spiritual thing. As the late Haida artist Bill Reid never tired of pointing out, "the highest morality is craftsmanship. One can give nothing more precious of oneself than something well-designed and well-crafted" (Bill Reid, personal communication to Richard Daly, in Daly, 2005: 45). This obviously takes art far beyond the merely decorative, shocking, or titillating functions it all too often serves in the modern Western world. Northwest Coast art is much more comparable to art in Europe in the medieval age of cathedrals and altarpieces. (The literature on Northwest Coast art is enormous. Some notable works that contain beautiful plates but also provide serious analysis include Barbeau, 1953; Duff, 1975; Holm & Reid, 1975; Jonaitis, 1981, 1986, 1988, 1991, 1999; Jonaitis & Glass, 2010; MacDonald, 1983; MacNair et al., 1984; McLennan & Duffek, 2000).

It therefore makes sense that carving was considered a powerful spiritual gift—from the Master Carver himself, in Haida thinking—and that many carvers were of high rank. Tsimshian artmakers were *gitsontk*, "people of the hidden studios"—they were highly respected people who worked in chambers or studios taboo to ordinary people and used for formal meetings as well as art (Garfield, 1939; Halpin, 1981a: 274). This sense that artists are specially gifted, and the corollary that high-born persons are drawn to that trade, is very much alive today. Studios are still power places for many, all up and down the coast.

Totem poles and other outdoor art weather away quickly in the Northwest, even if made of cedar. (They almost always are, because this wood is easy to work but relatively slow to weather away.) This disappearance is regretted, but seen as an inevitable consequence of the natural process. New poles are carved at need, or by request. One recalls medieval European attitudes toward art: wooden sculptures in the open, or days of work on exquisite detail far up in the church roof where no human eyes would ever see it. The art was for audiences that involved more than human eyes.

Daly (2005: 68) notes: "The art-appreciating public... forgets that its perceptions are informed by its own literate, written-down tradition... This public does not see past the rotting artifact to a culture alive and confident of its roots and its pedigree."

He is not in favor of making special efforts to preserve the poles, because the decay was part of the intention of the culture. The public could answer that culture is all very well, but there is also all humanity to consider. The species at large may have some moral claim on a truly beautiful object made for public display, and this is a moral claim recognized by all nations today (via UNESCO among other vehicles). Today the rule is still for the best poles to be in the open, to pass away with time, but also to carve replica poles for museums, if so desired.

Body, house, and cosmos are often equated. For the Puget Sound Salish, a house is a kneeling body, the house front being the face (Miller, 1999: 36). The hearth in the house could be the sun (heart?), the ridgepole stands for the backbone, river, and Milky Way all at once, and the four main supporting posts then become human limbs as well as the pillars that hold up the sky. Canoes were similarly represented. "In consequence, curing, bailing, and cleaning were all reflexes of a renewal of the world" (Miller, 1999: 36).

The Tsimshian, and Haida (Duff, 1981: 215), however, saw the house front as the whole front-facing side of an animal sitting on its haunches, with the result that the door is often the entrance to the womb. One famous Tsimshian housefront represented a female bear. The humans entered and exited through its vagina, thus either being born or returning to the womb whenever they passed. There is even a small face above the door where a clitoris would be.

Ronald Hawker (2003) has chronicled the "dark ages" of coast art, from 1921 to 1961. The first date represents the arrest of dozens of people and the confiscation of potlatch goods, including fine art objects, at a potlatch by the Cranmer family. This marked the beginning of harsh and general enforcement of the anti-potlatch law, previously less than seriously enforced in remote areas. The second date is the granting of citizen rights (including freedom of religion and thus of ceremonies) on the Canadian side. Hawker points out that art survived, but it was down to a slender thread (Hawker, 2003: 8). Still, the "dark ages" were less dark than cryptic. Art survived and flourished, but tended to be small-scale or concealed, because of the repression (see also Dawkins, 2019).

Hawker has some rather unfortunate things to say about those who speak of a "renaissance" after 1961. He was evidently not on the coast before 1961. If ENA may be permitted a bit of reminiscence, it was indeed possible, though difficult, to find good art then, though it was considered "mere craft" and a fine carving or basket would cost $10 or $20. Fine large pieces were hard to find. ENA's first contact was in 1960, when he found active basketmaking in many communities, and watched Mungo Martin carving a totem pole in Victoria. Having watched the renaissance, from 1960 onward, and interviewed many of the participants, ENA can testify that there was a quite incredible explosion of quantity and quality—and of price, as people of all ethnicities came to appreciate the work. Sordid lucre may be the worst way to evaluate such things, but it does give some measure of how much people value a tradition. There was indeed a renaissance.

A large number of excellent studies of recent art, with some pros and cons about the renaissance, are found in *Unsettling Native Art Histories on the Northwest Coast* (Bunn-Marcuse & Jonaitis, 2020). The chapters in this book deal with current questions of museums and repatriation, authenticity and tradition, change and resilience, and other current sociological issues in a glorious but contested art world.

Few artists continued to work in the 1920–1960 era. Charlie James in Alert Bay continued to carve "Kwakiutl" (Kwakwaka'wakw?) poles, with some government and business patronage, and taught his stepson Mungo Martin. The latter, who organized that first potlatch at the British Columbia Museum, was a genuinely great artist by any standards, European or Native. He was also a fine and patient teacher of art, culture, and ceremony. He taught a whole new generation, including the art historian Bill Holm, long-time curator at the Burke Museum at the University of Washington. Holm's classic analysis of Northwest Coast art as serious art rather than craft (Holm, 1965) was the opening shot of the revolution, with regard to public appraisal of and attention to the art. It remains a standard textbook, found in countless art studios throughout the coast. Holm, a highly capable artist himself, also knew other traditional forms. He was, for a while, the only person to know several classic Kwakwaka'wakw dances, and generously taught many young First Nations people. For this he received a Kwakwala name.

Meanwhile, art revived among the more northern First Nations. The Tlingit and Alaskan Haida benefited from the fact that potlatches and ceremonies had never been banned in Alaska (thanks to the First Amendment), though the missionaries tried their hardest to discourage them. In consequence, art had not shrunk so much there.

In any case, a rapid and enormous expansion of art followed legalization of the potlatch. Local White interest rapidly increased. Patronage by the University of British Columbia, as well as by museums and galleries, grew. Bill Reid became a leader of the Haida art renaissance, along with Robert Davidson and others. Like Mungo Martin, Reid and Davidson went far beyond mere copying of old forms. They were genuinely great creative artists (see, for instance, accounts and examples in Dawkins, 2019). Reid, who had a career in radio before turning to art, was also an excellent writer and speaker (De Menil & Reid, 1971; Reid, 2000; ENA's interviews with Bill Reid in 1984 and 1985 lie behind much of this chapter). Reid's dialogue on classic art pieces with Bill Holm (Holm & Reid, 1975) was a landmark in understanding the art as form and message. (For a fine example of Reid as both performer and artist watch: https://www.youtube.com/watch?v=5f_fkZ3tW3U.)

Another pioneer of the renaissance was Wilson Duff, long-time curator of art at the Royal British Columbia Museum in Victoria, later professor at the University of British Columbia. He analyzed the art in depth and with sensitivity and theoretical interest, trying to understand the mindsets behind it. His work was cut short by his early and tragic death, but stimulated serious thought as well as public interest (Abbott, 1981; Duff, 1975, 1981, 1996).

Wilson Duff, Bill Holm, and Bill Reid (1967) worked together on the famous "Arts of the Raven" exhibit at the Vancouver Art Gallery. (This is Vancouver's city art museum, not a private gallery.) The gallery was then directed by Doris Shadbolt, who deserves much credit for the show. Her husband Jack Shadbolt was a leading artist, who, although not Indigenous, drew on Indigenous themes and supported Indigenous art. The combination of Duff the anthropological theorist, Holm the structural analyst, and Reid the Indigenous artist was perfect. A major stated purpose of this exhibit was to get Native art taken seriously as fine art, rather than dismissed as "craft" and consigned to museums of natural history. As a result, the show succeeded brilliantly. Native art prices shot out of sight, at least for the best pieces, and many a career was launched. Also, the show more or less explicitly gave Native people the right to explore, expand on, and develop their traditions any way they saw fit, without being attacked for "unauthenticity" whenever they got beyond the pre-1921 forms. A later Legacy show (MacNair et al., 1984) highlighted new creations since the 1960s revival. The catalogue contains particularly fine photos and biographies.

We now have Native video, glassmaking art, mixed media, animation, and the art keeps diversifying and improving. More and more of the younger generation find emotional satisfaction as well as economic prospects in the art. Woodcarving, stonework, and jewelry (Dawkins, 2019) continue to follow old themes, but with constant increase in technical variety and virtuosity. New materials are also used, as in Tlingit artist Preston Singletary's magnificent glasswork (Post, 2009). Charles Edenshaw in the late nineteenth century played with non-Indigenous themes, often ironic cross-references; this was tracked in the "renaissance" by Bill Reid, who had some Edenshaw ancestry, and continued to expand these themes. Through all this, the spiritual knowledge that drives the art remains active and vibrant. Indigenous belief and art have reacted to Christianity and secular modernity, but those settler intrusions have strengthened the old concepts at least as often as they have weakened them. (This point derives largely from ENA's field work and interviews, but see e.g., Wyatt, 1994, 1999.)

The art is highly relevant to environmental ideology and practice. Above all, it encodes the spirit beliefs recorded above, putting them in dramatic visual form for all to see. (Among major collections, see e.g., Musée de l'Homme Paris, 1970). One major theme in Northwest art is the incorporation of other societies, including humans, within the forms, which demonstrates connectedness and relatedness in concrete form. Stonework was particularly difficult to make under aboriginal conditions, and thus particularly revealing, although it has been preserved for thousands of years, allowing us to see that Northwest Coast art developed slowly and steadily. A thorough collection of stone art, with all too brief introduction (the author died about the time the book was published), was provided by Wilson Duff (1975).

Totem poles serve as records of the kin groups that erect them. They display spirit forebears, human ancestors, and important events in the history of the group (recall Mr. Ross' dog, p. xxx). They usually represent major deities or spirit figures and various animals, usually the mythically powerful predators such as ravens, bears, eagles, and orcas. Smaller predators like wolves and seals are more common on masks and painted objects. Totem poles represent statements of ownership and history.

Masks are particularly important, because anyone wearing and dancing with a mask becomes the being or animal portrayed, or at least is inspired by its spirit. In the spectacular performances of the central and northern coast, huge masks were and are created, many of which could be changed during performances by strings attached to moving parts. Raven could open and close his beak, cranes could flap their wings, and supernaturals could open and close their eyes. Among the Tsimshian, fast shifting of masks was also practiced; men could shift from male to female and back by shifting masks, often accompanied by women doing the same (Halpin, 1981a). This partially explains one of the most striking works of Northwest Coast art: a set of two neatly fitting masks carved in hard stone by an unknown Tsimshian artist in the mid-nineteenth century (one was collected in 1879; Duff, 1975: 164). Apparently, metal tools were used. The inner mask has eyes wide open, with an alert look; the outer mask has its eyes closed and appears peacefully at rest. Duff (1975) and Marjorie Halpin (1981a) analyzed this pair, noting that they could be changed rapidly. In the flickering firelight of a winter ceremony, it would seem that a stone-headed being was opening and closing his eyes. Open eyes stand for alertness and perception. Halpin quotes a Tsimshian writer in 1863 describing religious conversion as having his eyes bored (Halpin, 1981a: 286). Tsimshian masks and other visual privileges are *naxnox* (spirit power vehicles), and often accompany societies of initiated persons of noble lineages, such as the Dog Eaters and Cannibals.

The Kwakwaka'wakw also have hereditary mask privileges that go with secret societies, including a supposedly cannibal society whose members can theoretically take bites out of people (but, in practice, fake it). The most famous cluster of initiate-related masks and dances concerns the monstrous man-eating birds known as the Hohokw, the Baqbaqwalanukhsiwe, the Crooked Beak of Heaven, and others. Masks of these beings are so large that only a very muscular person can dance with them; they can weigh 30 kg. The dances are highly athletic. Other classic masks include Moon, Sun, Wolf, Raven in various incarnations, *bukwus* (Wild Man of the Woods), *nuɬmaɬ* (a sinister "fool," an ominous clown), grizzly bear, sea creatures, and many more. Many of these are significantly ambiguous figures, like the terrifying but helpful monster birds and the humorous but dangerous *nuɬmaɬ*. The mix of danger with humor or help is an expression of strong spirit power, and the three together make up a basic underpinning of Kwakwaka'wakw philosophy, an armature on which to base ontological and moral thinking.

Art achieved its highest expressions in the winter dances, when spectacular performances were staged by powerful young men who could leap wildly around the hall while manipulating heavy masks, such as the Kluckwalle of the Makah and Nuu-Chah-nulth (Ernst, 1962). The masks represented spirit beings important in the stories of the social groups represented. To varying degrees, according to the local worldview, the dancers became the spirits and animal powers. At the very least, the performers were inspired by those beings. The interpenetration of human and nonhuman society was total. This totality was expressed in the most beautiful, compelling, intense, and dramatic ways humanly possible. Durkheim's view of ceremony as a way of engaging people in their cultural and religious world has never been more spectacularly confirmed. No one could miss the personhood of natural kinds, or fail to be moved emotionally and morally by the needs to respect and care for those kinds.

The visual arts, including dances, portrayed the myths and stories of the groups in question. Among images most frequently observed on totem poles are Raven, bears, orcas, frogs, eagles, beavers, and other frequent heroes of myth. The Kwakwaka'wakw winter dances include a whole cycle, danced largely by highborn persons, of myths about the human-eating Bakbakhwalanukhsiwe, Crooked Beak of Heaven, and Hohokw.

Surprisingly rare in Northwest Coast art are wolves, but the Nuu-chah-nulth are an exception, possibly because of the significance of the Kluckwalle ceremony (Ernst, 1962). Tracing relationships with these formerly common animals, they have complex dances using wolf masks. Otters are largely confined to shamanic art, because of their sinister magical implications. Black Oystercatchers are also associated with shamanism. Tales like the Bear Mother are often portrayed. Since the renaissance of the 1960s, art has expanded rapidly to portray any and every subject, and has gone off into abstract forms.

Conservation tales were often dramatized in the spectacular winter dance ceremonies. In 2010, ENA was privileged to see the ?Atla'gimma dance drama, a forest spirit ceremony owned by four Kwakwaka'wakw families from the Rivers Inlet area. It was performed at the 33rd Annual Conference of the Society of Ethnobiology, by kind courtesy of Kwaxsistalla (Adam Dick) and his family. It is a spectacular performance, involving some twenty dancers masked and costumed as fantastical spirit beings of the forest. The story is the standard one known from the Arctic to the Maya of Quintana Roo, Mexico, where ENA's friend Don Jacinto told it from his own experience: a young man goes hunting, takes more than he needs, and is caught by the spirits and warned never to do that again. The spirits instruct him in how to act properly and respectfully toward the forest and its living beings. A number of minor points of ecology appear incidentally; a young tree growing from a stump (a universal sight in the Northwest Coast) shows continuity and rebirth. The full performance is long and extremely dramatic, and is accompanied by many long speeches that thank, bless, and instruct all those present.

References

Abbott, D. (Ed.). (1981). *The world is as sharp as a knife: An anthology in honour of Wilson Duff*. British Columbia Provincial Museum.
Ackerman, L. A. (Ed.). (1996). *A song to the creator: Traditional arts of native American women of the plateau*. University of Oklahoma Press.
Barbeau, M. (1929). *Totem poles of the Gitksan, upper Skeena River, British Columbia*. National Museums of Canada.
Barbeau, M. (1950). *Totem poles*. National Museum of Canada.
Barbeau, M. (1953). *Haida myths illustrated with argillite carvings*. National Museum of Canada.
Boas, F. (1955). *Primitive art*. (Orig. 1927.) Dover.

References

Boas, F. (1995). *A wealth of thought: Franz Boas on native American art*. University of Washington Press.
Bunn-Marcuse, K., & Jonaitis, A. (Eds.). (2020). *Unsettling native art histories on the Northwest Coast*. University of Washington Press.
Daly, R. (2005). *Our box was full: An ethnography for the Delgamuukw plaintiffs*. University of British Columbia Press.
Dawkins, A. (2019). *Understanding Northwest Coast indigenous jewelry*. University of Washington Press.
De Menil, A., & Reid, B. (1971). *Out of the silence*. Harper & Row.
Duff, W. (1975). *Images: Stone: BC: Thirty centuries of Northwest Coast Indian sculpture*. Oxford University Press.
Duff, W. (1981). The world is as sharp as a knife: Meaning in northern northwest coast art. In D. Abbott (Ed.), *The world is as sharp as a knife: An anthology in honour of Wilson Duff*. British Columbia Provincial Museum.
Duff, W. (1996). *Bird of paradox*. Hancock House.
Duff, W., Holm, B., & Reid, B. (1967). *Arts of the raven*. Vancouver Art Gallery.
Ernst, A. H. (1962). *The wolf ritual on the Northwest Coast*. University of Oregon Press.
Garfield, V. (1939). Tsimshian Clan and Society. *University of Washington Publications in Anthropology, 7*, 167–340. University of Washington Press.
Garfield, V. E., & Forrest, L. A. (1961). *The wolf and the raven: Totem poles of Southeastern Alaska*. University of Washington Press.
Halpin, M. (1981a). 'Seeing' in stone: Tsimshian masking and the twin stone masks. In D. Abbott (Ed.), *The world is as sharp as a knife: An anthology in honour of Wilson Duff* (pp. 269–288). British Columbia Provincial Museum.
Halpin, M. (1981b). *Totem poles: An illustrated guide*. University of British Columbia Press.
Hawker, R. W. (2003). *Tales of ghosts: First Nations art in British Columbia, 1922-1961*. University of British Columbia Press.
Holm, B. (1965). *Northwest Coast Indian art: An analysis of form*. University of Washington Press.
Holm, B., & Reid, W. (1975). *Form and freedom: A dialogue on Northwest Coast Indian art*. Rice University Press.
Jonaitis, A. (1981). *Tlingit halibut hooks: An analysis of the visual symbols of a rite of passage* (Vol. 57). American Museum of Natural History.
Jonaitis, A. (1986). *Art of the northern Tlingit*. University of Washington Press.
Jonaitis, A. (1988). *From the land of the totem poles*. American Museum of Natural History.
Jonaitis, A. (Ed.). (1991). *Chiefly feasts: The enduring Kwakiutl Potlatch*. University of Washington Press.
Jonaitis, A. (1999). *The Yuquot whalers' shrine*. University of Washington Press.
Jonaitis, A., & Glass, A. (2010). *The totem pole: An intercultural history*. Douglas & McIntyre.
Lam, M., & Gonzalez-Plaza, R. (2007). Evolutionary universal aesthetics in ecological rationality. *Journal of Ecological Anthropology, 10*, 66–71.
MacDonald, G. F. (1983). *Haida Monumental Art*. University of British Columbia Press.
MacNair, P. L., Hoover, A. L., & Neary, K. (1984). *The legacy: Traditional and innovation in Northwest Coast Indian art*. University of Washington Press.
McLennan, W., & Duffek, K. (2000). *The transforming image: Painted arts of Northwest Coast First Nations*. University of Washington Press.
Miller, J. (1999). *Lushootseed culture and the shamanic odyssey: An anchored radiance*. University of Nebraska Press.
Musée de l'Homme Paris. (1970). *Chefs-d'oeuvre des art indiens et esquimaux du Canada*. Société des amis du Musée de l'Homme Paris.
Post, M. (2009). *Preston Singletary: Echoes, fire, and shadows*. University of Washington Press.
Reid, B. (2000). *Solitary raven: Selected writings of Bill Reid*. University of Washington Press.
Townsend-Gault, C., Kramer, J., & Ḵi-ḵe-in. (2013). *Native art of the northwest coast: A history of changing ideas*. University of British Columbia Press.

Wardwell, A. (1986). *Ancient Eskimo ivories of the Bering Strait*. Hudson Hills Press for American Federation of Arts.

White, D. (2013). 'Where mere words failed': Northwest coast art and law. In C. Townsend-Gault, J. Kramer, & Ḵi-ḵe-in (Eds.), *Native art of the northwest coast: A history of changing ideas* (pp. 633–676). University of British Columbia Press.

Wyatt, G. (1994). *Spirit faces: Contemporary masks of the Northwest Coast*. University of Washington Press.

Wyatt, G. (1999). *Mythic beings: Spirit art of the Northwest Coast*. University of Washington Press.

Chapter 13
Conclusions

Summary of What Has Gone Before

Among Indigenous Nations throughout America, ceremony and ritual are techniques developed to ensure that individuals retain their understanding of connections and responsibilities to all life forms (Pierotti, 2011, Chap. 5). Religion is famous for this, but successful conservation organizations have all figured out small ways to do this, from volunteer programs to beautiful calendars. Practice, such as those programs, and art, from calendars up to Northwest Coast winter dances, are clearly keys—essential to perpetuation of this ethic. Everything to encourage love of nature, respect for natural beings, and responsibility toward them is all to the good. For Elinor Ostrom's self-conscious rules, recorded in the introduction to this work, traditional cultures substituted emotional and religious engagement. These worked where rational appeal failed.

Emotional involvement and consequent responsibility are required to get people to self-police at the level required. The stunning effectiveness of a morality that prevents people from taking too many bow staves from a rare tree, from exterminating a tiny herd of caribou, and from "robbing a creek" shows what can be done.

Nancy Turner, leading authority on human ecology on the Northwest Coast, has compiled a list of lessons she wants the world to learn. She advocates "connecting with communities and places," celebrating generations and elders, recognizing relationships, being grateful and responsible while maintaining accountability, "valuing and supporting diversity," using different teaching styles, and using all these to re-create healthy, sustainably managed ecosystems (Turner, 2014: 2: 403, 404, 405–411).

The common threads that unite those many diverse groups are straightforward. The main points center on a few truths and a few design principles.

First, resources are always limited but can be managed sustainably. Doing so requires information on what can be sustained, but it also requires very wide-scale use of resources—ideally, using the total landscape. Concentration on one or a few resources is deadly. Also required is rapid feedback, so that users do not get trapped into a false sense of security by thinking benefits can be gained without costs. The dreadful story of oil is before us: nothing but benefits considered up into the twentieth century; then pollution recognized and partially controlled; but then the dreadful reality of global warming due to carbon emissions was discovered, too late to stop Big Oil from dominating the world. Such things were prevented in the Indigenous world by the rapidity of feedback: overhunting the game meant starvation next year.

Second, survival of communities over time depends on such management. Exhaustion or waste of resources leads to catastrophe. The "infinite substitutability" of formal economics is meaningless in the real world.

Third, managing resources, especially common-pool ones, is a social and community affair, and must be socially and culturally constructed.

Fourth, this requires firm rules and principles, including some sort of fairness (however structured by local hierarchy), clear limits on areas to be managed, reasonably clear limits on take, and constant monitoring of the state of populations and the reasonable yield (which will, note, always be less than the "maximum sustainable yield," since people are too prone to be overoptimistic and cheat or fudge on that).

Fifth, and this is where Northwest Coast people go beyond resource economists, *the fact that management is socially constructed means that it must be socially engaged. It must be part of an emotionally involving social responsibility.*

Sixth, this has come to mean, in the Northwest Coast and very widely from ancient Europe to South America, that people must feel close relationship with the nonhuman world, feel social responsibility toward it, and show full respect to its denizens as to other humans.

Seventh, emotional relationship with the wider world is represented and made intense and sharp by art forms: stories, songs, dances, visual art, and even gourmet feast food. Visual art especially has been elaborated to a world-class level on the Northwest Coast, and most of it is dedicated to animal and spirit beings that express and record human relations with the rest of the cosmos. These culminated in the spectacular winter ceremonies, which dramatized social relations in the human, ecological, and spirit worlds. The intense mobilization of group emotion drove home the messages of mutual care.

Eighth, constant emphasis on generosity and sharing—including the unashamedly competitive form in the potlatch—prevents the extreme concentration of wealth and power in the hands of a very few. Chiefs consolidated their rule, but only by being generous. This saved the Coast from the extreme inequality that allows the rich in the modern world to neglect the environment and avoid paying the price of destroying the resource base. It also forced everyone to consider others—at least other kinship groups and communities, if not every other human. Slavery and classism were major problems, with environmental as well as human costs.

Ninth, respect for all beings is absolutely essential. It is the constantly repeated key. The other R's, responsibility, relatedness, and reasonableness, follow from it.

Tenth, ideas that seem "irrational" to the settler may be totally rational in their implications, while ideas that seem the very soul of rationality to the settler may be literally insane from the long view. Some truths unquestioned in bioscience only a few decades ago, notably the Cartesian view of animals, have proved wrong. Some "religious" claims of Indigenous people, such as the relationships of beings in the wild and the consciousness of at least the higher animals, have proved scientifically correct. Pigeonholing views as "science" vs. "religion" or "tradition" is not helpful.

Eleventh, and the most important thing we personally are adding to the whole discourse, is that traditional management is heavily pragmatic and experience-based. It is neither the product of "noble savages in harmony with nature" nor of ignorant, isolated "premoderns." It is a solid, grounded practice based on continual interaction with the totality of the human and nonhuman world, as experienced by the people who are acting. It is based, often, on sober knowledge of the results of overuse, and of how to heal those results. Culture, management, and society all arise from ongoing interactive practice, culturally and socially constructed by thoughtful agents. The vital corollary is that people begin by responding to experiences that are similar across cultures: sheep, fish, trees. They perceive these according to cultural expectations, but cannot stray too far from reality if they are to make a living. But as they move farther from direct interactive experience, they create more and more abstract ideas, often inferred explanations. Everyone knows an earthquake when they feel one, but whether it is explained as dragons in the earth or subducting slabs of ocean floor depends on cultural inference and learning.

Combining consideration of all beings, respect for all beings, and loving concern for the wide world, with the basic design principles of resource economics is quite adequate to save resources, and with modern capacity for evaluating and preserving, it would be quite adequate to save the world. What we of the modern world lack is the concern for nature. Five thousand years of anti-nature ideology, from Sumer and Babylon to Rome to western Europe to the Americas, has implanted a deeply destructive ideology in settler colonialism. It is a long-standing cultural belief, holding over from long before capitalism, colonialism, and racism came along to bring out its very worst corollaries.

Colonialism and its economic theories, from Hobbes and Locke to neoliberalism, have propagated this view over the world. The basic concept was to destroy local nature and replace it with a totally culturally-built landscape, which is what England is today. This began with the wholly rationalized, gridded, monocropped irrigation agriculture systems of Babylon and Assyria. They have resulted in the rationalized, gridded, monocropped fields of central Oregon and eastern Washington. Through eliminating even the concept of sustainability, the attitude of destroying in hopes of building brought the sustainable resources of the Northwest almost to the vanishing point.

They could be restored, if we learned from the occupants of the land instead of from a wave of settlers who had not had time to develop either knowledge of or respect for the land, and who came with an ideology of destruction. Treating resources as if plants, animals, and people mattered, if forced by public opinion, would force managers to think of sustainability and diversity rather than instant drawdown in the hopes that the future will take care of itself.

This would allow the best of modern science and technology to be used in the service of conservation, rather than in the service of destruction in the hopes that something better could be done with a ruined land.

References

Pierotti, R. (2011). *Indigenous knowledge, ecology, and evolutionary biology*. Routledge.

Turner, N. J. (2014). *Ancient pathways, ancestral knowledge: Ethnobotany and ecological wisdom of indigenous peoples of northwestern North America* (p. 2). McGill-Queen's University Press.

Chapter 14
Appendix 1: Indigenous California

> ...We are constantly walking on herbs, the virtues of which no one knows.—Pastor, Chumash elder, as quoted by Fernando Librado Kitsepawit, Ventureño Chumash (Librado, 1979: 56; cf. Blackburn, 1975: 258, Hudson et al., 1977)

Trying to understand Indigenous California can seem confusing because so much damage was done to these peoples and their cultures by colonialism. Starting with the Spanish, we find ourselves dealing with deep and unresolved trauma. There was much ecological disruption. As an example, the Pomo, a people with strong connections to the sea, were forced into small rancherias in the interior. Much of the connection to the marine environment was forcibly severed. Even today, when California appears to be liberal and progressive, government agencies have to be subjected to popular pressure to acknowledging Indigenous rights. As a consequence, much of the old way has been lost, or at least deeply hidden, so Indigenous people are often less prone talk about issues, compared to those of the Pacific Northwest.

Another point to emphasize is that there were no early treaties, because of the Spanish efforts at colonialism, which included slavery and genocidal actions. Areas of Spanish influence were covered by the Treaty of Guadalupe Hidalgo of 1848, which recognized Spanish laws in such matters as Indigenous affairs. Local treaties were made later, but contemporary California has no treaties like the Washington tribes, that allowed the restoration of fishing and whaling rights. Finally, California's efforts on conservation have often been misdirected, so the results have turned many California Indigenous cultures into the equivalent of "conservation refugees" as described by Dowie (2009), who begins his book by discussing the removal of Miwok from Yosemite, because of attitudes advocated by (among others) famous "conservationist" and racist (a victim of his time), John Muir (though in fact the Miwok remained in the valley during the early decades of the park, being removed only in the 1930s).

Despite all of this, there are still glimmers of hope and the traces of the attitudes that almost certainly prevailed in California prior to the European invasion. Respect and responsibility were key ideas. Humans were related to other beings, regarding them to varying degrees as persons with their own consciousness and agency. People approached plants with reverence and respect. They felt the usual Native American kinship with the rest of the environment (Anderson, 2005: 57–59). Ceremonies for first fruits and seasonal foods bonded people to the resource base while simultaneously making them aware of the fact that resource could disappear if not treated carefully. Kat Anderson cites a Chukchansi Yokuts elder: "I've always wondered why people call plants 'wild.' We don't think of them that way. They just come up wherever they are, and like us, they are at home in that place" (Anderson, 2005: 41). She and Thomas Blackburn note: "Today, native peoples still retain a deep respect for the natural world, and retell stories that remind them of the absolute necessity for judicious harvesting. Elders are quick to tell younger gatherers, 'Do not take all—and leave the small ones behind'" (Blackburn & Anderson, 1993: 20). A more specific sense of genuine deeply-felt responsibility for conserving resources for the wider good was the basic attitudes of management. One difference from the Northwest Coast is that the Native Californians had fully developed agriculture, with standard southwestern crops or with native plants (Lawton et al., 1976).

Yet—whether because it is really lacking or because ethnography is so thin—there is little record of its being verbalized explicitly in a philosophic ideology, as is often found on the Northwest Coast. We suspect the ethnography is the problem. Research came somewhat later; more to the point, California was settled much earlier, with intensive missionary activity along the south and central coasts. There was more alteration of the cultures, and Californians were subjected to outright, explicit genocide (Madley, 2016), more bloody than anything north of southern Oregon. We thus found a sadly thin record of conservation narratives. Fortunately, California Native people survived, and are rapidly reviving much of their cultural heritage, including reverence and respect for the environment.

In the meantimes, the work of Kat Anderson (1999, 2005, 2017) and her associates (Anderson & Lake, 2013, 2017) has done much to document and preserve what is left. *California Indians and Their Environment* by Kent Lightfoot and Otis Parrish (2009) provides a thorough introduction to that subject. Basic reference works (Bean & Blackburn, 1976; Golla, 2011; Heizer, 1978; Kroeber, 1925; Lutkin, 2002) provide background. Bettinger (2015) describes political culture. California ethnobotany has been richly explored over several generations, but we have probably lost most of the old knowledge. Even so, what remains is stunning. Most of the southern California tribes have produced full ethnobotanical books (for the Cahuilla, Bean & Saubel, 1972; Chumash, Arnold, 1987, 2004; Gamble, 2008; Timbrook, 2007; Kawaiisu, Zigmond, 1981; Santa Ysabel Kumeyaay, Hedges & Beresford, 1986; Baja California Kumeyaay, Wilken-Robertson, 2018; Serrano, Lerch, 1981; others in manuscript) and northern California has not been neglected (e.g., Latta, 1977 on Yokuts; Welch, 2013 on Pomo; for Northern Paiute, just across the line in Nevada, Fowler, 1992, 2013; general, Gifford, 1957; Mead, 2003). Ethnozoology is less well covered (but see Timbrook and Johnson's Chumash ethnoornithology, 2013).

Old records remain important. J. P. Harrington's are the richest (see Timbrook, 2007). Records are still being discovered and made available.

An important exception to the lack of good documentation is the Klamath River region in the far north, where traditional culture continues to an appreciable degree. For that area we have not only a great deal of good ethnography, but a unique early book by a Native Californian—Lucy Thompson's *To the American Indian* (1991, orig. 1916), who points out that the Yurok carefully protected sugar pines, source of nuts and sugary sap (Thompson, 1991: 28ff). They conserved fish (Thompson, 1991: 178–179), and burned carefully and systematically (Thompson, 1991: 31–33). Her account stands in striking contrast to ethnographies by outsiders—she stresses the spiritual interaction with nature and its function in maintaining conservation. It seems highly likely that this was universal, and simply missed by early ethnographers.

Confirmation for the Yurok case comes from more recent work. Yurok spiritual teacher Harry Williams (Buckley, 2002; Burrill, 1993) tells us:

> I was with my grandfather, Charley Williams. We were walking on a dirt path down to the ocean. There was a bug crossing our path, and my grandfather told me, 'Reach down and help that bug on its way.' So I did. I reached down and helped the bug on the path to where it was going. 'Now, do you know what you have done?' Grandfather continued. 'You won't feel badly now, for perhaps a bird will someday eat the bug. But you must remember that the Creator created the bug for birds to eat. He didn't create them to get stepped on' (Burrill, 1993: 43; presumably the Creator is Wohpekumeu, the Yurok trickster-transformer; strictly speaking there is no Creator, since the Yurok teach that the universe has always existed).

Lest anyone think this is an exaggeration, ENA observed similar behavior among the Yucatec Maya based on similar teachings. Maya who picked up a bug to show ENA would always put it back on the path, unharmed, and headed the way it had been going (Anderson & Medina Tzuc, 2005).

Harry Williams' grandfather also said that rocks are living things, and that "the white man is like the wind. Nobody knows where he comes from." Williams also tells of a line he heard at a Native American conference: "Creator gave man two ears, and two eyes, but only one mouth. But the white man thinks he has five mouths, no ears, and no eyes. That must be why he talks so much" (Burrill, 1993: 106).

(Pierotti was raised in a similar way. He writes: Among the things my mother taught was that if I were stung by a bee or wasp it was almost certainly my fault, because I had upset a relative. Because of their highly social nature, such insects were considered close relatives to the Penateka, or 'wasp' band. [Pierotti's band within the Comanche nation.] One of my most profound experiences took place when the lawnmower I was riding flipped up a rock that dislodged a paper wasp (*Polistes exclamans*) nest. Given my upbringing, I was upset and ran to the nest, surrounded by flying, agitated wasps. When I reached to pick up the dislodged nest, the senior female of the group flew over and landed on my index finger. I was sure I was about to be stung when she bent her head and gently bit me on the finger. I put down the nest and she flew off. Later, after the wasps had calmed down, I picked up the nest and put it back in its original location. The next day the wasps had reattached the nest to the wall. Despite this potentially traumatic event, these wasps showed no increased aggression towards humans, even though humans passed within a few feet of their nest for the rest of their summer breeding season.)

For the Karuk, the "*Ikxareyavs* were old-time people, who turned into animals, plants, rocks, mountains, plots of ground, and even parts of the house, dances, and abstractions when the Karuk came to the country" (Harrington, 1932: 8; Lang, 1994). Many of the most feared of these beings turned into large and spectacular rocks, a story which the Karuk supposedly proved to Harrington by pointing out that you can still see the rocks. The *Ikxareyavs* who turned into abstractions remind us yet again of classical Greece, with its goddesses such as Sophia (Wisdom) and Nike (Victory), as well as the ancient Hindus, who visualized Time (*kali*) as a goddess.

The Hupa, culturally very close to the Karuk though linguistically unrelated, held that *Yinukatsisdai* "made all the trees and plants which furnish food for men.... If he sees food being wasted, he withholds the supply and produces a famine." If pleaded with, he may relent, and "then gives the food...in such bountiful quantities that acorns are found even under the pines" (Goddard, 1903: 77). The highly religious Hupa shared the Northwest Coast view of punishment for waste, but added a fascinating touch in the creator being capable or mercy, which nature typically was not. The corresponding masters of deer are the Tans, who withhold deer from hunters if deer are not treated with respect. As elsewhere in the wider Northwest and California (e.g., Rhoades, 2013: 81), respect means not only treating the dead deer respectfully but also totally refraining from waste, overhunting, or hunting without serious need for food. Yet other deities care for fish in similar ways (Goddard, 1903: 77–78). The fish, like the deer and other food animals, had been kept by entities understood to be supernatural, and had to be liberated by culture-heroes; there are many stories of these events, which are used in ceremonies to maintain the stock (Goddard, 1904). Less effective, but not unrelated, were prayers that birds (magpies, jays, and woodpeckers) and squirrels might not desire to eat, or cache, the acorn crop (in competition with the Hupa) (Goddard, 1903: 81).

The Yurok, Karuk and Hupa joined in enormous ceremonies intended to preserve and renew the world and its resources (Keeling, 1992a, b). These ceremonies almost died out, but managed to survive, and are now once again celebrated regularly. The belief in the need of humans to hold elaborate and active rituals to keep the game, fish, and plant foods productive has itself been preserved and renewed, especially since these groups have seen the result of modern Californian indifference to conservation. The Hupa even regarded trails as sacred persons. Many myths detail the origin of these ceremonies and the need for them, but specific teachings describing directly conservationist behavior are rather limited. However, it seems clear that a general reverence and spiritual concern for the landscape has a preservationist effect. People will not thoughtlessly waste resources that are personally and spiritually important (Anderson, 1996).

Among the Lake Miwok, as in several other parts of North America, "game animals were believed to be immortal and under spirit control, and it was believed that animals sometimes transformed themselves into other species" (Callahan, 1978: 272). The Yuki held that deer are immortal, their souls living in a mountain under care of a Deer Guardian who is second only to the Creator and who controls availability of obsidian as well as deer. This belief in animal souls within a mountain occurs widely in North America, and is found as far south as *Huitepec*, the sacred mountain near San Cristobal in Chiapas. It is more than likely that the immortality and protectedness of spirits of game was universal.

The Northern Pomo had a concept of

> *xa*, manifestations of the supernatural spirits...left on earth from the beginnings and investing certain peculiar objects with supernatural attributes.... *Xa* is the genius of procreation, acquisition, alien to human activities...but a spiritual concomity of men whose aid may be engaged through prayer and possession of its symbols.... *Xa* is summoned by sexual contact, is the mystery of conception and gestation, leaves its stamp on the buttocks of new born till erased by cognoscence; places an indelible mark on the skin of a favored mortal.... *Xa* is the inspiration of song..., the rhythmic impulse of song-dance ceremonies, the buoyancy of regalia...and the stimulus of fingers tapping upon the flute. It is the celestial, beneficent influence as opposed to the terrestrial demon of diaster... (notes of John Hudson, ca. 1900, from Welch, 2013: 169).

Xa resides in hawk and falcon and eagle feathers, in crests and red tails of birds such as woodpeckers, and in omens and apparitions. Five plants have *xa:* trail plant (*Adenocaulon bicolor*), angelica, sweet cicely (*Osmorhiza* sp.), Fendler's meadowrue (*Thalictrum fendleri*), and leather root (Welch, 2013: 169). As elsewhere in California, tobacco and Jimson weed (*Datura wrightii*) were sacred in a different way: yielding direct visions and healings. All this probably fed back on resource management, but no early record of this seems to have survived.

Equally revealing is a story related by Fernando Librado (1979: 113): A man was trapping rats in a pitfall trap; an old Santa Rosa Chumash told him: "'You are polluting our mother, *Xutash!* [The Chumash earth being]. Remove this at once. If you defile our Mother, she will give us nothing.'"

The Cocopa, who lived in a landscape of abundance in the fertile Colorado River delta, lived simply and never worried much about food—though famine threatened if a drought led to the river being very low. William Kelly made careful enquiries about religious beliefs connected with fertility, harvest, wild foods, agriculture, and the whole suite, and found:

> Harvest festivals...were...religious in nature; yet their function, explicit and implicit, was in connection with group life and social organization, and they were neither related to the harvest as such nor a mechanism aimed at increasing effort or diligence in farming (Kelly, 1977: 44).

This seems general throughout California (the Cocopa live today in Baja California). The only rule related to such issues that is ecological in function was the universal Native American rule that a boy could not eat his own first kills (Kelly, 1977: 45), which teaches responsibility and reciprocity. In North America such practices are part of teaching the boy respect for both the game and the human social group that shares the kill. The Cocopa tabooed doves but apparently for totemic, not environmental, reasons. As agricultural people, they probably had a different take on myth, with Coyote a creator of good crops and transformer of insulted ones into bitter wild foods, rather than a producer of good wild foods for people to gather.

Gary Nabhan (2013) provides a brilliant, incisive analysis of this contrast in southwestern mythology.

We are, however, surely missing a great deal. The Northern Paiute, whose territory included the northeast corner of California, offered prayers to slain game animals. They left the tail tip of a hunted deer under a rock with the prayer "'Deer, thank you, and come again.' A similar offering was made for bighorn sheep" (Fowler, 1992: 181). The deer will be reincarnated, and will again offer itself to the hunter if treated with respect—a universal North American belief. Even harvesting roots required an offering or prayer. Eagle feathers were harvested without killing the eagle (ibid.). A number of prayers, to the Sun and other powers (see below), were given daily or frequently. Ceremonies insured continued production of food resources. Yet Fowler does not record conservation myths either.

Taboos may also have had a conservation effect. The Yana, for instance, did not allow salmon to be eaten with deer meat, small game, or roots taken from gopher burrows (Sapir, 1910: 156). The salmon would cease to present themselves for harvest if this taboo was violated. Possibly they did not like to associate with prototypically "land" foods, but more likely it was inappropriate to mix foods meant to be taken in different seasons (Brody, 1982: 196–197; Pierotti, 2011: 37). There is no evident conservation here, but at least some respect for all these foods was apparent. The Wintu and related peoples tabooed a large number of things, including most birds of prey and predatory animals (Du Bois, 1935). The Nutuwich Yokuts even tabooed bear and deer, being thus reduced to eating rabbits for meat, a most unusual degree of forbearance (Gayton, 1948: 166). Taboos this extensive would have a major ecological effect. They preserved the hunted species indirectly, by preserving keystone species in the ecosystem. Heizer (1978) cites a number of taboos and rules from around the state that affect land use.

The Chilula (Athapaskan) culture is barely known from a very few elderly people just after 1900. One of them "was a medicine woman for troubles caused by the deer gods" (Goddard, 1914: 379). That is all we know of Chilula animal religion, outside of a few generic myths, which imply existence of spirit guardians of the game such as we know from most of Native America.

Attested from one end of the state to the other are harangues by chiefs telling people to work hard and diligently at hunting, gathering, and food production in general. (The best description is for the Atsugewi [Garth, 1978], but the custom is attested all the way to the Mohave [Kroeber, 1972] and Cocopa [Kelly, 1977: 66].) Many groups had a special designated Orator as well, who could do this. People could ignore if they chose, but they would then be subject to major criticism and be shunned, and hard work was a strong value everywhere from the northwest corner (e.g., Buckley, 2002, Kroeber, 1972 for the Yurok) to just beyond the southeast one (Cocopa: Kelly, 1977: 23). This would rather tend toward overexploitation of the resources than conservation of them, and thus may have much to do with the archeological evidence noted above of local over-harvest.

Attitudes and Representations: Philosophy.

Pierotti states:

A common general philosophy and concept of community appears to be shared by all of the Indigenous peoples of North America, which includes: 1) respect for nonhuman entities as individuals, 2) the existence of bonds between humans and nonhumans, including incorporation of nonhumans into ethical codes of behavior, and 3) the recognition of humans as part of the ecological system (Pierotti, 2011: 198–199).

This statement, and all it implies is fully true for California.

Concerning Respect: The full panoply of North American Native conservation attitudes is reflected in an astonishing prophecy that Cora Du Bois recorded from Kate Luckie, a Wintu shaman, in 1925:

> When the Indians all die, then God will let the water come down from the north. Everyone will drown. That is because the white people never cared for land or deer or bear. When we Indians kill meat, we eat it all up. When we dig roots, we make little holes. When we build houses, we make little holes. When we burn grass for grasshoppers, we don't ruin things. We shake down acorns and pine nuts. We don't chop down the trees. We only use dead wood. But the white people plow up the ground, pull up the trees, kill everything. The tree says, 'Don't. I am sore. Don't hurt me.' But they chop it down and cut it up. The spirit of the land hates them... The Indians never hurt anything, but the white people destroy all. They blast rocks and scatter them on the earth. The rocks says, 'Don't! You are hurting me.' But the white people pay no attention. When the Indians use rocks, they take little round ones for their cooking. The white people dig deep long tunnels. They make roads. They dig as much as they wish. They don't care how much the ground cries out. How can the spirit of the earth like the white man? That is why God will upset the world—because it is sore all over (Du Bois, 1935: 75–76; cf. Heizer, 1978: 650).

Luckie continued to say that water could not be permanently hurt, because it eventually runs to the ocean, and that it would thus survive to destroy the current world by flood, a perhaps astonishingly accurate prediction, concerning climate change, and cultural memories of changes in sea level over the last 15,000 years. The Wintu share the widespread North American belief that there have been four worlds of people so far, and we are in the fifth; which will be destroyed by flood, according to Wintu tradition. Another Wintu shaman commented that "the gold feels sorry" for the Indian people because they were driven from their homes by men seeking that metal (Du Bois, 1935: 76). (It is worth noting that Du Bois had no special interest in ecology or environment, but was meticulous about documenting shamans and all they said; hence these unique recordings.)

The California Native peoples shared the widespread Native American belief that disrespect of powerful animals brought danger. A story has made it from the Chumash consultant Juan Justo to early ethnographer John Peabody Harrington, thence to Chumash expert John Johnson, and then into Lynn Gamble's book on the Chumash—a typically indirect route:

> ...Juan's uncle began to laugh and shout and make fun of [a rattlesnake].... The other man advised Juan's uncle to be quiet,... but Juan's uncle made all the more noise....whereupon the other man left him and went on alone. When he was alone, Juan's uncle looked around a saw a whole pile of *guicos* [alligator lizards] with their mouths open towards him and their tongues out.... he... shut his eyes and went jumping and climbing to break through the lizards, and when he opened his eyes there was nothing there (Gamble, 2008: 216, quoting John Johnson's edited version of a Harrington text).

The lizards were probably warning Juan's uncle that if he teased a snake again he would suffer, probably through a bite from the snake.

General respect for plants and animals and their spirits and spirit guardians existed. Respect guided conservation generally. "One took what he needed and expressed appreciation.... Without these attitudes the California Indians could have laid waste to California long before the Europeans appeared" (Heizer, 1978: 650). There was a very general sense that plants needed human care, a sentiment backed up by experience, but going well beyond the facts of Native management and care. Plants and animals *need* to be used, as a mark of respect. Neglecting them wounds their spirits, and they decline and become weedy, poorly grown, and despondent-looking (Anderson, 2005; Blackburn & Anderson, 1993). California's useful plants do respond to care; basketry plants put out long straight shoots, nut trees crop more, and so on (Anderson, 2005; Benedict et al., 2014).

Such attitudes survive today in areas where something like traditional plant uses can exist. Michael Wilken-Robertson, interviewing Kumeyaay people in Mexico just south of the California border, heard from elder Teodora Cuero Robles:

> This I can assure you, the ancient ones never damaged a tree, no, never; they loved them as something very sacred. They would tell us not to go breaking the branches of the pines, not to play there, nor to climb up on any small tree, they said that they were almost just like humans; 'They are watching us, they are taking care of us, they give us our food. Don't go around damaging them don't be shouting, none of that,' they would say, 'take special care of them,' for this reason we know very well that we must take care of these trees. Also the medicinal herbs, those they especially charged us to care for, we shouldn't just go out and cut for no reason, go out and cut them and throw them away to dry up, no. They told us many things, that we should even care for the rocks, just imagine! The rocks, the sand, the springs, the water flowing, all these things they said we must respect (Wilken-Robertson, 2004: 49; see also Wilken-Robertson, 2018: 231–232).

From the Paipai, a group that moved from Arizona to northern Baja California about 300 years ago, comes a story told by Eufemio Sandoval. The Mexican government forbid them cutting juniper posts because the junipers were getting rare. Sandoval commented "we have never cut the plant to the root, but rather it has been a form of pruning that we carry out. We just take what is useful as a post and leave the rest to keep growing and developing" (Wilken-Robertson, 2004: 53–54, reprinted Wilken-Robertson, 2018: 236). Sandoval held that this was better conservation than pure neglect. Recall Wilke's findings on Great Basin junipers.

There is little reference to animals letting themselves be taken if they are respected, but apparently the belief existed. One Mohave did say that the Creator gave hunting to the desert tribes but not the Mohave, so when the Chemehuevi "see game, the animals cannot run fast, or they sit down...they want to be caught. The same with the Walapai. But if Mohave go to hunt, the animals run swiftly away" (Kroeber, 1972: 84). A Wintu hunter who failed to get deer would say "The deer don't want to die for me anymore" (Heizer, 1978: 651). Many stories around the state imply that animals not respected will not let themselves be killed. Conversely, they might go away. Elsewhere in North America, some groups have noted that game disappears as white settlers fill up the landscape, and suspect this is because

Native hunting is outlawed and the game is offended and leaves. Dramatic declines of deer have taken place since hunting has been banned in settled areas, but in fact the reasons are drought (first of all), suburbanization, and introduced diseases. Still, the Native view has its merits; failing to keep the game alert, and failing to weed out sick and slow individuals, has its costs.

As we have indicated, the belief that wild plants and animals, and even rocks, must be treated with *respect* is shared all over North America (Anderson, 1996; Pierotti, 2011). All have spirits, which are ever-present; and they deserve respect as elders, helpers, friends, and possible self-sacrifices to the human hunter or forager. As a result, there are absolute rules against overhunting, overcollecting, and waste, and these are observed in the remotest areas. The widespread North American taboo against a youth eating his first kill was probably general. Among the Chumash one report says that a hunter or fisher could *never* eat his own kill, on pain of never succeeding again (Grant, 1978: 512, citing Z. Engelhardt; cf. Grant, 1965), even though this ignores what happens when an individual kills to save themselves from starvation. This is evidently part of the North American complex of respect for animals. The Chumash are known to have prayed to the swordfish to drive whales on shore, and, as a result, had a swordfish dance and also revered other powers of the sea (Blackburn, 1975; Gamble, 2008).

California tribes also had the idea, general in western North America, that the bones of an animal should be treated carefully and respectfully, because such things as breaking a bone would mean the animal would reincarnate with a broken or missing leg; at least this is attested in myth, if not in actual practice (Blackburn, 1975: 131, in a myth recorded by Harrington). Another version of the same myth has Momoy—*Datura* personified as an old woman—protesting against a young man killing unnecessarily: "Have you no sense at all? You are just killing for the sake of killing" (Blackburn, 1975: 147). Occurrences like this, in Harrington's thorough materials on the Chumash, make it seem very likely that California Indians did not lack the usual western North American values; however, some ethnographers simply failed to record them.

Among the Monache, careful enquiry indicated otherwise: "No special ritual precautions accompanied the hunting of deer or bear. Animals were not addressed before, during, or after the kill" (Spier, 1978: 428). This is in marked contrast to the situation in the Northwest Coast and many other areas, where respectful addresses to the animals were required, and were part of a conservation-related ideology. Generally, nothing is reported either way for California peoples; "absence of evidence is not evidence of absence," but the lack is suggestive given Spier's report.

Despite these negative views (Spier may not have asked the right questions), California groups thought of the natural world as closely allied to the human one, almost always with actual kin relations or equivalent social bonds. Many Californian groups, especially in the center and south, had lineages and moieties with animal emblems. Moieties named Coyote and Wildcat (Bobcat) were found widely, as among the Cahuilla and Serrano. The Yokuts named them West and East, but divided the animals among them; "the Tachi assigned Eagle, Crow, Killdeer, Raven, Antelope, and Beaver to the…West moiety and Coyote, Prairie Falcon,

Ground [Burrowing] Owl, Great Horned Owl, Skunk, Seal" (species unspecified), and other beings to the East (Wallace, 1978: 453, from E. W. Gifford's data; reference to Valley Yokuts, but all Yokuts groups apparently had more or less the same). Anna Gayton (1976) points out that people really felt close to their lineage animals. Eagle and Coyote were the lead animals, respectively. Within the moieties were animal-named patrilineages; the Eagle lineage supplied chiefs. The Coyote moiety had its own chiefs, however (Wallace, 1978: 454). Moieties sometimes owned certain foods, and feasted each other with their respective foods (Gayton, 1976: 84).

The Miwok seem to have reached an extreme, extending human society to the entire cosmos by classifying everything (at least everything they noticed) into either the Land or the Water Moieties; the former included mostly up-country beings, the latter not only aquatic but also lowland creatures (Gifford, 1916, summary in Kroeber, 1925: 455; it is worth noting that Gifford's now-obscure studies of Miwok society are one of the more amazing achievements in the history of kinship studies, being far ahead of their time in almost every way; Gifford had a high-school education and was a true autodidact). These had nothing to do with individuals' spirit power animals or other less global social symbolism. The Monache had lineages named after birds or sometimes other animals; a lineage's namesake was called its "dog" in the sense of "pet animal" (Spier, 1978: 433). The Eagle lineage was the chiefly one; messengers came from Roadrunner and Dove lineages. The Yokuts used their Dove lineages for this purpose, and Magpie lineage members as criers.

Humans as Part of the Ecological System

This phrase understates the powerful, deep, and complex emotional attachments to the land. For every Californian group, their land is home—not just their personal home but the home of their people since the time of creation (or at least of transformation into our recognizable world). Lynn Gamble and Michael Wilken-Robertson, in a recent, particularly sensitive account of Native Californian relationships with the land (2009), describe "...a landscape that is permeated with symbolic and ritual meanings [,] that embraces mythical histories, ancestral pasts, and moral messages that overlay a landscape where economic resources, such as foods and medicines, abound."

Related to this ideational landscape are the themes of landscape as memory and landscape as identity. Specific places are reminders of a social past that was filled with "triumphs and disasters" and other stories. People remember "not just a boulder, but the significant events associated with the boulder" (Gamble & Wilken-Robertson, 2009: 148; see also Basso, 1996). Every stream and hill, as well as every sizable boulder, has its stories, and even individual trees are often important landmarks. Often the historic associations of these landmarks blend into myths and origin stories (as elsewhere in California; see, e.g., Woiche, 1992). Other, related groups in Baja California maintain similar ties with the land, including long-lost homes on the US side of the border, where they have kin and other social relations (Garduño, 2016). This refers specifically to the people of the Tijuana River basin in Baja California, but it could be said with equal truth of every group in California, or for that matter the whole of the North American Pacific coast, or even all of North America. Every ethnographer who has written much about the ideational culture of Native Californians has emphasized their extreme attachment to and concern for the land. Directions—including up toward the sky and down toward the lower world—as well as places have enormous significance ritually and culturally.

The Yumans of the Colorado River drainage—close linguistic relatives of the Tijuana River people— speak of "'Coyote Law,'...the law of the land—sometimes capricious and unreasonable like Coyote himself—but nevertheless, the way things are. [Their] tales tell about Coyote Law" (Hinton & Watahomigie, 1984: 6; Harney, 1995 for the Shoshone).

This train of investigation leads to two broad conclusions. First and most important, views of the land and its resources are impassioned. Native people are not sizing up "resources" with the cold eye of the economic planner. Instead, they are looking at their home. For California's people, the whole land is not only their family home, but the home of their entire people since the beginning of time. We may understand the latter clause as meaning "since the beginning of the group as an identifiable cultural and social entity," but that does not diminish its psychological force. The land is loved, but the emotional involvement is much more than that; it is a total personal involvement, the sort of interaction that Emmanuel Levinas regarded as literally infinitely important, because it makes us who and what we are (Levinas, 1969).

Second, knowledge is derived from interactive practice; knowing the land comes from living on it and making a living from it. Knowledge of plants and animals comes from working with them in the field, handling them and depending on them for sustenance or health. Rural Anglo-Americans in our youth learned the same way, and contrasted it quite sharply with "book learning." They knew that interactive practice is far better for learning actual life skills and work skills, whatever it may cost in knowledge of grand theory. However, their wider knowledge of the world was book-learning (or TV-learning). For traditional Californians, *all* knowledge came from interactive practice. People grew up knowing the local ecology from personal experience; they knew what the fish ate, which plants grew together, what was needed for a healthy ecosystem. Research comparing Native and White rural folk in the northern Midwest is relevant here; even Whites who knew the outdoors as intimately as the Natives thought very differently, seeing species as separate and relatively isolated rather than part of a great web that included humans (Medin et al., 2006).

Awareness of the possibility of overpopulation—too much population pressure—is found in the story, reported for every well-studied Californian group, of Coyote or some similar creature bringing death because the world would become too crowded if people lived forever. Where Coyote is the death-bringer, his own child is usually the first to die, and he regrets his choice (Pierotti, 2011; Pierotti & Fogg, 2017;

Smith, 1993: 3 where it is the very first tale presented). A similar account is provided by the Mohave shaman *Nyavarup* had "the small lizard" as deathbringer; who says "'I wish people to die. If they all keep on growing, there will be no room. There will be no place to go; if we defecate, the excrement will fall on someone's foot'" (Kroeber, 1972: 6).

Mythic animals were most conspicuously predators: Coyote, Wolf, Fox, Eagle, Falcon, Condor, because hunter-gatherers identify with predators, who they recognize as similar to themselves (Pierotti, 2011, Chap. 3). Many game animals seem to have been merely game animals even in mythic time. Among the Wappo, Elk was a humanoid pre-animal in mythic time, but hunted deer, which were ordinary game animals (see tales in Radin, 1924). Deer are rather rarely seen as having been humanoid in mythic time (though the Karuk have several Deer stories; Kroeber & Gifford, 1980). Among the Chemehuevi, Coyote and Fox hunted rats and mice (Laird, 1976).

Southern Californian groups had long and complex origin cycles, involving creation by heroic individuals. In the Serrano song cycle for mourning ceremonies, the first song spoke of the earth, the second *chukiam*, "all growing things" (Lerch, 1981: 11). This refers to plants and animals, but apparently to plants above all; it included a passage about the *Datura* plant, ritually used as a halluncinogen in puberty rites and in medicine. Cognate words such as *chukit* are known among other southern Californian Shoshonean groups. The Serrano creator died in Big Bear Valley, which, as a result, has an enormous variety of plants, many of them endemic. It is interesting that "[t]he Serrano "were not only aware of the phenomenon, they had an explanation for it in their cosmology" (Lerch, 1981: 14). The mourners turned to pines, which still stand in ranks around the valley (due, in modern terms, to the layered rock outcrops). The related Cahuilla have a long cycle of creation myths centering on Mukat, a human-like figure who brought agriculture among other useful plant and animal management strategies.

By contrast, the far northwest of the state had no origin myths; the cosmos always existed. However, creator-like beings had altered it greatly and made it suitable for humans, who appear after the time of such beings as the Karuk *ikxareyavs* (see below).

The Yuki and Kato had a high god (Taikomol in Yuki, Nagaitcho to the Kato; Goddard, 1909) who was above Coyote and his fellow creatures, though in Yuki and Kato creation these animals did the final tune-ups. Taikomol created the universe by song and speech.

In a beautiful Kato telling of the creation story by Bill Ray in 1906, Nagaitcho and the dog he has created end by rejoicing in their world:

"My dog, come along behind me and look."
   Vegetation had grown, fish had come into the creeks.
   Rocks had become large....
"Walk fast, my dog."
   The land was good. Valleys had appeared....
   Water had begun to flow. Springs had come....
"I made the land good, my dog,

Walk fast, my dog."
Acorns were growing, pine cones were hanging,
Tarweed seeds were ripe, chestnuts were ripe,
Hazelnuts were good, manazanita berries were getting white,...
Buckeyes were good, peppernuts were black-ripe,
Bunch grass was ripe, grasshoppers were growing,
Clover was with seed. Bear-clover was good. Mountains had grown.
Rocks had grown, different foods were grown.
"My dog, we made it good."
Fish had grown that they will eat.
"Waterhead Place we have come to now."
Different plants were ripe. They went back, they say,
His dog with him.
"We will go back," he said.
(Goddard, 1909: 93–94; slightly rewritten for comprehensibility. Nagaitcho and his dog return to the north, whence they came, and leave us this beautiful world.)

This hymn to all the wonderful foods of the north coast ranges—and indeed, they are excellent eating—is only a tiny part of a very long creation story that mentions virtually every plant, animal, fish, and geological feature in Kato habitat. It is all very reminiscent of Psalm 104, but far more richly detailed. It also reveals something of Bill Ray's personal narrative style, which included long chanted lists of plants, animals, and geographic features of the environment, alternating with narrative that is largely spoken and is so telegraphic as to be dreamlike. The combination is powerful enough to make Ray one of the more distinctive and poetically gifted California myth-tellers. Such lists are a widespread stylistic feature in myths in many languages of north-central California and elsewhere in the world (think of Hesiod's *Theogony*).

The Athapaskan original takes full advantage of the exquisite beauty and potential for sound-poetry of the Athapaskan languages. The above is rhythmical and rhymed poetry in the original, rhyming with the repeated chorus-line word *kwanang* ("they say"). Note that the mountains and rocks grow; they are living things in California belief. Like Hanc'ibeyjim's creation story (Shipley, 1991), Ray's is one of the greatest religious poems in world literature, and it deserves more than languishing in a forgotten monograph. Here are the first few lines, in Kato, simplified for easier reading from Goddard's linguistic transcription:

E lot, shiit la, nan dal,
  O dut t ge ka la e kwanang,
  To nai nas de le kwanang.
  Sha na ta se gun cha ge kwanang.
  N gun sho ne kwanang.
  Kakw chqal yani kakw ko winyal, e lots ul chin yani ne n gun sho ne kwanang....

Bill Ray's descendents are maintaining the Kato language and reviving it, with the help of Goddard's documentation, and anthropologist Sally Anderson has compiled a dictionary from this material (Anderson, 2018; emails to ENA from Kitty Lynch and S. Anderson, 2021).

Many other groups had Earthmaker and Coyote or Wolf and Coyote as creators. Earthmaker or Wolf was the senior, more responsible and sober one, the stereotypic elder sibling. Coyote was young and wild, everybody's crazy kid sibling. Everyone respected Earthmaker but loved Coyote. A particularly beautiful and moving version of this story is William Benson's Pomo version (Benson, 2002), where Coyote is called by his Pomo name or alternative incarnation Marumda. A more poignant story is told by the Paiute, in which Wolf loses patience with his creation (in the story *Tracks of the Creator*):

> It is said that the (Paiute) Creator, Gray Wolf (*Numuna*), burnt everything in the old world...Gray Wolf talked with the Sun, "There should be a flood"...and the Mountains were covered with Water. Then Gray Wolf went with his woman far across the water.
>
> After a time, it is said, the water began to dry up, the mountains appeared, there were banks and shores again. Then the Sun said to Gray Wolf, "You should make children." "Yes," said Gray Wolf, and he created pine trees, juniper trees, aspen trees, cottonwood trees, willows, springs, deer, otter, beaver, trout, buffalo, horses, mountain sheep, bears.
>
> When Creation was finished, Gray Wolf's children began to do wrong; they fought amongst themselves...Gray Wolf, became angry, and kicked them all out. He decided to go south; he said, "My children are not going to see me again!" Then his wife cried, "But my children are here!" But they went down to the water anyway, it is said, and walked away over its surface.
>
> Gray Wolf and his wife came to a tall mountain, with a pine-covered summit. He said, "I am going in there; afterwards my children will see my tracks going in. Here I have come and left my tracks; Nuhmuhnuh will see them and so will white men." So it was. (Ramsey, 1977, p. 231)

Some say this is why wolves are no longer found in Oregon and Northern California.

The Pomo had a range of creation stories, often conflicting, because of their diversity and the diversity of peoples around them; they were influenced by the Yuki to the north, the Guksu Cult of the Patwin to the east, and so on (Barrett, 1933). Other fine stories are Hanc'ibeyjim's Maidu one (Dixon, 1912; Shipley, 1991; this is one of the greatest pieces of oral literature ever recorded in the Americas), Chemehuevi stories (Laird, 1976, 1984), and stories from the Kiliwa (Mixco, 1983) and Kawaiisu (Zigmond, 1981).

One exception is in an astonishing Wappo text, "The Chicken-Hawk Cycle," recorded from Jim Tripo (Radin, 1924: 87–147). This is one of the most impressive and striking mythic texts ever recorded in California, and deserves better than to languish obscure in a forgotten volume. Tripo spins it out to 60 pages; more typical is a Karuk version of the same story that manages only two pages (Kroeber & Gifford, 1980: 250–252.) Much of the plot turns on the anger of Moon, a captious old man, at the Hawk chief and his son harvesting pine cones by breaking off the branches of Moon's pine trees. Moon is not portrayed in a sympathetic light, but it is clear that such a damaging way of getting pine nuts was genuinely bad by Wappo standards.

Songs were basic to activities and teachings. They gave and expressed power, and had power themselves. A good explanation was given by a California basketmaker, Mrs. Mattz. Richard West reports that she:

"...was hired to teach basket making at a local university. After three weeks, her students complained that all they had done was sing songs.... Mrs. Mattz, taken aback, replied that they were learning to make baskets. She explained that the process starts with songs that are sung so as not to insult the plants when the materials for the baskets are picked. So her students learned the songs..... Upon their return to the classroom, however, the students again were dismayed when Mrs. Mattz began to teach them yet more songs. This time she wanted them to learn the songs that must be sung as you soften the materials in your mouth before you start to weave.... The students protested.... Mrs. Mattz...patiently explained the obvious to them: "You're missing the point," she said, "a basket is a song made visible." (Quoted Wilkinson, 2010: 382).

This account reveals a major difference in perception between Indigenous and settler midnsets, which may have influenced the way California Indigenous cultures have been characterized.

The most important conservation mechanism of all for California Indigenous peoples was burning, although all those other techniques were important as well. Fire was the chief way of managing the environment, even though the record is somewhat confusing. There is no question that California Native peoples set fires everywhere that would burn, and that these very substantially altered the vegetation over vast areas of the state (Anderson, 1999, 2005; Anderson & Rosenthal, 2015; Lewis, 2003; Lightfoot & Parrish, 2009, with major review of literature; Pyne, 2004; Timbrook, 2007). This was to be expected, for all Native American peoples except those in non-combustible environments (basically, Arctic and high-alpine areas and sand deserts) burned regularly (Pyne, 2004; Stewart et al., 2002), with major effects on the vegetation.

Juan Crespí's diary from 1769 (2001; see Gamble, 2008) is particularly revealing. He noted not only widespread deliberate burning, but also that the vegetation in many areas was short annual pasture rather than the chaparral and coastal sage scrub that are now, or recently found in those locations. The reviews by M. K. Anderson and by Lightfoot and Parrish list hundreds of sources covering dozens of groups.

Much of the state is affected by dry lightning, which in the mountains can be an almost daily phenomenon in late summer, and that other sources of ignition exist. These might range from volcanism to spontaneous combustion in animal nests. California's bone-dry summers and highly inflammable vegetation combine to guarantee natural fires on a cyclic basis. California would burn sooner or later, indigenous people or no (see also Sugihara et al., 2006).

Chaparral and some California forest formations are characterized by large numbers of species that evolved to burn: they dry out in summer and contain resins, waxes, and other compounds that are highly inflammable. These species all either stump-sprout aggressively after fire, have fruits that need fire to open them, or have seeds that need fire to germinate. Some authorities think that these plants evolved to eliminate competition and maximize their own dominance by this aggressive route.

For instance, California's most distinctive pine groups, the closed-cone and knobcone pines, have cones that normally do not open unless burned. They live in chaparral, grow and fruit rapidly, and are designed to burn on 20-to-50-year cycles. I have lived to see the knobcone pine forest on the San Bernardino Mountains go through two cycles and get well into a third. Obviously, they did not evolve in the last few centuries, and thus it is clear that California has burned since long before the Native Americans perfected their management systems.

Be that as it may, Native Californians burned chaparral regularly, to increase edible plant, mammal, and even insect resources (Anderson, 2005; Jones & Codding, 2020). M. K. Anderson and Jeffrey Rosenthal (2015) report, for instance, that caterpillars, as well as grasshoppers, were managed by fire, which causes rapid regrowth of the tender new shoots on which they feed. These authors describe the values of each stage of regrowth after fire. Burning also opened the brush, making travel possible; a stand of mature chaparral is impenetrable, or at best very slow going. Annual plants often produce more seeds (they need heavy seeding to survive) and greens than perennials do.

Even fishery management could be helped. Michelle Stevens and Zelazo (2015) point out that burning in summer opened up floodplains that flooded in fall, winter, and spring. Fish that spawned in those areas, including many important ones endemic to the central part of the state, were increased. Another benefit was increase in number and quality of stems of plants used to make fishnets, such as Indian hemp (*Apocynum cannabinum*) and milkweed. These plants grow in moist areas and produce longer, straighter stems after burning.

Richard Minnich (2008) has established that the bunchgrass prairies of California's interior valleys were nonnatural, and indeed many of them were purely mythical—early mappers' overgeneralizations. The potential vegetation of most of the dry San Joaquin Valley is saltbush and other brush. Minnich, formerly somewhat skeptical about the extent of Indigenous burning, has qualified his stand (Minnich, 2008 and pers. comm. to ENA, 2009–2010; Minnich & Franco-Vizcaino, 2002). It appears that chaparral and even desert vegetation can be burned much more often than it would naturally do. This has made him aware of the importance of Native American burning as a landscape shaper.

Californians were careful fire managers; they made very small fires for their own use. J. W. Powell, writing on the Paiute, says: "…an Indian never builds a large fire…and expresses great contempt for the white man who builds his fire so large that the blaze and smoke keep him back in the cold" (Powell, 1971: 53; this confirms a very widespread American folk observation of comparative fire-culture).

In short, the evidence is unequivocal. They certainly managed well-populated parts of the state by burning. On the other hand, their ability to reshape the vast lightning-prone mountains of the state seems limited. M. K. Anderson (2005, and pers. comm, Feb. 4, 2014) finds that they maintained and expanded the mountain meadows and coastal prairies of the state. These are now rapidly growing up to forest, in spite of lightning strikes; but deliberate fire suppression and the current years of drought (which favor trees over meadow grass) are involved in this.

Another equivocal case is oak woodlands. Oak seedlings die when burned. Frequent burning of oak groves would eliminate them. On the other hand, oaks survive burning when they grow large enough to have thick bark. Coast liveoaks sprout rapidly back from the very hottest fires. It takes about ten years for a live oak to reach fire-withstanding age. Thus, rarer burning—once a new generation of oaks had grown up—would eliminate fungal and insect pests, thin out the competition, and maintain the groves.

A problem for everyone trying to reconstruct Californian vegetation as of 1700 is that Europeans replaced deliberate burning with deliberate fire suppression. The Chumash were already seeing this as a major hardship, and complaining about it, by 1800 (Gamble, 2008; Timbrook, 2007). The Achomawi, later, complained and regretted the ruin of the forests (Rhoades, 2013: 112). The only possible conclusion is that human-set fire profoundly affected areas near large population centers, minimally affected remote mountain and desert areas, and affected to an unknown and probably unknowable degree the vast in-between zone. Today, fire suppression has led to buildup of fuels, which with the drought and heat now following global warming is leading to massive destruction of forests and permanent altering of vegetation (Minnich, 2008; Wigglesworth, 2021).

Finally, in the contemporary world, the image of much of rural California is massive out of control forest fires. It is unknown whether Indigenous burning might have prevented these events from happening, but it seems almost certain that the situation would have been much less sever and destructive, because as we pointed out above, the purpose of Indigenous burning was to minimize the chances of such fires becoming massive and out of control. It would be highly beneficial for the state to consult with its native peoples about how to manage a combustible environment in a rapidly changing world.

References

Anderson, E. N. (1996). *Ecologies of the heart*. Oxford University Press.
Anderson, M. K. (1999). The fire, pruning, and coppice management of temperate ecosystems for basketry material by California Indian tribes. *Human Ecology, 27*, 79–113.
Anderson, M. K. (2005). *Tending the wild: Native American knowledge and the management of California's natural resources*. University of California Press.
Anderson, M. K. (Ed.). (2017). *Special issue: California geophytes* (Vol. 44). Fremontia.
Anderson, S. (2018). *Naanessh Kwineeshe': Cahto Dictionary*. Cahto-Dictionary WedEdition 2018.pdf.
Anderson, M. K., & Lake, F. (2013). California Indian ethnomycology and associated forest management. *Journal of Ethnobiology, 33*, 33–85.
Anderson, M. K., & Lake, F. (2017). Beauty, bounty, and biodiversity: The story of California Indians' relationship with edible native geophytes. *Fremontia, 44*(3), 44–51.
Anderson, E. N., & Medina Tzuc, F. (2005). *Animals and the Maya in Southeast Mexico*. University of Arizona Press.
Anderson, M. K., & Rosenthal, J. (2015). An ethnobiological approach to reconstructing indigenous fire regimes in the foothill chaparral of the Western Sierra Nevada. *Journal of Ethnobiology, 35*, 4–36.
Arnold, J. E. (1987). *Craft specialization in the prehistoric Channel Islands, California* (p. 18). University of California, Publications in Anthropology.
Arnold, J. E. (2004). *Foundations of Chumash complexity* (Vol. 7). Cotsen Institute of Archaeology, Perspectives in California Archaeology.

Barrett, S. A. (1933). *Pomo Myths*. Public Museum of the City of Milwaukee, Bulletin 15, pp. 1–608.

Basso, K. (1996). *Wisdom sits in places: Landscape and language among the Western apache*. University of New Mexico Press.

Bean, L., & Blackburn, T. C. (Eds.). (1976). *Native Californians: A theoretical retrospective*. Ballena Press.

Bean, L. J., & Saubel, K. S. (1972). *Temalpakh: Cahuilla Indian knowledge and usage of plants*. Malki Museum Press.

Benedict, M., Kindscher, K., & Pierotti, R. (2014). Learning from the land: Incorporating indigenous perspectives into the plant sciences. In C. Quave (Ed.), *Strategies for teaching in the plant sciences* (pp. 135–154). Springer.

Benson, W. R. (2002). "Creation." Tr./ed. Jaime de Angulo. In H. Luthin (Ed.), *Surviving through the days: A California Indian reader* (pp. 260–310). University of California Press.

Bettinger, R. L. (2015). *Orderly anarchy: Sociopolitical evolution in aboriginal California*. University of California Press.

Blackburn, T. (1975). *December's child: A book of Chumash oral narratives*. University of California Press.

Blackburn, T., & Anderson, M. K. (Eds.). (1993). *Before the wilderness: Environmental management by native Californians*. Ballena Press.

Brody, H. (1982). *Maps and dreams*. Pantheon Books.

Buckley, T. (2002). *Standing ground: Yurok Indian spirituality, 1850-1990*. University of California Press.

Burrill, R. (1993). *Protectors of the land: An environmental journey to understanding the conservation ethic*. Anthro Company.

Callahan, C. (1978). Lake Miwok. In R. L. Heizer (Ed.), *Handbook of North American Indians* (California) (Vol. 8, pp. 264–273). Smithsonian Institution Press.

Crespí, J. (2001). *A description of distant roads: Original journals of the first expedition into California, 1769-1770*. San Diego State University Press.

Dixon, R. (1912). *Maidu Texts* (p. 4). Publications of the American Ethnological Society.

Dowie, M. (2009). *Conservation refugees: The hundred year conflict between global conservation and indigenous peoples*. MIT Press.

Du Bois, C. (1935). *Wintu ethnography* (Vol. 336, pp. 1–148). University of California Publications in American Anthropology and Ethnology.

Fowler, C. (1992). *In the shadow of fox peak: An ethnography of the cattail-eater northern Paiute people of Stillwater marsh*. U. S. Department of the Interior, Fish and Wildlife Service, Stillwater National Wildlife Refuge, Cultural Resource Series #5.

Fowler, C. (2013). The *Kasaga'yu:* An ethno-ornithology of the Cattail-Eatern Northern Paiute people of Western Nevada. In M. Quinlan & D. Lepofsky (Eds.), *Explorations in ethnobiology: The legacy of Amadeo Rea* (Vol. 1, pp. 154–178). Society of Ethnobiology, Contributions in Ethnobiology.

Gamble, L. (2008). *The Chumash world at European contact*. University of California Press.

Gamble, L. H., & Wilken-Robertson, M. (2009). Kumeyaay cultural landscapes of Baja California's Tijuana River watershed. *Journal of California and Great Basin anthropology, 28*, 127–152.

Garth, T. R. (1978). Atsugewi. In A. L. Kroeber (Ed.), *Handbook of North American Indians* (Vol. 8, pp. 236–248). Smithsonian Institution.

Garduño, E. (2016). Making the invisible visible: The Yumans of the U.S.-Mexico transborder region. *Human Organization, 75*, 118–129.

Gayton, A. (1948). Northern foothill Yokuts and Western mono. *University of California Anthropological Records, 10*, 143–302.

Gayton, A. (1976). Culture-environment integration: External references in Yokuts life. In L. J. Bean & T. C. Blackburn (Eds.), *Native Californians: A theoretical retrospective* (pp. 79–97). Ballena Press.

References

Gifford, E. W. (1916). *Miwok moieties* (Vol. 12, pp. 139–194). University of California, Publications in American Anthropology and Ethnology.

Gifford, E. W. (1957 [1936]). California balanophagy. In R. Heizer (Ed.), *The California Indians: A source book* (pp. 237–241). University of California Press.

Goddard, P. E. (1903). *Life and culture of the Hupa* (Vol. 1, p. 1). University of California Publications in American Archaeology and Ethnology.

Goddard, P. E. (1904). *Hupa texts* (Vol. 1, p. 2). University of California Publications in American Arcaeology and Ethnology.

Goddard, P. E. (1909). *Kato texts* (Vol. 5, pp. 65–238). University of California Publications in American Arcaeology and Ethnology.

Goddard, P. E. (1914). *Chilula texts* (Vol. 10, pp. 289–379). University of California Publications in American Archaeology and Ethnology.

Golla, V. (2011). *California Indian languages*. University of California Press.

Grant, C. (1965). *The rock paintings of the Chumash: A study of a California Indian culture*. University of California Press.

Grant, C. (1978). Eastern coastal Chumash. In R. L. Heizer (Ed.), *Handbook of north American Indians* (California) (Vol. 8, pp. 509–519). Smithsonian Institution Press.

Harney, C. (1995). *The way it is: One water . . . One air . . . One mother earth*. Blue Dolphin.

Harrington, J. P. (1932). *Tobacco among the Karuk Indians of California*. United States Government, Bureau of American Ethnology, Bulletin 94.

Hedges, K., & Beresford, C. (1986). *Santa Ysabel ethnobotany*. San Diego Museum of Man, Ethnic Technology Notes 20.

Heizer, R. (Ed.). (1978). *California. Handbook of North American Indians* (Vol. 8). Smithsonian Institution Press.

Hinton, L., & Watahomigie, L. (1984). *Spirit Mountain: An anthology of Yuman story and song*. University of Arizona Press.

Hudson, T., Blackburn, T., Curletti, R., & Timbrook, J. (Eds.). (1977). *The eye of the flute: Chumash traditional history and ritual as told by Fernando Librado Kitsepawit to John P. Harrington*. Santa Barbara Natural History Museum.

Jones, T. L., & Codding, B. F. (2020). The native California commons: Ethnographic and archaeological perspectives on land control, resource use, and management. In L. R. Lozny & T. H. McGovern (Eds.), *Global perspectives on long term community resource management* (pp. 255–280). Springer.

Keeling, R. (1992a). *Cry for luck: Sacred song and speech among the Yurok, Hupa and Karuk Indians of northwestern California*. University of California Press.

Keeling, R. (1992b). Music and culture history among the Yurok and neighboring tribes of northwestern California. *Journal of Anthropological Research, 48*, 25–48.

Kelly, W. H. (1977). *Cocopa ethnography*. University of Nevada Anthropology Papers 29.

Kroeber, A. L. (1925). *Handbook of the Indians of California*. U. S. Government, Bureau of American Ethnology, Bulletin 78.

Kroeber, A. L. (1972). *More Mohave myths*. University of California Anthropological Records 27.

Kroeber, A. L., & Gifford, E. W. (1980). *Karuk myths*. University of California Press.

Laird, C. (1976). *The Chemehuevis*. Malki Museum Press.

Laird, C. (1984). *Mirror and pattern*. Malki Museum Press.

Lang, J. (1994). *Ararapikva, creation stories of the people: Traditional Karuk Indian literature from northwestern California*. Heyday Books.

Latta, F. F. (1977). *Handbook of Yokuts Indians* (2nd ed.). Bear State Books.

Lawton, H. W., Wilke, P. W., DeDecker, M., & Mason, W. M. (1976). Agriculture among the Paiute of Owens Valley. *Journal of California Anthropology, 3*, 13–51.

Lerch, M. (1981). *Chukiam (all growing things): The ethnobotany of the Serrano Indians*. B.A. thesis, University of California, Riverside, Department of Anthropology.

Levinas, E. (1969). *Totality and infinity*. Trans. by A. Lingis (Fr. orig. 1961). Duquesne University Press.

Lewis, H. (2003). Review of fire, native peoples, and the natural landscape. *Journal of Ethnobiology, 23*, 161–163.

Librado, F. (1979). *Breath of the sun*. Malki Museum Press.

Lightfoot, K. G., & Parrish, O. (2009). *California Indians and their environment: An introduction*. University of California Press.

Lutkin, H. W. (Ed.). (2002). *Surviving through the days: A California Indian reader*. University of California Press.

Madley, B. (2016). *An American genocide: The United States and the California Indian catastrophe, 1846-1873*. Yale University Press.

Mead, G. R. (2003). *The ethnobotany of the California Indians*. E-Cat Worlds.

Medin, D., Ross, N. O., & Cox, D. G. (2006). *Culture and resource conflict: Why meanings matter*. Russell Sage Foundation.

Minnich, R. (2008). *California's fading wildflowers: Lost legacy and biological invasions*. University of California Press.

Minnich, R. A., & Franco-Vizcaino, E. (2002). Divergence in Californian vegetation and fire regimes induced by differences in fire management across the U.S. Mexico boundary. In L. Fernandez & R. T. Carson (Eds.), *Both sides of the border: Transboundary environmental management issues facing Mexico and the United States* (pp. 385–402). Kluwer.

Mixco, M. J. (1983). *Kiliwa texts: "When I have donned my crest of stars"*. University of Utah Press.

Nabhan, G. (2013). The wild, the domesticated, and the coyote-tainted: The trickster and the tricked in hunter-gatherer versus farmer folklore. In M. Quinlan & D. Lepofsky (Eds.), *Explorations in ethnobiology: The legacy of Amadeo Rea* (pp. 129–140). Society of Ethnobiology, Contributions in Ethnobiology, 1.

Pierotti, R. (2011). *Indigenous knowledge, ecology, and evolutionary biology*. Routledge.

Pierotti, R., & Fogg, B. R. (2017). *The first domestication: How wolves and humans coevolved*. Yale University Press.

Powell, J. W. (1971). *Anthropology of the Numa: John Wesley Powell's manuscripts on the Numic peoples of Western North America 1868-1880*. Smithsonian Institution Press.

Pyne, S. J. (2004). *Tending fire: Coping with America's wildland fires*. Island Press (Shearwater Books).

Radin, P. (1924). *Wappo texts*. University of California Publications in American Archaeology and Ethnology 19.

Ramsey, J. (1977). *Coyote was going there: Indian literature of the Oregon country*. University of Washington Press.

Rhoades, L. (2013). *Lela Rhoades, pit river woman. As told to Molly Curtis*. Heyday Books.

Sapir, E. (1910). *Yana texts* (Vol. 9, pp. 1–235). University of California, Publications in American Archaeology and Ethnology.

Shipley, W. (1991). *The Maidu Indian Myths and Stories of Hanc'ibyjim*. Heyday Books.

Smith, A. M. (1993). *Shoshone tales*. University of Utah Press.

Spier, R. F. G. (1978). Monache. In R. L. Heizer (Ed.), *Handbook of North American Indians* (Vol. 8, pp. 426–436). Smithsonian Institution Press.

Stevens, M. L., & Zelazo, E. M. (2015). Fire, floodplains and fish: The historic ecology of the lower cosumnes River watershed. In P.-Y. Yu (Ed.), *Rivers, fish, and the people: Tradition, science, and historical ecology of fisheries in the American west* (pp. 155–187). University of Utah Press.

Stewart, O. C., Lewis, H., & Anderson, M. K. (2002). *Forgotten fires: Native Americans and the transient wilderness*. University of Oklahoma Press.

Sugihara, N. G., van Wagtendonk, J. W., Shaffer, K. E., Fites-Kaufman, J., & Thode, A. E. (Eds.). (2006). *Fires in California's ecosystems*. University of California Press.

Thompson, L. (1991 [orig. 1916]). *To the American Indian: Reminiscences of a Yurok woman*. New edition, edited and with foreword by P. Palmquist; introduction by J. Lang. Heyday Books.

References

Timbrook, J. (2007). *Chumash ethnobotany: Plant knowledge among the Chumash peoples of Southern California*. Santa Barbara Museum of Natural History and Heyday Books.

Timbrook, J., & Johnson, J. R. (2013). People of the sky: Birds in Chumash culture. In M. Quinlan & D. Lepofsky (Eds.), *Explorations in ethnobiology: The legacy of Amadeo Rea* (Vol. 1, pp. 179–218). Society of Ethnobiology, Contributions in Ethnobiology.

Wallace, W. J. (1978). Southern Valley Yokuts. In R. L. Heizer (Ed.), *Handbook of North American Indians* (Vol. 8, pp. 448–461). Smithsonian Institution Press.

Welch, J. R. (2013). *Sprouting Valley: Historical ethnobotany of ther northern Pomo from Potter Valley*. Society of Ethnobiology, Contributions in Ethnobiology, 2.

Wigglesworth, A. (2021, June 5). As wildfires decimate the giant sequoia, California faces unprecedented loss. *Los Angeles Times*, B1.

Wilken-Robertson, M. (2004). Indigenous groups of Baja California and the environment. In M. Wilken-Robertson (Ed.), *The U.S.-Mexican border environment, tribal environmental issues of the border region* (pp. 31–48). San Diego State University Press.

Wilken-Robertson, M. (2018). *Kumeyaay ethnobotany: Shared heritage of the Californias*. Sunbelt Publications.

Wilkinson, C. (2010). *The people are dancing again: The history of the Siletz tribe of Western Oregon*. University of Washington Press.

Woiche, I. (1992). *Annikadel: The history of the Universe as Told by the Achomawi Indians of California*. Recorded and edited by C. Hart Merriam. University of Arizona Press.

Zigmond, M. L. (1981). *Kawaiisu ethnobotany*. University of Utah Press.

Chapter 15
Appendix 2: Wider Connections

A Circumpolar Worldview

The Northwest Coast worldviews are special instantiations of a more general worldview that has been called the "Old Northern" view (Anderson, 2014). This concept is based on Waldemar (Vladimir) Bogoras' (1929) idea of a circumpolar culture area, which links clearly related items of material culture, such as snowshoes, dogsledding, and fishing techniques, with common folktales and religious themes, thus identifying a single vast cultural realm. It may have evolved as humans established and refined their coevolutionary relationship with wolves, which eventually became domesticated animals, i.e., dogs (Pierotti & Fogg, 2017). Like wolves becoming domesticated, this general way of life probably had no single place of origin; it developed through constant feedback, over thousands of years, between the multitude of societies that share it and used it to dominate other cultural traditions, including Neandertals (Shipman, 2011, 2015a, b). Thanks to continued contact, migration, and information flow around the circumpolar land masses, it has remained rather uniform. A steady diffusion of beliefs and practices has tended to homogenize beliefs about subsistence and environment.

The various traits that define it were probably invented in different parts of the region, and have coalesced since being widely shared, evolving to fit local needs. Any new idea or modification, wherever it started, could spread all over the region fairly fast, given the high mobility and frequent trade among the majority of groups. Siberia is central, and therefore a logical point of origin for some traits, but that is not to say Siberia is the origin point. Our first records of the more widespread folktales are from Greece, but there is no reason to believe they originated in Greece; the Greeks just wrote things down before the Irish, let alone the Tsimshian, which allows them to claim credit among written cultural traditions. Consider snowshoes, which probably began as flat boards or vegetable fiber mats tied to people's feet. This could have been invented anywhere that gets snow, once humans realized this made walking more efficient. They evolved in the east (east Siberia and Native North America) into snowshoes, but in Europe they evolved into skis (Bogoras, 1929).

Animal spirit beliefs are even harder to localize, being widely shared outside the Old Northern area. In the Old Northern worldview and its descendants and relatives, humans are people-in-nature, not people separate from nature, and certainly not people opposed to nature or (as Mao Zedong puts it) "struggling against nature" (This theme has been the major theme of Pierotti's research during the twenty-first century; summarized in Pierotti, 2011). The wider way of thinking is typical of almost all peoples who can claim the title of Indigenous, i.e., born to, or of, a place. This has ethical implications, many of which are stressed in recent scholarship, and allows scholars to generalize without have to recite the tedious mantra, "of course all cultures are unique," lest we be accused of "pan-Indianism." Cultures are unique, and adapted to local places; but, like species, they have shared roots, if one looks deeply enough (Pierotti, 2011; Pierotti & Wildcat, 2000).

The Old Northern worldview (like related worldviews south of it) sees humans and animals, and often plants, mountains, and streams, as spiritual beings, linked by spirit contacts and interactions. It is an instantiation of the "animistic" type of ontology as defined by Philippe Descola (2013). This is a hunters' world (Brody, 2000; Descola, 2013). If humans depend on animals; they must show respect, care for the animals' souls, hunt carefully, and communicate with animals via rituals and visions (Pierotti, 2011).

The Old Northern worldview, as defined here, was shared across the Old World and New World arctic and subarctic zones, and as humans radiated east and south, it can be seen throughout the hunting-gathering societies of North America and the hunters and pastoralists of northeast Asia and high central Asia, all of which may have survived, and thrived, because of their relationships with wolves (Pierotti & Fogg, 2017; Shipman, 2015a). This perspective seems to become progressively diluted as it enters pastoral and agricultural societies, and erodes further with the coming of cities, mass trade, and industrial production, and increased hostility towards the predatory species that originally defined these cultures (Pastoureau, 2007). It eventually disappears as an identifiable worldview in Europe and its colonial outposts, including the USA and Canada, however, it survives in cultural relicts, e.g., the nature consciousness of the Celtic peoples, animistic beliefs about tree spirits and *fengshui* in Chinese folk culture, and the survival of beliefs in animal companion spirits (*naguales*) in modern Mexico. It is a fuzzy set, with no definable boundaries in space or time.

South America's indigenous cultures often show Old Northern origin or influence, though they vary greatly (see above; also, Balée, 1994; Kohn, 2013; Lentz, 2000; Sullivan, 1988). Some have something close to the North American sense of interpenetration with the natural (Kohn, 2013; Reed, 1995, 1997; Reichel-Dolmatoff, 1971, 1976, 1996; Sullivan, 1988; Ventocilla et al., 1995). One group, for instance, believes that humans are reborn as peccaries and vice versa, leading to an alternation of generations. The peccary the hunter kills is his relative sacrificing himself, or herself, to feed the human family (Blaser, 2009; Descola, 2013). This clearly resonates with widespread Old Northern reincarnation concepts, such as the Haida belief that men reincarnate as killer whales and vice versa.

Recently, the Mayanist Vernon Scarborough and his colleagues (2020) have generalized the broadly animist view to characterize New World cultures in general, in particular its great civilizations, giving the familiar examples of apologizing to trees that are felled, regarding animals and mountains as people, and, in general, the broad list provided above in Chap. 7. They use it as part of the back story of the ecologically integrated civilizations of the New World: less destructive of nature, and also less technologically oriented than those of the Old.

Some astonishing parallels occur between societies at the far ends of the continuum. The medieval Irish and the Haida were both island societies on the northwest edges of continents. Old Northern links—indirect but real—led them to learn similar stories and create similar images. The wren is a powerful bird for both Irish and the Northwest (including Haida, Salish, and Quileute), because although it is tiny, it sings in the worst winter storms, and its song rings loud and clear above the howling of the wind. In societies that share a belief that song gives spirit power, this makes the bird a truly powerful being (Lawrence, 1997 for the Irish; ENA's research on Haida; Adamson, 2009: 31–32 for Salishan groups). Among the Nuu-chah-nulth the wren is the source of reliable words, "One Who Always Speaks Rightly" (Atleo, 2004, 2011: 98). The wrens are slightly different species (Common or Eurasian Wren in Ireland, Pacific Wren in the Northwest), but so similar as to be indistinguishable to nonexperts; they were considered one species until very recently.

The belief in the magic power of song is shared all across Europe and Siberia, along with several folktales. Also similar are a number of religious practices, including reverence for tree spirits. Moving to Scotland, we may add the concept of a clan crest, closely parallel to Northwest Coast concepts. Norway has or had many of the same cultural concepts, such as the sea-raider lifestyle. Both indirect connection, via intellectual exchanges around the Arctic, and being ecologically situated on rainy coasts with rich salmon resources, combine to produce similar cultures.

Animism?

This was a hunters' world, in which human and other-than-human persons are intertwined and can change into one another. Healers and religious officiants can go into trance, visit the spirit lands, and find out how to deal with human problems ranging from sickness to bad hunting luck. Typically, the underworld, this world, and the upper world are linked by a sacred tree. This became the ash Yggdrasil in Scandinavia, the birch in the northern Uralic cultures, the white spruce in parts of the Canadian subarctic, and so on, and has gone as far as Yucatan, where the Maya still revere the ceiba as the world-pole tree. The world tree is known beyond the Old Northern area, and may even be humanity-wide; it surfaces as the banyan figs of India, the date palm in Arabia, and even the metaphorical "tree of knowledge of good and evil" of Genesis 2, which was widely portrayed as a world-tree of Tree of Life in European and west Asian art (Cook, 1974).

The spectacular cave art of Europe, which goes back some 35,000 years (Clottes, 2008), implies that this worldview existed at that time. Beautifully drawn animals, along with human-animal hybrid figures, were drawn in remote, dark caves (*not* in the inhabited, sunlit caves beloved of cartoonists).

Some aspects of the Old Northern view may be survivals of the original worldview of humans from our earliest origins, e.g., the desire for large wolflike dogs as companion animals. Joseph Campbell traced many similarities between the most remote and distinctive surviving humans, the San of South Africa, and hunter-gatherers elsewhere. He inferred a basic worldview, the "way of the animal powers" (Campbell, 1983). Most scholars caution that he overstated, but he may have had a case, especially as evidence accumulates that modern humans originated in northeast Africa a mere 150,000 years or so ago, and probably did not get far from home until less than 100,000 years ago. They have been in Australia for 50–60,000 years, in Europe 40,000 years, in northeast Siberia a mere 30,000 or less, in the Americas around 20,000. In more recent millennia, there has been constant migration, exchange, and mutual learning, leading to such widespread homologies as the animal-soul beliefs of the Old Northern system. Campbell points not only to animal spirits but to trance-healing, pictograph art showing animals, the ritual power of song and dance, and the belief in a skyworld for spirits and an underworld for the dead, as well as the mythic importance of hunters (cf. Fontenrose, 1981) and their association with transformers. Animals, the world tree, waterfalls, snakes, and sacred mountains occur in religious symbolism everywhere (as of course do images of mother, father, and child), and qualify as archetypes.

Belief in, and reverence for, the spirits and deities in all things is more or less what E. B. Tylor (1871) called "animism." Tylor viewed it as "primitive," which led to the term "animism" being shunned for a while, but it is now being rehabilitated by scholars who realize it is not some sort of primitive holdover. It is as well developed and sophisticated a religious view as any other (Descola, 2013; Harvey, 2006; Hunn & Selam, 1990). Animism is based on the tendency to assign spirits or personhood to living things, or sometimes to everything. Chinese folktales assign willful, agentive spirits to rags, discarded bits of paper, and old roof tiles.

Humans in animism are clearly different from animals and other nonhuman beings, but each is considered able to transform into the others and return. However, the very fact that this is a subject for myth shows that it is not expected to be routine. Similarly, culture is different from nature. Claude Lévi-Strauss (1958, 1962, 1963, 1964–1971) was right: most groups (but perhaps not all) seem to make a distinction between culture and nature. Many human groups worldwide have a belief in "natural humans" like Bigfoot—hairy "savages" (from Latin *homo sylvaticus* "forest person"), people in their "natural state," lacking culture. These stories may be locally inspired by bears or apes and monkeys, but clearly depend heavily on people projecting a view of humans-without-culture onto the world, and imagining such beings (Bartra, 1994).

Shared from the Northwest Coast to pre-civilization Europe and most of the Indigenous New World are beliefs in the universality of animal souls, the full personhood of animals, the need to respect them, the idea that they sacrifice themselves to the hunter, many ideas of "power animals" and plants and places, and other details (Hayden, 2013).

Paul Radin (1927, 1957) early documented the striking range of difference in belief that existed within one small Native American group. Indeed, Brian Hayden notes that, in traditional hunter societies, typically 10–20% of individuals do not believe in spirits at all, even if they accept the importance of nonhuman lives (Hayden, 2013: 495; cf. Hayden, 2003). The belief that people in traditional cultures share all their cultural norms is an armchair scholars' mistake. It is important to keep in mind that Ridington (1988: 95) reports a traditional "dreamer" addressing this issue by pointing out that people who do not sing and dance, i.e., participate in ceremonies, get lost on their way to heaven, and then have bad luck which leads to dissolution of marriages and even death. Every society has its questioners and its nonbelievers. Thus, these are widely, but not universally, shared ideas.

Throughout the Old North, souls are considered to be highly mobile—they can and often do leave the body. They can inhabit other venues throughout the process. This implies that soul travel, soul exchange, dream journeys, and ghosts are all regular features of life. A given soul can inhabit animal or human bodies, and shape-shifting is common.

Trees and mountains are persons and have their own spirits. High regard for trees and their spirits is one of the main beliefs binding the Old Northern world into a single worldview. In the Northwest, even fire can be a person, conscious and verbally fluent (Jacobs & Jacobs, 1959: 93–95).

The Siberian Udmurt (Shutova, 2006), and Khanty (Wiget & Balalaeva, 2011), show rather similar regard for trees to that reported from Northwest Coast peoples (Turner, 2005). Rivers are important, and dramatic water features such as waterfalls and sinkholes are revered—again all the way from the Khanty (Wiget & Balalaeva, 2011: 114) to the Northwest Coast, and on south to the Maya. Dramatic rocks and prominences are also revered. This leads to cultural traditions that reinforce conservation; Tibet's Pure Crystal Mountain, for one, maintains its forests (Huber, 1999). Conservation throughout Tibet was, before Han Chinese invasion and destruction, very much on the Old Northern pattern (Salick & Moseley, 2012).

When animals are not interacting with people, they often live in villages, look like humans, and see their world in humanlike terms (e.g., Ridington, 1988). This view can be considered a way of reinforcing the idea that nonhumans are persons (Pierotti, 2011, Chap. 2), and is universally attested in the literature from ancient Scandinavian and Celtic culture right across Siberia and throughout Native North America. It is also very widespread in South America (see, e.g., Descola, 2013; Lloyd, 2007: 145). American Native views probably developed from Old Northern views brought across the Bering Strait sometime in the last 20,000 years. There was been continual contact and mutual reinforcement of such concepts, until Europeans arrived in the sixteenth century, emphasizing the domination and conquest of nature, which eroded the Indigenous perspective, especially through conversion to contemporary monotheistic religions, which reinforce separation.

Animal people are not quite the same as human people, even when in their humanoid forms. They still act like themselves. Thus Descola (2013) sees "animism" as postulating similarity in material form but not in essence. Yukaghir reindeer people in their humanoid form grunt instead of talking and eat lichen instead of meat (Willerslev, 2007: 89–90). One Siberian Yukaghir hunter told Willerslev of a visit he 'experienced' to a village of wild reindeer in their human form (Willerslev, 2007: 89–90), and Ridington (1988: 58–59) reports that, during his vision quest, Japasa lived with Silver Foxes in their home, and also encountered Snowshoe Hares "wearing clothes like people." The stress of the vision quest may have affected his perception. Willerslev's hunter source admitted he might have "dreamed it all," but was fairly convinced it had been real, which depends on what you mean by "real."

Among the gods and goddesses who protect animals are Artemis/Diana in Europe, Cybele in Asia Minor, and Prajapati in India. These are related to the "Keepers of the Game" of each animal species in much of the New World. In Georgia (Caucasus), the worship of pre-Christian deities who served as "Masters of the Game" persisted into the twentieth century. The Georgians revere Ochopintra in this role (Tuite, 1994: 73, 134). The closely related Svan people worshiped the goddess Dael or Dali (Tuite, 1994: 18, 45–47, 126, 141–143), equivalent to the Roman Diana. She protected game, and punished those who hunted improperly, wastefully, and destructively. Significantly, these deities, and many related beliefs associated with plants and animals, survived 1700 years of Christianity. On the other hand, the European separation of nature from culture is present—though not extreme—in Georgia, and very possibly goes back to ancient times (as it does among the Greeks).

The associated reverence for animal persons is perhaps most famously shown in a widely-known Georgian ballad about a woman whose son died killing a leopard (or tiger; Tuite, 1994: 37–41). The woman sings to the leopard's mother:

> I will go, yes I will see her,
>   And bring her words of compassion.
> She will tell her son's story
> And I will tell her of mine.
> For he too is to be mourned
> Cut down by a merciless sword.

Human and animal realms interpenetrate to a dangerous degree. The "werewolf" is a common type in European cultures, but the "bear doctor" is commoner. "Bear doctor" is a Californian term, used by Native Californians speaking English (see Blackburn, 1975). Such an idea can be found from America to Scandinavia, where the *berserkers* represent the same type of were-being; *berserk* means "bear shirt." In historic times, berserkers were ordinary humans who fell into a battle-frenzy, but they may have been bear transformers in the mythic past. The Celtic battle-frenzy (made famous by Caesar's accounts of the Gauls and by Irish medieval epics) must have been similar, though no bear stories survive about it.

Common, if not universal, are myths in which a trickster-transformer god or chief spirit makes things what they are today. The head transformer is Coyote in most of western North America, Raven in the northwest and in northeast Siberia, and various humanoid or composite beings in most of the rest of the circumpolar realm. The cognate European and Chinese figure is the fox, but he has degraded to a mere folklore figure. So has the Japanese trickster, the raccoon-dog, also known as the mangut (its Evenki name), tanuki, or neoguri (*tanuki* is generally mistranslated "badger" in English, thanks to some early translator now remembered only for zoological incompetence).

Tricksters are sometimes female, but the dominant Trickster-Transformer deity is almost always male, a fact emphasized by his frequently randy adventures. (Many a Coyote story was translated partly in Latin in the old days.) Such figures are powerful and mischievous, often foolish, and easily seduced by women or by greed. From the Finnish Väinamöinen to the East Asian raccoon dog to the Raven and Coyote, he does the same general sorts of things: changing his own shape, changing supernaturals into harmless forms, turning animals from humanoid to animal form, providing light and water and song, all the while having a run of salacious adventures. Even the long-Christianized Irish retain many Transformer stories, often—ironically—told of St. Patrick (see Dooley & Roe, 1999). He did not transform himself, but he transformed the giants of old into ordinary men, drove the snakes and almost all other wildlife out of Ireland, and changed the landscape.

Widespread or universal Old Northern stories draw on this belief system. Naturally, given the hunting-dominated lifestyle of most ancient Old Northern societies, the hero and/or heroine are very often hunters. The abundance of hunters in Greek myths is one of the proofs that ancient Greece was solidly Old Northern.

These stories extend all the way from Ireland and Greece to North America. One is the swan maiden story (see above, p. xxx). Another is the Orpheus myth (widespread in North America, often with Coyote as the Orpheus figure; see Adamson, 2009: 9–29, but especially 293–303 for a long, dramatic, terror-filled version; Ballard, 1929: 128–133; Hunn & Selam, 1990: 243–246; Archie Phinney, 1934: 268–285, discussed in Ramsey, 1977; Sapir, 1990: 133–143; compare, in California, Blackburn, 1975: 172–175; and see almost any large collection of American myths). The Orpheus myth, in ancient Greece as in much Native America, usually had a tragic ending, but some versions allow Coyote or his equivalents to regain his lost wife, as in the Coquille form (Seaburg, 2007: 221–227), and in the beautiful ballad version from the Shetland Islands (Child, 1882: 215–217; Coffin, 1976: 75). The tragic ending we know today was set firmly in place by Virgil in the *Georgics* (Fantham, 2006). Some Native American versions, including Archie Phinney's text collected from his Nez Perce grandmother, and Louis Simpson's text collected by Sapir, are as tragic as Virgil's. One proof that human minds run in the same channels is the astonishing similarity not only in plot but in emotional portrayals and character development between Phinney's shattering and heartbreaking text and Rainer Maria Rilke's incomparable retelling of Virgil's story (Rilke, 1964: 142–151). Among the Nehalem Tillamook, the story becomes one of a brother saving his brother; other than the gender, it follows the successful-version storyline (Jacobs & Jacobs, 1959: 96–98).

Another is the transformation of animals into stars. The Great Bear was originally a literal bear, with an unusually long tail, in stories told throughout the Old Northern region (Schaefer, 2006). The resemblance of the constellation to a bear is far from obvious; this implies that the story actually spread from one source, rather than being independently invented. Also widespread are a tale of two sisters who married stars (Thompson, 1955–1958), tales of persons transformed into animals by witchcraft (as in the European story of the princess kissing the frog), and countless stories of animal creators. The earth-diver myth, in which the land is created from a bit of mud retrieved by a diving animal from the bottom of the cosmic waters is closely linked with this northern worldview, but is found far outside its borders. Even Northwest Coast mnemonic lists of trees' value for firewood (Adamson, 2009: 162) find Celtic parallels (Anderson, 2014, documenting survival from medieval Celtic to modern English-language versions).

In this worldview, people are not subsumed into nature or mere pawns of the environment. Quite the reverse; rugged, dramatic individualism is characteristic. It is especially visible in Celtic epics and ballads, but also in Northwest Coast hero tales, Korean and Finnish epics, and other traditional vocal forms. There is an emphasis on the value of individuals, even ordinary ones, as in the Scottish ballads with their everyday heroes and heroines. This theme also appears in the many tales of the lone shaman battling dark forces in an alien world (e.g., Nowak & Durrant, 1977).

The Old Northern view, with its tales of heroic ordinary people, is a major source of Greek individualism. The writings of Homer and the myths retold by Hesiod show that it comes from the earliest stratum of Greek thought. The ancient Greeks invented the "individual" (the man or woman alone against the world and the gods), and also invented dramatic tragedy, through the dramas of Aeschylus and Sophocles; they made heroes of ordinary men and women standing tall in the face of terrible fates. The idea of the ordinary person dauntlessly facing destiny, though alone and beset on every side, appears here for the first time in history; earlier heroes like Gilgamesh had been semi-divine and rather stereotyped. The Northwest Coast, however, had known such stories for a long time. Particularly common from Europe to Native North America is the story of the poor or handicapped boy who secretly develops his strength and becomes a hero, often saving his people and winning wealth, which he then has to give away.

Many of the beliefs make sense primarily in relation to hunting. Hunters must stay pure and avoid smelling like humans, since game animals are acutely aware of such things. Apparently universal in the Old Northern world is a prohibition against sex before hunting, especially with a menstruating woman; the smell is too human (see Willerslev, 2007: 84). This may be purely psychological, but it may have roots in actual game-animal alertness. Hunting is often sexualized (Willerslev, 2007: 110, who notes worldwide comparisons).

Also extremely widespread is avoidance of talking about one's hunting plans, or naming animals one intends to hunt. They might hear and be offended. Worse still is to speak the name of a predator such as a bear. Bears are thus referred to by various euphemisms. In fact, the English words "bear" and "bruin" go back to a Germanic root cognate with "brown," apparently used because the Indo-European root derivative came to be avoided. (The Indo-European root is itself interesting. A philologist working from such forms as Latin *ursus* and Greek *arctos* reconstructed it as, approximately **rtko* or **ughrtko*—close enough to *grrhaahr* to suggest where that root came from! See Watkins 2000: 72). A Yukaghir hunter talking about having seen the tracks of "a Russian in felt boots" (apparently with badly untrimmed toenails; Willerslev, 2007: 100)). Another commented, "We'll pay him a visit tomorrow." Of course, the felt boots—no doubt mentioned in a significant tone of voice—were the giveaway here. The Siberian Khanty, especially when out hunting, call him "forest master." They have many ceremonial practices for dealing with a slain bear. "These very careful attempts to maintain the integrity and dignity of the bear's remains were aimed at ensuring the subsequent rebirth of the animal..." but trying also to hide the hunter's identity so the reborn bear will not kill the hunter (Filtchenko, 2011: 187). The Khanty share a belief in masters and mistresses of the forests and game with many Northwest Coast groups. In a similar manner, tigers are referred to by a whole flock of euphemisms all over east and south Asia, including their formerly extensive range in Siberia.

Conservation Ethics

Of major concern to our purposes here is the universal conservation ideology of the Old Northern view, especially in regard to animals. In Europe, gods or goddesses like Artemis—Masters or Mistresses of Animals—protect game and nongame. In North America, the masters are animals, exceptionally large and powerful ones or spiritually gifted ones. Trees, or at least some individuals or species, are sacred everywhere, and one must ask permission for taking bark or wood, and must be respectful of them. Water is widely regarded as sacred, and rules against polluting it are correspondingly widespread among cultures that have some shortage of it, especially in Siberia and Mongolia (see posting "Water" on ENA's website, www.krazykioti.com). Mountains are sacred (at least some of them), and often the sacred mountains are taboo for hunting and getting resources, others are taboo for all but the lowest-impact uses.

On the other hand, many societies within this vast area have retained only a little of what probably was once a more conservationist ideology. The extremely thorough ethnographies of the Siberian Koryak (Jochelson, 1908, 1926) and Chukchee (Bogoras, 1904–1909, 1919) show almost none, though creation stories hint at more animist beliefs in the past. For example, A Chukchee scholar writes of the early times of his people through a dialog:

Things you understand, have a life of their own, regardless of people naming them and measuring them...But of all the living creatures only humans have words"... So where did the other gods, like the four winds come from?"..."there aren't any gods...People just made them up because they feared mystery. When you can't be bothered to use reason to understand mystery, that's when you make up gods. As many gods as there are mysteries in the world. You can blame everything on gods... These days people even say their own powers are gods-given. Disgraceful!"..."We do revere the whales though"... "A whale is not a god,"... He is simply our ancestor and brother. He lives beside us, always ready to come to our aid. That's all there is to it." (Rytkheu, 2019: 39).

These groups lived largely by reindeer herding (though some lived earlier on by sea mammal hunting) and managing reindeer herds and of dogs (ritually very important) had taken over the role of respect and care of wild animals. Rytkheu goes on to describe how Chukchee culture was destroyed by greed and lust for power, so that his people were forced into reindeer herding after the sea mammals abandoned them because of their selfish actions. The attitudes of reverence and concern are directed to these domesticates (or semi-domesticates; the reindeer were still quite wild).

Jochelson also reports no conservationist stories for the Aleut (Jochelson, 1990). This is in the very center of the Old Northern distribution, but has been subjected to intensive missionary contact for 250 years, which may explain the lack of stories. As one leaves the arctic and subarctic, the Old Northern respect for nature falls off rapidly in some areas. The general attitudes of respect and care for all nature extend throughout East Asia and most of North America, but is much more developed among some societies than others. Even within the Northwest Coast, groups including the Tsimshian and Secwépemc are much better documented in this regard than the Oregon cultural groups, an artifact of poor knowledge of the latter. The changes are likely due to colonial and settler-society influences and sheer cultural loss over time since contact.

All the participants in this worldview have strong indigenous traditions of conservation, or at least of sustainable resource management. A particularly noteworthy account of a Siberian shamanistic society and its conservation practices is provided by a Tuvan writer (Mongush Kenin-Lopsan, 1997). Hundreds of sources deal with indigenous management in northwestern and western North America (Anderson, 1996).

In some areas the old beliefs persist with enthusiastic support, and contribute to modern political debate (see, e.g., Kendall, 1985 for Korea; Metzo, 2005 for Mongolia). In others, including Japan and most of Europe, they are largely displaced by later-coming belief systems, but they are still there (for Japan, see esp. Blacker, 1986; for Scotland, Carmichael, 1992). As an example, wolves used to be respected and even revered in Japan, until they had livestock introduced by Europeans, at which point they quickly turned against wolves and exterminated them (Walker, 2005). This reinforces the idea that nothing pushes a culture away from conservation more than acquiring domestic ungulates (i.e., chattel: Pierotti, 2011, Chap. 2). This may explain why the Northwest has retained its conservationist tradition: they lack domesticated ungulates.

It should not surprise us if reality is often short of ideal. Traditional people, like all people, frequently fail to live by their ecological morals. (For Native American society, see Kay & Simmons, 2002; for other areas of the world, Kirch, 1994, 1997,

2007; cf. comparable problems in the Mediterranean world described by McNeill, (1992) and Ponting (1991)). Plains Indians, when they got horses and guns and were forced out of older homelands onto the high plains, lost a good deal of their resource management practice. (See e.g., Black Elk (DeMaillie, 1984; Lame Deer & Erdoes, 1972)). Reports of wasteful hunting by Interior Salish people is probably a result of the fur trade (Teit & Boas, 1927: 8: 103, 268–269; cf. Martin, 1978), but the near-universal presence of stories about overhunting, as seen in the main text (and see Boas, 1918: 167–171), show that conservation was learned the hard way, e.g., Pierotti (2011, Chap. 8) argues that Indigenous Americans probably enforced conservation only after seeing the megafauna disappear around 10,000 ybp.

However, throughout North America, the strong conservation ethic had its effect. Native American views, at least in North America, stem from this worldview and carry forward its close relations with nature. For areas outside the Northwest, see Bernard and Salinas Pedraza (1989), Davis (2000), Haly (1992), Neihardt (1932), Pierotti (2011), Tanner (1979). Further documentation includes Alberta Society of Professional Biologists (1986), Callicott (1994), Martinez (2009), Lopez (1992), Scott (1996).

For the Great Lakes area, we have a considerable literature; notable is *American Indian Environmental Ethics* by Baird Callicott and Michael Nelson (2004), a rigorous and excellent study by two ethical philosophers, based on actual Native American texts recorded by a traditional Native American ethnographer, William Jones of the Sauk and Fox. There are also many short accounts; particularly good collections include Mary Evelyn Tucker and John Grim's (Tucker & Grim, 1994). One could point to many other works (e.g., Davis, 2000; Medin et al., 2006).

Native California has been less well served, probably because the cultures were severely impacted by foreign settlers before anthropologists went among them, but there are several important texts available (Anderson, 2011). For other areas, to the classics of Black Elk (DeMaillie, 1984; Neihard, 1932) and Lame Deer and Erdoes (1972) we can now add Pierotti's review (Pierotti, 2011). Work rarely linked to Native Americans, although it should be, given that the author grew up in South Dakota, running trap lines with local Lakota are Paul Errington's classics (Errington, 1967, 1973) on hunting and conservation, which argue that predators improve, rather than harm prey populations. Errington was Aldo Leopold's graduate student, and was the one who enlightened Leopold on the role of predation.

One problem common to most sources is a lack of grounding in study of the ethical systems of the traditional and local cultures under study. There are, in fact, few detailed studies of traditional Native American ethical systems. However, a thorough reader should examine V. F. Cordova's *How It Is* (Cordova, 2007) and David Martinez's *Dakota Philosopher: Charles Eastman and American Indian Thought* (2009), as well as books by Charles Eastman himself (Eastman, 1902, 1904). Two classic studies carried out by philosophers of ethics, rather than anthropologists, may be selected for consideration. They do not deal with classic Old Northern societies, but with societies influenced by town and city traditions from pre-Columbian Mexico. These are Richard Brandt's *Hopi Ethics* (1954) and John

Ladd's *The Structure of a Moral Code* (1957) about the Din'e. Both were stimulated by anthropological work, at Chicago and Harvard respectively.

Richard Brandt, a young man starting his career at that time, went on to a utilitarian ethicist. It is fairly clear that his Hopi experiences influenced him. (If one compares pp. 319–336 of Brandt (1954) with his philosophy, one sees a direct carry-over from Hopi tribal teachings to an effects-based morality.) His work is noteworthy for following the grand utilitarian tradition of Mill and Sidgwick.

Brandt's account of the Hopi begins with the moral terms (Brandt, 1954: 91). He stresses the basic importance of *hopi*—properly human, "the Hopi way"—versus *kahopi* (we now spell it *qahopi*), "not Hopi." However, Brandt notes that the moral implications of this term are not simply that custom or tradition is ideal (Brandt, 1954: 94). Most social customs are mere conventions, not especially moral. Conversely, individualist behavior can be very moral. (Ladd notes the same for the Navaho and for morals in general.) More common, but less revealing, are oppositions of *anta* "good, morally right" and *ka-anta* "bad," and *loloma* "good, good-looking, nice" and *kaloloma*. The equation of the good and the beautiful in that last word is also found in Navaho, a topic, of course, much discussed and debated in ancient Greece.

The problems of studying traditional ethics were beautifully summed up by Clyde Kluckhohn in his foreword to Ladd's book (Kluckhohn, 1957: xiv). As he reports it, he asked their mutual friend and frequent consultant Bidaga (a.k.a. Son of Many Beads) why he—Kluckhohn—had never gotten such ethical details. Bidaga gave him "a long lecture which culminated with the sentence: 'I have been trying to explain these things to you for thirty years, but you never asked me the right questions.'"

In Ladd's work, the famous Din'e (Navaho) word *hozhon* (now spelled *hozhǫǫ*) "good, beautiful, harmonious, proper" and its opposite *hochoon* (*hochǫǫ*) "bad" are discussed; subsequent discussions (Farella, 1984; Witherspoon, 1977) are more detailed. From these later accounts we realize that the Din'e are much more self-consciously and socially moral than Ladd thought, and that their concepts of hozhon and hochon refer to the whole universe—divinities, humans, nature. The universe is a fundamentally moral place; the Din'e would reject Hume's claim that one cannot deduce an ought from an is. This also goes a long way toward adding a more Kantian side to the utilitarian and Aristotelian threads that Brandt and Ladd describe.

Hozhon is a concept that environmentalists should adopt. The idea is that beauty, goodness, harmony, health, peace, and smooth function are aspects of one general reality. The Navaho, significantly, also realize that ugliness, disharmony, death, and other bad things have a necessary place in the world. As Ladd notes, the monster-slaying hero-twins were persuaded to leave four monsters—hunger, violence, old age, and death—so that the world would not become static and overpopulated. If there were always harmony, there would be no change, progress, innovation, or valuable conflict (Farella, 1984; Witherspoon, 1977). The world must maintain a dynamic tension between peaceful harmony and dynamic destruction. Coyote, the source of mischief in Navaho cosmology, is a necessary and even a revered animal.

Related to such attitudes is a high order of social and personal morality, enforced by spiritual agents in the natural world. A Navaho ritual singer told the great pioneer ethnographer Washington Matthews: "Why should I lie to you? I am ashamed before the earth; I am ashamed before the heavens; I am ashamed before the blue sky; I am ashamed before the darkness; I am ashamed before the sun; I am ashamed before that standing within me which speaks with me. Some of these things are always looking at me, I am never out of sight. Therefore, I must tell the truth. That is why I always tell the truth. I hold my word tight to my breast." The Navaho are rather Aristotelian in their morality, based on cultivating individual virtues to avoid troubles as well as social sanctions and shame or guilt before the world (Ladd, 1957).

Traditional Native American worldviews typically go with a particular educational strategy (see, e.g., Armstrong, 2005a, b, 2020; Cajete, 1994; Goulet, 1998; Sharp, 2001). Children learn by doing, with relatively little verbal instruction. They copy what their elders do, without much correction, and learn by their mistakes as they try to imitate. Verbal teaching, if any, is done by personal stories. What little correcting of the young is done verbally is done by telling stories on oneself: "When I was young, I..." is a typical formula starting off a story of obvious—but completely tacit—application to the younger hearers present. (See also Greenfield (2004) on such learning among the Tzotzil Maya and Hunn (2008) on early learning among the Zapotec.) Once again, rural Anglo-Americans, including Hedrick's ranchers, converge on Native Americans in these matters. They teach by the book in schoolhouses and churches, but in the field or on the job, they teach by example and personal story, as is true of rural European and American tradition as well.

Shamanism

The Old Northern complex is associated with shamanism (see above, p. xxx), but many of the relevant groups have ancestor cults and other religious manifestations rather than the shamanistic system. True shamanism is characterized by rather specific ideas of the cosmos. The shaman goes into trance and goes through a long dark passage, or climbs the world tree, takes a lengthy flight, or otherwise takes a long and arduous "shamanic journey" to reach the lands of spirits above or below. Aids to trance such as flashing lights, hypnotic drumming, dancing, and hyperventilation are used.

Shamans sometimes competed in showing off their powers. Folktales in shamanic societies often include descriptions of competitions between rivals, to see who can assume the most powerful shape, foiling the rival (see Bland, 2012). Such stories endure even in modern Europe, e.g., the Child ballad of "The Twa Magicians," a Scottish tale that survived into the twentieth century (its protagonists are mere shape-changing magicians, but the idea has shamanic roots at some distant point).

The shaman is an individual performer, but socially recognized, and shamans may share secrets quite widely, so there is something like an organized religion here in spite of the lack of a "church." In fact, today, Siberian shamanism is well organized and once again becoming a major religion in Siberia.

Alternatively, many Native American religions are as complex, elaborated, and self-conscious as the Asian ones, but involve a more complex vision questing in which all individuals seek spirit power and are expected to get it, but only some go on to higher or more specialized powers that are comparable to the shamanic ones of Asia (Benedict, 1923 and many subsequent works).

The tendency to use "shamanism" for all Native American (let alone all "noncivilized"!) spiritual beliefs seems too sloppy and vague to be acceptable. I Agree There is a huge debate on how wide to extend the term, and how to define it, in the literature (Harvey, 2006; Humphrey & Onon, 1996; Kenin-Lopsan, 1997; Nowak & Durrant, 1977; Willerslev, 2007).

There is every reason to believe that spirit mediumship developed from shamanic religion in China, Japan (Blacker, 1986), and Korea (Kendall, 1985, 1988). Similarly, ancient Greek religion is clearly a solidly Old Northern religion, influenced—more and more heavily over time—by Mesopotamian, Levantine, and Egyptian religions (cf. Dodds, 1951; Fontenrose, 1974, 1981).

As so often, insightful comparisons with medieval Ireland can be made. The unique, fascinating, and complex story of "mad Sweeney" (Heaney, 1984) gets into realms of shamanic soul travel. It is a Christian epic; Sweeney offends an early missionary, who curses him by making him go "mad" and flee from a battle—an unbearable shame for an Irish chief. Sweeney wanders through Ireland. He finally finds peace in his last days in a Christian community. It is quite obvious that Sweeney is not so much "mad" as engaging in ancient shaman-like practices, though the anonymous author also describes strikingly well the effects of war-induced terror. Sweeney was evidently a practitioner of a shamanic spiritual discipline that is demonized by Christianity. He flies, takes great leaps, shifts shape, foretells the future, talks with wolves, diagnoses mental states, and otherwise acts like a classic northeast Asian shaman. The anonymous author, though Christian, has an obvious sympathy with Sweeney and his "madness." The result is one of the most amazing "inside" accounts of soul travels in all world literature, and it is very well worth comparing with Jay Miller's and others' Northwest Coast "shamanic odysseys" and with Caroline Humphrey's classic work with Urgunge Onon (Humphrey & Onon, 1996). We have this work in a rather late and somewhat corrupt version, but the language, medieval Irish Gaelic, includes a great deal of extremely beautiful poetry. (Readers of Irish can consult J. G. O'Keeffe's excellent bilingual edition of O'Keeffe, 1913).

Another Irish mythic figure, King Conare, was part bird in origin and was protected by birds who could shed their feather mantles and appear as humans. From them he got power but had to observe certain rules (*geasas*) to keep it. He was constantly being put in positions of having to break these rules, and thus eventually died (Gantz, 1981: 60–106). The shape-shifter spirits granting power, the rules attending that, and the myth turning on loss of power through having to break the rules are exactly the same in Northwest Coast mythology.

References

Adamson, T. (2009). *Folk-tales of the coast Salish*. (New edn. with introduction by W. R. Seaburg and L. Sercombe; original, New York: American Folklore Society, Memoir 27, 1934). University of Nebraska Press.

Alberta Society of Professional Biologists. (1986). *Native people and renewable resource management*. Alberta Society of Professional Biologists.

Anderson, E. N. (1996). *Ecologies of the heart*. Oxford University Press.

Anderson, E. N. (2011). Yucatec Maya botany and the 'nature' of science. *Journal of Ecological Anthropology, 14*, 67–73.

Anderson, E. N. (2014). *Caring for place*. Left Coast Press.

Armstrong, J. (2005a). En'owkin: Decision-making as if sustainability mattered. In M. K. Stone & Z. Barlow (Eds.), *Ecological literacy: Educating our children for a sustainable world* (pp. 11–17). Sierra Club Books.

Armstrong, J. (2005b). Okanagan education for sustainable living: As natural as learning to walk or talk. In M. K. Stone & Z. Barlow (Eds.), *Ecological literacy: Educating our children for a sustainable world* (pp. 80–84). Sierra Club Books.

Armstrong, J. (2020). Living from the land: Food security and food sovereignty today and into the future. In N. J. Turner (Ed.), *Plants, people and places: The roles of ethnobotany and ethnoecology in indigenous peoples' land rights in Canada and beyond* (pp. 36–50). McGill-Queen's University Press.

Atleo, E. R. (2004). *Tsawalk: A Nuu-chah-nulth worldview*. University of British Columbia Press.

Atleo, E. R. (2011). *Principles of Tsawalk: An indigenous approach to global crisis*. University of British Columbia Press.

Balée, W. (1994). *Footprints of the forest*. Columbia University Press.

Ballard, A. C. (1929). Mythology of southern Puget Sound. *University of Washington Publications in Anthropology, 3*, 31–250.

Bartra, R. (1994). *Wild men in the looking glass: The mythic origins of European otherness*. University of Michigan Press.

Benedict, R. F. (1923). *The concept of the Guardian spirit in North America*. American Anthropological Association. Memoir 29.

Bernard, H. R., & Salinas Pedraza, J. (1989). *Native ethnography. A Mexican Indian describes his culture*. Sage.

Blackburn, T. (1975). *December's child: A book of Chumash Oral narratives*. University of California Press.

Blacker, C. (1986). *The catalpa Bow: A study of shamanism in Japan* (2nd ed.). George Allen & Unwin.

Bland, R. L. (2012). Bernard Fillip Jacobsen and three Nuxalk legends. *Journal of Northwest Anthropology, 46*, 143–166.

Blaser, M. (2009). The threat of the Yrmo: The political ontology of a sustainable hunting program. *American Anthropologist, 111*, 10–20.

Boas, F. (1918). *Kutenai tales*. United States Government, Bureau of American Ethnology, Bulletin 59.

Bogoras, W. (1904–1909). *The Chukchee*. Memoir of the American Museum of Natural History, XI; Report of the Jesup North Pacific Expedition, VII.

Bogoras, W. (1919). *Chukchee mythology*. Memoir of the American Museum of Natural History, XII; Report of the Jesup North Pacific Expedition, VIII.

Bogoras, W. (1929). Elements of culture of the circumpolar zone. *American Anthropologist, 31*, 597–601.

Brandt, R. B. (1954). *Hopi ethics: A theoretical analysis*. University of Chicago Press.

Brody, H. (2000). *The other side of Eden: Hunters, farmers, and the shaping of the world*. Northpoint Press.

Cajete, G. (1994). *Look to the mountain: An ecology of indigenous education*. Kivaki Press.

Callicott, J. B. (1994). *Earth's insights: A multicultural survey of ecological ethics from the Mediterranean Basin to the Australian outback.* University of California Press.
Callicott, J. B., & Nelson, M. P. (2004). *American Indian environmental ethics: An Ojibwa case study.* Pearson Prentice Hall.
Campbell, J. (1983). *The way of the animal powers.* A. van der Marck Editions.
Carmichael, A. (1992). *Carmina Gadelica.* Lindisfarne Books.
Child, F. J. (1882). *The English and Scottish popular ballads* (Vol. I). Houghton, Mifflin Harcourt.
Clottes, J. (2008). *Cave art.* Phaidon.
Coffin, T. (1976). *The singing tradition of child's popular ballads.* Princeton University Press.
Cook, R. (1974). *The tree of life.* Avon Books.
Cordova, V. F. (2007). *How it is: The native American philosophy of V.F. Cordova.* University of Arizona Press.
Davis, T. (2000). *Sustaining the forest, the people, and the Spirit.* SUNY Press.
DeMaillie, R. J. (1984). *The sixth grandfather: Black Elk's teachings given to John G. Neihardt.* University of Nebraska.
Descola, P. (2013). *Beyond nature and culture.* Trans. by Janet Lloyd (French original 2005). University of Chicago Press.
Dodds, E. R. (1951). *The Greeks and the irrational.* University of California Press.
Dooley, A., & Roe, H. (1999). *Tales of the elders of Ireland: A new translation of Acallam na Senórach.* Oxford University Press.
Eastman, C. A. (1902). *Indian boyhood.* McClure, Philips & Co.
Eastman, C. (1904). *Red hunters and the animal people.* Harper & Bros.
Errington, P. (1967). *Of predation and life.* Iowa State University Press.
Errington, P. (1973). *The red God's call.* Iowa State University Press.
Fantham, E. (2006). Introduction. In *Georgics*, by Virgil. Trans. by P. Fallon (Latin orig. ca. 38 BC) (pp. xi–xxxiii). Oxford University Press.
Farella, J. (1984). *The Main stalk: A synthesis of Navaho philosophy.* University of Arizona Press.
Filtchenko, A. (2011). Landscape perception and sacred places amongst the Vasiugan Khants. In P. Jordan (Ed.), *Landscape and culture in northern Eurasia* (pp. 178–197). Left Coast Press.
Fontenrose, J. (1974). *Python: A study of Delphic myth and its origins.* Biblio and Tannen.
Fontenrose, J. (1981). *Orion: The myth of the hunter and huntress.* University of California Press.
Gantz, J. (1981). *Early Irish myths and sagas.* Dorset.
Goulet, J. A. (1998). *Ways of knowing: Experience, knowledge, and power among the Dene Tha.* University of Nebraska Press.
Greenfield, P. M. (2004). *Weaving generations together: Evolving creativity in the Maya of Chiapas.* School of American Research Press.
Haly, R. 1992. On becoming a mountain: Or why native Americans have animals for souls. In *Paper, American Anthropological Association, Annual Meeting.*
Harvey, G. (2006). *Animism: Respecting the living world.* Columbia University Press.
Hayden, B. (2003). *Shamans, sorcerers and saints: A prehistory of religion.* Smithsonian Books.
Hayden, B. (2013). Hunting on heaven and earth. *Current Anthropology, 54*, 495–496.
Heaney, S. (1984). *Sweeney astray.* Faber & Faber.
Huber, T. (1999). *The cult of pure crystal mountain.* Oxford University Press.
Humphrey, C., & Onon, U. (1996). *Shamans and elders: Experience, knowledge, and power among the Daur Mongols.* Oxford University Press.
Hunn, E. (2008). *A Zapotec natural history: Trees, herbs, and flowers, birds, beasts and bugs in the life of San Juan Gbëë.* University of Arizona Press.
Hunn, E., & Selam, J. (1990). *Nch'i-Wana, the big river.* University of Washington Press.
Jacobs, E. D., & Jacobs, M. (1959). *Nehalem Tillamook Tales.* University of Oregon Books.
Jochelson, W. (1908). *The Koryak.* Memoir of the American Museum of Natural History, X; Reports of the Jesup North Pacific Expedition, VI.
Jochelson, W. (1926). *The Yukaghir and the Yukaghirized Tungus.* Memoir of the American Museum of Natural History, XIII; Reports of the Jesup North Pacific Expedition, IX.

References

Jochelson, W. (1990). *Unangam Ungiikangin Kayux Tunusangin, Unangam Uniikangis Ama Tunuzangis, Aleut Tales and Narratives*. Alaska Native Language Center, University of Alaska-Fairbanks.
Kay, C. E., & Simmons, R. T. (Eds.). (2002). *Wilderness and political ecology: Aboriginal influences and the original state of nature*. University of Utah Press.
Kendall, L. (1985). *Shamans, housewives, and other restless spirits: Women in Korean ritual life*. University of Hawaii Press.
Kendall, L. (1988). *The life and hard times of a Korean shaman*. University of Hawaii Press.
Kenin-Lopsan, M. B. (1997). *Shamanic songs and myths of Tuva*. Akadémiai Kiadó.
Kirch, P. V. (1994). *The wet and the dry: Irrigation and agricultural intensification in Polynesia*. University of Chicago Press.
Kirch, P. V. (1997). Microcosmic histories. *American Anthropologist, 99*, 31–42.
Kirch, P. V. (2007). Hawaii as a model system for human ecodynamics. *American Anthropologist, 109*, 8–26.
Kluckhohn, C. (1957). Foreword. In J. Ladd (Ed.), *The structure of a moral code* (pp. i–xiv). Harvard University Press.
Kohn, E. (2013). *How forests think: Toward an anthropology beyond the human*. University of California Press.
Ladd, J. (1957). *The structure of a moral code*. Harvard University Press.
Lame Deer, J. F., & Erdoes, R. (1972). *Lame deer, seeker of visions*. Simon & Schuster.
Lawrence, E. A. (1997). *Hunting the wren: Transformation of bird to symbol*. University of Tennessee Press.
Lentz, D. L. (Ed.). (2000). *Imperfect balance: Landscape transformations in the Precolumbian Americas*. Columbia University Press.
Lévi-Strauss, C. (1958). *Anthropologie structurale*. Plon.
Lévi-Strauss, C. (1962). *La pensée sauvage*. Plon.
Lévi-Strauss, C. (1963). The sorcerer and his magic. In *Structural anthropology* (Chap. 9, pp. 167–185). Trans. by C. Jacobson and B. Grundfest Schoepf. Basic Books.
Lévi-Strauss, C. (1964–1971). *Mythologiques* (4th edn). Plon.
Lloyd, G. (2007). *Cognitive variations: Reflections on the unity and diversity of the human mind*. Oxford University Press.
Lopez, K. L. (1992). Returning to fields. *American Indian Culture and Research Journal, 16*, 165–174.
Martin, C. (1978). *Keepers of the game*. University of California Press.
Martinez, D. (2009). *Dakota philosopher: Charles Eastman and American Indian thought*. Minnesota Historical Society.
McNeill, J. R. (1992). *Mountains of the Mediterranean world*. Cambridge University Press.
Medin, D., Ross, N. O., & Cox, D. G. (2006). *Culture and resource conflict: Why meanings matter*. Russell Sage Foundation.
Metzo, K. R. (2005). Articulating a Baikal environmental ethic. *Anthropology and Humanism, 30*, 39–54.
Neihardt, J. (1932). *Black Elk Speaks*. William Morrow.
Nowak, M., & Durrant, S. (1977). *The tale of the Nišan Shamaness: A Manchu folk epic*. University of Washington Press.
O'Keeffe, J. G. (1913). *Buile Suibne (the Frenzy of Suibhne), being the adventures of Suibhne Geilt: A middle-Irish romance*. Irish Texts Society.
Pastoureau, M. (2007). *The bear: History of a fallen king*. Belknap Press of Harvard University Press.
Phinney, A. (1934). *Nez Percé Texts* (pp. 1–497). Columbia University Contributions to Anthropology.
Pierotti, R. (2011). *Indigenous knowledge, ecology, and evolutionary biology*. Routledge.
Pierotti, R., & Fogg, B. R. (2017). *The first domestication: How wolves and humans coevolved*. Yale University Press.

Pierotti, R., & Wildcat, D. (2000). Traditional ecological knowledge: The third alternative. *Ecological Applications, 10*, 1333–1340.

Ponting, C. (1991). *A green history of the world: The environment and the collapse of great civilizations*. Penguin.

Radin, P. (1927). *Primitive man as philosopher*. Appleton.

Radin, P. (1957). *Primitive religion*. Dover. (Orig 1937; this has a new preface.).

Ramsey, J. (1977). *Coyote was going there: Indian literature of the Oregon country*. University of Washington Press.

Reed, R. K. (1995). *Prophets of agroforestry: Guaraní communities and commercial gathering*. University of Texas Press.

Reed, R. K. (1997). *Forest dwellers, Forest protectors: Indigenous models for international development*. Allyn & Bacon.

Reichel-Dolmatoff, G. (1971). *Amazonian cosmos: The sexual and religious symbolism of the Tukano Indians*. University of Chicago Press.

Reichel-Dolmatoff, G. (1976). Cosmology as ecological analysis: A view from the rain Forest. *Man, 11*, 307–316.

Reichel-Dolmatoff, G. (1996). *The Forest within: The world-view of the Tukano Amazonian Indians*. Themis, imprint of Green Books, Foxhole, Dartington, Totnes, Devon.

Ridington, R. (1988). *Trail to heaven: Knowledge and narrative in a northern native community*. University of Iowa Press.

Rilke, R. M. (1964). Orpheus. Eurydice. Hermes. In *New poems* (pp. 142–147). Translated by J. B. Leishman. New Directions.

Rytkheu, Y. (2019). *When the whales leave*. Trans. by I. Y. Chavasse. Milkweed Editions.

Salick, J., & Moseley, R. K. (2012). *Khawa Karpo: Tibetan traditional knowledge and biodiversity conservation*. Missouri Botanic Garden.

Sapir, E. (1990). In W. Bright (Ed.), *The collected works of Edward Sapir. VII. Wishram texts and ethnography*. Mouton de Gruyter.

Scarborough, V. L., Isendahl, C., & Fladd, S. (2020). Environment and landscapes of Latin America's past. In L. R. Lozny & T. H. McGovern (Eds.), *Global perspectives on long term community resource management* (pp. 213–234). Springer.

Schaefer, B. E. (2006, November). The origin of the Greek constellations. *Scientific American*, 96–101.

Scott, C. (1996). Science for the west, myth for the rest? The case of James Bay Cree knowledge construction. In L. Nader (Ed.), *Naked science: Anthropological inquiry into boundaries, power, and knowledge* (pp. 69–86). Routledge.

Seaburg, W. R. (Ed.). (2007). *Pitch woman and other stories: The Oral traditions of Coquelle Thompson, upper coquille Athabaskan Indian*. University of Nebraska Press.

Sharp, H. (2001). *Loon: Memory, meaning and reality in a northern Dene Community*. University of Nebraska Press.

Shipman, P. (2011). *The animal connection*. W. W Norton & Sons.

Shipman, P. (2015a). How do you kill 86 mammoths? Taphonomic investigations of mammoth megasites. *Quaternary International, 359–360*, 38–46.

Shipman, P. (2015b). *The invaders: How humans and their dogs drove Neandertals to extinction*. Harvard University Press.

Shutova, N. (2006). Trees in Udmurt religion. *Antiquity, 80*, 318–327.

Sullivan, L. (1988). *Icanchu's drum: An orientation to meaning in south American religions*. Macmillan.

Tanner, A. (1979). *Bringing home animals*. St. Martin's Press.

Teit, J., & Boas, F. (1927–1928). *The Salishan tribes of the Western Plateaus* (pp. 25–395). Bureau of American Ethnology, Annual Report.

Thompson, S. (1955–1958). *Motif-index of folk-literature*. Indiana University Press.

Tucker, M. E., & Grim, J. E. (Eds.). (1994). *Worldviews and ecology: Religion, philosophy, and the environment*. Orbis Books.

Tuite, K. (1994). *An anthology of Georgian folk poetry*. Associated University Presses.
Turner, N. J. (2005). *The Earth's blanket*. Douglas and MacIntyre, University of Washington Press.
Tylor, E. (1871). *Primitive culture*. John Murray.
Ventocilla, J., Herrera, H., & Nuñez, V., with editorial assistance of H. Roeder. (1995). *Plants and animals in the life of the Kuna*. Trans. by E. King. University of Texas Press.
Walker, B. (2005). *The lost wolves of Japan*. University of Washington Press.
Watkins, C. (2000). *The American heritage dictionary of Indo-European roots*. Houghton Mifflin.
Wiget, A., & Balalaeva, O. (2011). *Khanty: People of the Taiga*. University of Alaska Press.
Willerslev, R. (2007). *Soul hunters: Hunting, animism, and personhood among the Siberian Yukaghirs*. University of California Press.
Witherspoon, G. (1977). *Language and art in the Navajo universe*. University of Michigan Press.

Chapter 16
Appendix 3: The "Wasteful" Native Debunked

At some point (we leave it to the end!), it is necessary to deal with two old chestnuts that are always raised when anyone speaks of Native American conservation practices.

First, buffalo jumps. Native Americans on the plains stampeded bison over cliffs, thus killing a number at a time. This has been condemned as "wasteful" by generations of white writers (the same people that shot tens of millions of bison for the hides and tongues or solely to starve the Indians out, leaving the carcasses of the bison to rot). The idea seems to be that the Indians drove a million buffalo over a cliff every time they wanted a light lunch (Krech, 1999, who should know better, still spins this story, based on a single incident of mass bison death, which may not even have involved humans).

In fact, the vast majority of buffalo jumps were used only once, and the rest rather rarely. There were not very many sites in all, considering the enormous amount of space and time involved (Bamforth, 2011). The busiest, the Head-Smashed-In Jump in Alberta, was used only once a generation during its peak years (according to displays at the site, now a World Heritage Site). The total number of bison killed in all the jumps in the plains, through the 16,000 or so years of human occupancy, was probably in the low millions.

There was, however, undeniably some waste—too many buffalo killed for the population to use, which of course assumes that anything not consumed by humans represents waste.

> It may well be true that waste occurred, if you define "waste" solely in terms of human use. Given the number of other predators (wolves, bears, eagles, ravens, crows, magpies, etc.) with which Indigenous human hunters shared the West, I find it unlikely that much meat went uneaten... at least until Europeans arrived and worked hard to exterminate these other predators along with bison and Indians (Pierotti, 2011: 168).

This is especially true if it is considered that many Plains peoples learned how to use buffalo jumps from working with wolves who were full participants in the activity.

In Blackfoot cosmology, beavers constructed the world in which we now live, but wolves were the first to learn how to utilize the most precious food resource in the world that beavers built: bison. Blackfoot learned to live together and hunt bison in social groups from wolves, and refer to their ancestors as "the wolf people." Blackfoot reportedly called wolves their brothers and protected them (McClintock, 1910: 434, 476). Wolves were associated with cunning and traveling long distances, and wolf-songs were used in hunting and in war (McClintock, 1923: 210, 281).16 These stories coexist with iinisskimm (Buffalo Jump) stories, and draw attention to an important ecological fact: wolves drive bison (From Barsh & Marlor, 2003: 581).

The problem is that bison, unlike cattle, are not domesticated to the point of stupid obedience. You cannot pick out one when driving them. If you start a herd, you cannot stop them, and they may—but, in fact, rarely did—all go over the edge. Moreover, the extremely high numbers killed at places like Head-Smashed-In were apparently not killed just for the local population. Grease and dried meat were prepared in quantity, for storage and extensive trade (Bamforth, 2011). Wolves were also full participants: "Wolf-Also-Jumped pisskan is a reminder than humans and wolves hunted bison alongside each other" (Barsh & Marlor, 2003: 578). There were mouths other than human to be fed, as described by George Bent (Cheyenne):

> (During) winter buffalo hunts...when all the tribe was on foot (i.e., before they had horses). A herd of buffalo was surrounded by the people (and the "dogs" (wolves) and driven into deep drifts...If a buffalo got away the dogs would set on it and quickly drive it back to the deep drifts...After the buffalo are skinned (and butchered) the (wolves dragged) the bundles of meat over the ice...As soon as the camp was reached, the dogs were loosed, and *at once the whole pack* rushed back... to the [kill site, where]... they feasted on the parts that had been thrown aside [during butchering]...mothers who had puppies in camp *would run to the [site], gorge themselves with meat, and then run back to camp and disgorge part of the meat for the puppies to feed on.* Sometimes a mother would make several trips to get enough meat for her litter of young ones (from Pierotti & Fogg, 2017: 146; emphasis and parenthetical elements added).

Getting them to go at all is the hard part, which is where the wolves came in, because as mentioned above; wolves drive bison. Unlike domestic cattle, bison do not herd easily, and are too intelligent to stampede randomly over a cliff. Arranging a jump involved considerable planning—watching the bison routes, building cairns that looked like warriors posted along the path to the cliff, organizing the work force, and so on. Then when the time came, the bison had to be stampeded such that the ones in back forced the ones in front over the cliff; otherwise, the ones in front would simply stop and not go over. This involved serious danger, since the bison would naturally tend to turn when they got to the cliff, and run back the other way—right over the waiting Native hunters. This had to be discouraged by such measures as lighting smoky fires, or by charging wolves. If the bison did fall over the cliff, they could fall on other people waiting to dispatch them. That is how Head-Smashed-In got its name; one poor soul got too close to the landing point (according to interpretive materials at the site).

Tribes learned to employ this method from working with and observing wolves take bison using this technique (Barsh & Marlor, 2003; Fogg et al., 2015). Western scholarship emphasizes only the human/bison interaction and argues that if more bison are killed than humans can consume, this results in waste (as argued by Krech, 1999). Native people observed that this dynamic process involved much more than the humans and the bison. It included wolves, coyotes, crows, ravens, magpies, and other species (Barsh & Marlor, 2003). Even if the humans could not use all the meat, many other organisms obtained food, and all the bison killed were consumed. From the perspective of the ecological community, there was no waste. This explains why, in the story of the Great Race told by the Tsitsista (Cheyenne) people, every species in the community participated in the race to determine whether humans would eat bison or vice versa. Humans were recognized by birds to be more cooperative than bison, thus the birds helped the humans to win (Grinnell, 1923; Pierotti, 2020).

The other chestnut is the reputed Native American extermination of the Pleistocene megafauna, originally claimed by Paul S. Martin in the 1960s (Martin, 1967, 1984, 2002, 2005; Martin & Klein, 1984). His only evidence was that the Native Americans appeared in North America about the same time that several large species—including all the largest ones—became extinct.

At least the buffalo jumps were real, and a lot of buffalo died that way. But the extinction of the megafauna has not been connected with Native Americans by much direct evidence. There are very few mammoth and mastodon kill sites from the early millennia. Reputed kill sites for horses and other extinct creatures have not held up well on analysis. Evidence is all in the other direction: Mathew Stewart et al. (2021) showed that there is no detectable correlation, at least with present data, of human arrival and megafaunal extinction. The correlation is with the sudden onset of intense cold in the Younger Dryas period, around 12,500 years ago, following a long warming time that led to expansion of megafaunal populations and their northward spread. Neither the megafaunal species group as a whole nor individual large species show a correlation with human activity. The present authors do not rule out some real effects by humans, however; data are still thin, and correlations thus difficult and somewhat inconclusive.

Martin's original position was that humans entered North America about 11,000 years ago, moved very fast and increased their numbers very fast (3% per annum), and wiped out the megafauna in a sudden "blitzkrieg." This version of the overkill hypothesis is not credible—such increases would have been impossible in a difficult environment filled with predators, new diseases, and such. But it now appears that people were in the Americas much earlier, at least from 20,000 years ago. Also, mastodons and some other species lasted longer than was once thought. All this makes a slower and more believable process possible. John Alroy (2001a, b) has created a simulation showing that people could indeed cause the extinction under such circumstances.

The deadly Clovis points that are actually found in mammoth and bison carcasses did not develop till many of the megafaunal species were already extinct (or at least are no longer found in the paleontological record; Wolverton et al., 2009). Clovis flourished around 11,000 years ago (Waters & Stafford, 2007), while most of the megafauna disappear around 12,000–14,000 years back. For instance, the main collapse of the megafauna in what is now Indiana was between 14,800 and 13,700 years ago, long before Clovis (Gill et al., 2009). Indeed, there is no evidence for humans in Indiana at that time. If there were any, they were exceedingly few. Extinction in California came later, mainly around 12,000 years ago, when more people were present but when climate was also dramatically changing.

The only real evidence for Martin's position is that not only in North America, but also in Australia, New Zealand (Worthy & Holdaway, 2002), and countless smaller islands, the coming of modern humans coincided with the extinction of the larger and slower species of wildlife, and usually with most of the ground-nesting birds. This is indeed very good circumstantial evidence. It also suggests some other possibilities beyond hunting. All scholars now seem to agree that the very widespread use of fire to manage the landscape is the main thing driving extinctions in Australia. Rats and cats (introduced by humans) ate eggs and even adults, which apparently drove most of the extinctions of flightless island birds. People hunted moas in New Zealand, but otherwise most of the extinct bird species on oceanic islands are tiny, obscure, and rare, not the sort of thing people hunt. The problem is that moas and dodos (large flightless pigeons of Mauritius) are island species, passing for charismatic megafauna on islands. Disease has certainly wiped out much island life, as has climate change; Hawaii's native birds are being slowly but surely exterminated by recently introduced avian malaria. Fire, other invasive animal species, and diseases could all have had a hand in the American extinctions. Fire was surely involved, given both the Australian case and our current knowledge of the universality of fire in Native American management.

Martin's idea has some merit. The phenomenon of pioneering peoples sweeping through an island or continent without thought for the morrow is apparently universal among humans. We have seen it in the colonial histories of the Americas, Australia, and other areas. Indeed, nonhuman species can explode when they enter a new area, only to stabilize later as they eat themselves out of a home and as diseases and parasites build up. The earliest Native Americans thus might have hunted with no thought of the morrow until game grew scarce enough to force them to change. Regardless of cause, Pierotti (2011) has argued that the disappearance of megafauna may have signaled to Indigenous Americans that conservation was needed.

Martin argued that American fauna, like animals on small isolated islands, were unused to predators like humans, and thus not adapted to predation. However, America's megafauna included lions, wolves, sabretooths, giant bears, and other carnivores. Martin argued that humans are somehow different, but any experience with game animals shows that this is not the case. Animals exposed to other predators quickly learn to be wary of humans too.

There is a paper that argues that naivete of moose when encountering reintroduced wolves or grizzly bears (Berger et al., 2001) shows that prey species are naïve when encountering new predators. For some reason the editors of the journal *Science* decided this should be published even though this study demonstrated unequivocally that the prey involved—moose (*Alces alces*) and American elk (*Cervus elaphus*)—appeared unfamiliar with wolves or bears at first, but once individuals of these species observed wolves or bears take a member of their species, they learned immediately to respond to them through avoidance and defense. They also taught their offspring to avoid these predators, Pierotti has argued:

> The result of the Berger et al. (2001) study, which to me clearly does not support the Pleistocene Overkill hypothesis, amazingly is cited by Paul Martin (2005) as follows: "Berger's [sic] recent findings suggest that *at first contact* the American fauna would have lacked behavioral defenses against humans including the fear and alarm response necessary to inspire potential prey to fight or flee... Very likely it would have taken longer for potential prey to learn to fear the new human predators than it did for the moose to learn to fear the wolves, *which their ancestors had known to be dangerous*" (Martin, 2005: 139; emphasis added). This is the response of a person who studies fossils rather than living creatures—the behavioral assumptions implicit within it bear a closer resemblance to science fiction than to science. It also shows that not only the public, but even scientists are often prepared to uncritically believe what a scientist tells them, rather than to critically evaluate the evidence themselves.
>
> It is interesting to note that Martin considers moose more likely to learn quickly than mammoths or mastodons, who like most proboscideans are much more "intelligent," or at least behaviorally complex, than cervids such as moose and elk. Even more interesting is Martin's apparent invocation of a sort of ancestral "memory" in moose, while ignoring the fact that mammoths and mastodons also had ancestral experience of human hunters in Asia, where they had coexisted for as long as 40,000 years, as well as the well-established ability of proboscideans to pass knowledge across generations (Payne, 2000; McComb et al., 2001; Goebel et al., 2008).

Similar areas in the Old World—Europe, China, southeast Asia, and so on—did not see their megafauna disappear when modern humans came (see, e.g., Louys et al., 2021). In fact, China still has a few elephants, rhinos, tigers, pandas, and so on, and maintained a quite rich megafauna until the rise of the Chinese empire. Africa lost considerable megafauna over time, but gradually with drying of the continent (Faith et al., 2018), and of course much of the megafauna survived until recently.

When they do show destruction, it is unmistakable in the archeological record. Contrasting with the lack of evidence in North America are the massive boneyards on Sicily and Cyprus, where humans unquestionably exterminated the local dwarf elephants and hippos (Simmons, 2007; displays in Sicily's historical museum at Syracuse, observed by ENA, Jan. 1, 2009). On Cyprus, one site alone has the bones of over 500 pigmy hippos (Simmons, 2007: 231)—couple that with post-Pleistocene drying and heating, and there is no question why that species went extinct! This is exactly what we do *not* find anywhere in early North America. It is simply not credible that there was better preservation on a couple of Mediterranean islands than in the whole North American continent. There is also the fact that the vast terminal-Pleistocene boneyards we do have, such as the La Brea tar pits, contain few or no human kills.

The other main candidate for causing the megafaunal extinctions is climate change. We know that much earlier glaciations ended with comparably huge extinction events, although no humans were around to affect this (Finnegan et al., 2011). The cold peak 18,000 years ago was extreme—more severe than any other well-known ice age in the last several million years—and the warm-up after that was rapid. By 14,800 years ago, when Indiana began to lose megafauna, the climate was still very cold; changes in vegetation, and the coming of frequent fire, were far in the future (Gill et al., 2009). However, by 13,000 years ago, times were warm and dry. Then a sudden, dramatic reversal (the Younger Dryas event) froze the continent again around 12,500 years ago. Cooling took perhaps only a few centuries. It was followed by an extremely rapid shift to hot, dry conditions. It is difficult to imagine large, slow-breeding animals tracking all this, and indeed much less dramatic swings in earlier millennia all led to massive extinction events, as in the late Eocene, late Miocene, early Pleistocene, and other times. Stewart et al. (2021) found a very marked correlation between the Younger Dryas and the great extinction. They point out that monogastric megafauna (the proboscideans and sloths) might have had a harder time adjusting than bison, elk, moose and other cud-chewers. More significant is the observation that:

> Pleistocene overkill requires that the tiny founding populations of New World peoples *swiftly hunted to extinction 35 separate genera of large mammals*, while the substantial harvest pressure and *severe impacts of the enormous late Holocene human populations required millennia and failed to cause the extinction of a single species* (Broughton, 2002: 67, emphasis added, as cited in Pierotti, 2011: 171–172).

After Europeans arrived there was a rolling wave of extinctions of both mammals and birds: Steller sea cow, Passenger Pigeon, Great Auk, Carolina Parakeet, and, almost, the bison and the native wolf species.

Shorter term climate change resulting from rapid drying and heating before and after the Younger Dryas could also be significant. In 2001–2002, 2006–2007, and 2020–2022, almost no rain fell throughout the entire southwestern quarter of North America. A two- or three-year period like that would have exterminated the megafauna. Such a period is quite likely for the Altithermal, the hot dry period that peaked around 7000–8000 years ago. The lack of major change in Indiana at the early dates of decline makes human hunting more imaginable as an explanation (Johnson, 2009), but there were few (if any) humans there, and no evidence of a deadly hunting technology till much later.

Further proof that climate did much of the work is the extinction of several large waterbirds, such as the La Brea stork (Grayson, 1977), and water mammals like the capybara. This is clearly climate-driven. These birds and mammals did not depend on the megafauna, but rather on wet Pleistocene conditions. Martin dismissed Donald Grayson's birds as "cowbirds" that depended on the megafauna, despite the fact that close relatives elsewhere are not cattle-followers. He also argues, repeatedly (see all cited sources), that the *lack* of evidence proves his point, since the wipeout happened so fast! This brings us close to the conspiracy theorists of the popular press, who prove their theories by arguing that the absence of evidence shows how successful "the government" or "the Trilateral Commission" has been in hiding it.

Climate change after the glaciation was extremely rapid and disruptive. Similar rapid and dramatic extinction events occurred at the ends of previous glaciations, such as the Ordovician-Silurian event (Finnegan et al., 2011), when humans were obviously not involved. North America 18,000–20,000 years ago was probably the coldest it has ever been. It was hot and dry by 12,000, but then the Younger Dryas event dropped temperatures back to Ice Age levels around 11,000 years ago. This in turn reversed, and an extremely hot and dry period set in by about 6–7000 BCE. The changes were extremely rapid.

The arguments presented by Clovis Overkill advocates serve as good examples of the subjectivity of some issues in science. Numerous sites in North America, as well as several in South America, have been dated as significantly older than the Clovis site in New Mexico. Every time a new site is identified, its validity is questioned, usually by C. Vance Haynes, emeritus professor of geosciences at the University of Arizona, member of the National Academy of Sciences, and not coincidentally, the individual responsible for dating the Clovis site in the first place (Pierotti, 2011: 188). Haynes appears to have appointed himself defender of the Clovis site as the first major human site in North America, even though it is several thousand miles from the Bering Strait. By extension Haynes also serves as one of the primary defenders of the Bering Strait argument. Paul Martin, who developed the overkill hypothesis and serves as its chief advocate, is also an emeritus professor of geosciences at the University of Arizona. This means that the two individuals largely responsible from pushing Clovis arguments are long-term colleagues who have steadfastly resisted attempts to establish evidence suggesting that humans have been in the Americas much longer than their findings suggest.

Prominent non-Native scholars examining the evidence have described the Clovis hypotheses as follows:

> Martin's recent writings suggest to us that he is no longer trying to approach this issue within a scientific framework. As we have noted, he explicitly maintains that the North American overkill position does not require supporting evidence. He is unconcerned that archaeologists "wash their hands" of his ideas. He criticizes the search for pre-Clovis sites in the New World as "something less than serious science, akin to the ever popular search for 'Big Foot' or the 'Loch Ness Monster." As one of us has observed elsewhere, *Martin's position has become a faith-based policy statement rather than a scientific statement about the past*, an overkill credo rather than an overkill hypothesis." (Grayson & Meltzer, 2003: 591, emphasis added; cited in Pierotti, 2011: 192–193).

Overall, it seems possible but probable that humans contributed to the extinction of the mammoths and mastodons, and perhaps also did in the giant ground sloths. Pierotti addressed this issue (2011: 190):

> Why did mammoths and mastodons go extinct, whereas tropical elephants survived? Admittedly there are a few mastodon kill sites, of which Martin makes a great deal. He suggests that early American humans may have specialized on these large, social, intelligent, and potentially dangerous mammals. In my opinion, because of their intelligence, strength, and tendency to act as an organized group, elephants are much less likely to have been regular targets of hunters than ungulates such as bison, moose, wapiti (elk), deer, caribou and pronghorn, all of whom persist to this day.

Ground sloths were sluggish, and unable to defend themselves against missiles, but they survived longer than other megafauna—until around 7000 years ago, well after the collapse of the Clovis culture. Other extinct megafauna includes a number of horse species, a huge camel, a llama, southerly species of ox and mountain goat, and several smaller forms. These seem to have disappeared before the rise of human hunters, or at least there is no credible association of them with humans. They are the sort that would not make it through rapid climate change. Relatives of three of them—the llama, the capybara, and a flat-headed peccary—not only survive but locally flourish in South America, despite extremely heavy hunting pressure today.

Many carnivorous mammals and birds died off, including all the largest, but almost everyone agrees these went because their large-herbivore prey disappeared. In short, humans were likely to have been involved in finishing off the last few megafauna; especially if they were concentrated around water holes by a drought like that of 2006–7. As argued above, these disappearances may have provided motivation to develop conservation ethics. We share the view of Barnosky et al. (2004), that the situation is unclear, but it seems hard to question "the intersection of human impacts with pronounced climatic change" (2004: 70; see also Kay & Simmons, 2002, Surovell et al., 2016).

Although this common-sense middle ground exists, the argument has remained polarized, with True Believers and total skeptics. The sociology of the controversy is possibly as interesting as the fate of the megafauna. Most of the scientifically-trained believers appear to be biologists, many of whom clearly have a very sour view of humans in general, at least as managers of nature (e.g., Ward, 1997). Others are mere popularizers who have no scholarly competence in this regard, and at least a few are outright racists who simply hate Native Americans. On the other hand, anthropologist Raymond Hames (2007), who has studied hunting peoples in the field, has recently provided a well-argued, well-reasoned defense of the overkill hypothesis.

Skeptics include Vine Deloria Jr. (1988, See Pierotti, 2011, Chap. 9 for an analysis); Donald Grayson (1977, 1991, 2001), Grayson & Meltzer, (2002, 2003), Shepard Krech (1999), in spite of his exaggeration of the buffalo jumps; and many others. Steven Wolverton et al. (2009) provide a particularly sophisticated review of the skeptical position. Skeptics seem to be either Native Americans or anthropologists, rather than biologists.

Perhaps the most interesting thing about all this is a point made by Deloria: any overhunting by people 12,000 years ago has very little to do with anything Native Americans may be doing now. Pioneers everywhere are notorious for overrunning their newly settled lands and wasting resources lavishly; they learn better soon. The Anglo-American settlement of North America, with its wave of deforestation followed by a wave of reforestation, shows this clearly enough. 12,000 years is a long time. Cultures do change, and archaeology shows that the old story of the changeless, timeless Indian was preposterous (e.g., Pryce, 1999).

The present book should be enough to point out that (1) Native Americans have a very lively sense of conservation, and observe it in practice, but (2) they teach it through highly circumstantial stories of overhunting that led to starvation. This should more or less settle the case. Yes, Native American conservation is real, but, no, Native Americans are no different from anyone else; they learn through experience, and some of the experience is hard-won. They probably did contribute somewhat—we will never know how much—to the extinctions.

References

Alroy, J. (2001a). A multispecies overkill simulation of the end-pleistocene megafaunal mass extinction. *Science, 294*, 1893–1895.
Alroy, J. (2001b). Response. *Science, 294*, 1459–1460.
Bamforth, D. (2011). Origin stories, archaeological evidence, and Postclovis Paleoindian bison hunting on the Great Plains. *American Antiquity, 76*, 24–40.
Barnosky, A. D., Koch, P. L., Feranec, R., Wing, S. L., & Shabel, A. (2004). Assessing the causes of Late Pleistocene extinctions on the continents. *Science, 306*, 70–75.
Barsh, R. L., & Marlor, C. (2003). Driving bison and blackfoot science. *Human Ecology, 31*, 571–593.
Berger, J., Swenson, J. E., & Persson, I. L. (2001). Recolonizing carnivores and naïve prey: Conservation lessons from quaternary extinctions. *Science, 291*, 1036–1039.
Broughton, J. M. (2002). Pre-Columbian human impact on California vertebrates: Evidence from old bones and implications for wilderness policy. In C. E. Kay & R. T. Simmons (Eds.), *Wilderness and political ecology: Aboriginal influences and the original state of nature* (pp. 44–71). University of Utah Press.
Deloria, V., Jr. (1988). *Custer died for your sins: An Indian manifesto*. University of Oklahoma Press.
Faith, J. T., Rowan, J., Du, A., & Koch, P. L. (2018). Plio-Pleistocene decline of African megaherbivores: No evidence for ancient hominin impacts. *Science, 362*, 938–942.
Finnegan, S., Bergmann, K., Eiler, J. M., Jones, D. S., Fike, D. A., Eisenman, I., Hughes, N. C., Tripati, A. K., & Fischer, W. W. (2011). The magnitude and duration of late ordovician-early Silurian glaciation. *Science, 331*, 903–906.
Fogg, B. R., Howe, N., & Pierotti, R. (2015). Relationships between Indigenous American peoples and wolves 1: Wolves as teachers and guides. *Journal of Ethnobiology, 3*, 262–285.
Gill, J. L., Williams, J. W., Jackson, S. T., Lininger, K. B., & Robinson, G. S. (2009). Pleistocene megafaunal collapse, novel plant communities, and enhanced fire regimes in North America. *Science, 326*, 1100–1103.
Goebel, T., Water, M. R., & O'Rourke, D. H. (2008). The Late Pleistocene dispersal of modern humans in the Americas. *Science, 319*, 1497–1502.
Grayson, D. K. (1977). Pleistocene avifaunas and the overkill hypothesis. *Science, 195*, 691–693.
Grayson, D. K. (1991). Late pleistocene mammalian extinctions in North America: Taxonomy, chronology, and explanations. *Journal of World Prehistory, 5*, 193–231.
Grayson, D. K. (2001). Did human hunting cause mass extinction? *Science, 294*, 1459.
Grayson, D. K., & Meltzer, D. J. (2002). Clovis hunting and large mammal extinction: A critical review of the evidence. *Journal of World Prehistory, 16*, 313–359.
Grayson, D. K., & Meltzer, D. J. (2003). A requiem for north American overkill. *Journal of Archaeological Science, 2003*, 1–9.
Grinnell, G. B. (1923). *The Cheyenne Indians: Their history and way of life*. Yale University Press.
Hames, R. (2007). The ecologically noble savage debate. *Annual Review of Anthropology, 36*, 177–190.
Johnson, C. (2009). Megafaunal decline and fall. *Science, 326*, 1072–1073.
Kay, C. E., & Simmons, R. T. (Eds.). (2002). *Wilderness and political ecology: Aboriginal influences and the original state of nature*. University of Utah Press.
Krech, S. (1999). *The ecological Indian: Myth and reality*. W. W. Norton.
Louys, J., et al. (2021). No evidence for widespread animal extinctions after post-pleistocene hominin arrival. *Proceedings of the National Academy of Sciences, 118*, e2023005118.
Martin, P. S. (1967). Pleistocene overkill. *Natural History, 76*, 32–38.
Martin, P. S. (1984). Prehistoric overkill: The global model. In P. S. Martin & R. Klein (Eds.), *Quaternary extinctions: A prehistoric revolution* (pp. 354–405). University of Arizona Press.

Martin, P. S. (2002). Prehistoric extinctions: In the shadow of man. In C. E. Kay & R. T. Simmons (Eds.), *Wilderness and political ecology: Aboriginal influences and the original state of nature* (pp. 1–27). University of Utah Press.

Martin, P. S. (2005). *Twilight of the mammoths: Ice age extinctions and the rewilding of America.* University of California Press.

Martin, P. S., & Klein, R. (Eds.). (1984). *Quarternary extinctions: A prehistoric revolution.* University of Arizona Press.

McClintock, W. (1910). *The old north trail; or, life, legends, and religion of the Blackfeet Indians.* MacMillan.

McClintock, W. (1923). *Old Indian trails.* Houghton Mifflin.

McComb, K., et al. (2001). Matriarchas as Repositories of social knowledge in African elephants. *Science, 292,* 491–494.

Payne, K. (2000). *Silent thunder: In the presence of elephants.* Viking Press.

Pierotti, R. (2011). *Indigenous knowledge, ecology, and evolutionary biology.* Routledge.

Pierotti, R. (2020). Learning about extraordinary beings: Native stories and real birds. *Ethnobiology Letters, 11,* 253–260.

Pierotti, R., & Fogg, B. R. (2017). *The first domestication: How wolves and humans coevolved.* Yale University Press.

Pryce, P. (1999). *Keeping the lakes' way.* University of Toronto Press.

Simmons, A. H. (2007). *The Neolithic revolution in the near east: Transforming the human landscape.* University of Arizona Press.

Stewart, M., Carleton, W. C., & Groucutt, H. (2021). Climate change, not human population growth, correlates with late quaternary megafauna declines in North America. *Nature Communications, 12,* 965.

Surovell, T. A., Pelton, S., Anderson-Sprecher, R., & Myers, A. (2016). Test of Martin's overkill hypothesis using radiocarbon dates on megafauna. *Proceedings of the National Academy of Sciences, 113,* 886–891.

Ward, P. (1997). *The call of distant mammoths: Why the ice age mammals disappeared.* Springer.

Waters, M. R., & Stafford, T. W., Jr. (2007). Redefining the age of Clovis: Implications for the peopling of the Americas. *Science, 315,* 1122–1126.

Wolverton, S., Lyman, R. L., Kennedy, J. H., & La Point, T. W. (2009). The terminal pleistocene extinctions in North America, hypermorphic evolution, and the dynamic equilibrium model. *Journal of Ethnobiology, 29,* 28–63.

Worthy, T., & Holdaway, R. N. (2002). *The lost world of the moa: Prehistoric life of New Zealand.* University of Indiana Press.

Printed in the United States
by Baker & Taylor Publisher Services